NANO-INNOVATIONS IN FOOD PACKAGING

Functions and Applications

NANO-INNOVATIONS IN FOOD PACKAGING

Functions and Applications

Edited by

Shiji Mathew, PhD

E. K. Radhakrishnan, PhD

AAP | APPLE ACADEMIC PRESS

First edition published 2023

Apple Academic Press Inc.
1265 Goldenrod Circle, NE,
Palm Bay, FL 32905 USA

760 Laurentian Drive, Unit 19,
Burlington, ON L7N 0A4, CANADA

CRC Press
6000 Broken Sound Parkway NW,
Suite 300, Boca Raton, FL 33487-2742 USA

4 Park Square, Milton Park,
Abingdon, Oxon, OX14 4RN UK

Library and Archives Canada Cataloguing in Publication

Title: Nano-innovations in food packaging : functions and applications / edited by Shiji Mathew, PhD, E.K. Radhakrishnan, PhD.
Names: Mathew, Shiji, editor. | Radhakrishnan, E. K. (Edayileveettil Krishnankutty), editor.
Description: First edition. | Includes bibliographical references and index.
Identifiers: Canadiana (print) 20220205191 | Canadiana (ebook) 20220205221 | ISBN 9781774639726 (hardcover) | ISBN 9781774639733 (softcover) | ISBN 9781003277422 (ebook)
Subjects: LCSH: Food—Packaging. | LCSH: Nanocomposites (Materials) | LCSH: Polymers. | LCSH: Nanotechnology.
Classification: LCC TP374 .N36 2023 | DDC 664/.09—dc23

Library of Congress Cataloging-in-Publication Data

CIP data on file with US Library of Congress

ISBN: 978-1-77463-972-6 (hbk)
ISBN: 978-1-77463-973-3 (pbk)
ISBN: 978-1-00327-742-2 (ebk)

About the Editors

Shiji Mathew, PhD

Shiji Mathew, PhD, has worked as a microbiology tutor at Sri Rajiv Gandhi Dental College, Bangalore, Karnataka, India. She has published papers based on her research work in various international peer-reviewed journals. She has also contributed more than five book chapters in international book volumes from publishers such as Elsevier, Wiley, and Cambridge Scholars Publishers. Her work has been cited over 305 times, and she has an h-index of 8 and i10-index of 7. Her area of research is medical microbiology and nanotechnology and is specialized in development of nanotechnology-based polymer composites, their characterization, and specialized applications in food packaging and in the medical sector. Dr. Mathew completed her doctoral degree in Medical Microbiology at Mahatma Gandhi University, Kottayam, Kerala, India. Her doctoral thesis was titled, "Antimicrobial effects of nanoparticle conjugates." She also obtained her MPhil in Biosciences and her MSc and BSc degrees in Medical Microbiology from Mahatma Gandhi University.

E. K. Radhakrishnan, PhD

E. K. Radhakrishnan, PhD, is an Assistant Professor at the School of Biosciences; Director, Business Innovation and Incubation Center; and Joint Director of the Inter University Centre for Organic Farming and Sustainable Agriculture, Mahatma Gandhi University, Kottayam, India. During his 12 years of research, he has published over 100 research publications, many book chapters, and several review papers. He edited two books with Springer Nature and Elsevier, and six books are in progress. His work has been cited almost 2254 times, and his h-index is 24 and i10-index is 56. To date, he has delivered over 40 invited talks at various national and international conferences and seminars. He has completed several research projects for various funding agencies and has five ongoing projects as PI.

His research areas include plant microbe interactions, microbial natural products, microbial synthesis of metal nanoparticles, and development of polymer-based nanocomposites with antimicrobial effects for food packaging and medical applications. He completed his doctoral degree at Rajiv Gandhi Centre for Biotechnology, Thiruvananthapuram, Kerala, India, and his postdoctoral studies at the University of Tokyo, Japan.

Contents

Contributors

Gauravi Agarkar
Department of Biotechnology, SGB Amravati University, Amravati, Maharashtra, India;
E-mail: gauravi7agk@gmail.com

Sony George
Department of Food Technology, Saintgits College of Engineering, Kottayam, Kerala, India;
E-mail: sojen143@gmail.com

M. Girilal
Department of Food Technology, Saintgits College of Engineering, Kottayam, Kerala, India;
E-mail: girilal@gmail.com

Nishar Hameed
Factory of the Future, Swinburne University of Technology, Hawthorn, VIC, Australia;
E-mail: nisharhameed@swin.edu.au

K. S. Joshy
School of Energy Materials, Mahatma Gandhi University, Kottayam 686560, Kerala, India;
E-mail: joshyk.s@gmail.com

Sourabh Suresh Kale
Dairy Technology Department, SMC College of Dairy Science, Anand Agricultural University, Anand, 388110, Gujarat, India; E-mail: sourabhkale502@gmail.com

Gemechu Berhanu Kerorsa
College of Agriculture and Veterinary Medicine, Dambi Dollo University, Ethiopia;
E-mail: gemechuberhanu@yahoo.com

Luiza Helena Da Silva Martins
ISPA (Institute of Animal Health and Production) Federal Rural University of Amazonia Avenue Presidente Tancredo Neves, Terra Firme, Belém, Pará, Brazil; E-mail: luhelemarte@gmail.com

Shiji Mathew
School of Biosciences, Mahatma Gandhi University, Kottayam 686560, Kerala, India;
E-mail: shiji_jan23@yahoo.co.in

Mahendra Pal
Narayan Consultancy on Veterinary Public Health and Microbiology, Anand, Gujarat, India;
E-mail: palmahendra2@gmail.com

Jyotishkumar Parameswaranpillai
Department of Science, Faculty of Science & Technology, Alliance University, Chandapura-Anekal Main Road, Bengaluru 562106, Karnataka, India;
E-mail: jyotishkumarp@gmail.com

Suneeta Pinto
Dairy Technology Department, SMC College of Dairy Science, Anand Agricultural University, Anand, 388110, Gujarat, India; E-mail: suneetavpinto@aau.in

G. L. Praveen
Wimpey Laboratories, Ras Al Khor, Industrial Area 2, Dubai; E-mail: emailtopgl@gmail.com

E. K. Radhakrishnan
School of Biosciences, Mahatma Gandhi University, Kottayam 686560, Kerala, India;
E-mail: radhakrishnanek@mgu.ac.in

Mahendra Rai
Department of Biotechnology, SGB Amravati University, Amravati, Maharashtra, India;
E-mail: mahendrarai7@gmail.com

Nisa Salim
Swinburne University of Technology, Faculty of Science, Engineering and Technology, Hawthorn, VIC, 3122, Australia; E-mail: vsnisa@gmail.com

Aiswarya Sathian
School of Energy Materials, Mahatma Gandhi University, Kottayam, India;
E-mail: sathianaiswarya@gmail.com

K. Senthilkumar
Center of Innovation in Design and Engineering for Manufacturing (CoI-DEM),
King Mongkut's University of Technology North Bangkok, Wongsawang, Bangsuc,
Bangkok 10800, Thailand; E-mail: kmsenthilkumar@gmail.com

Suchart Siengchin
Department of Mechanical and Process Engineering, The Sirindhorn International Thai–German Graduate School of Engineering (TGGS), King Mongkut's University of Technology North Bangkok, Bangsue, Bangkok 10800, Thailand; E-mail: suchart.s.pe@tggs-bangkok.org

Sabu Thomas
School of Chemical Sciences, Mahatma Gandhi University, Kottayam 686560, Kerala, India;
E-mail: sabuthomas@mgu.ac.in

C. Vibha
Department of Mechanical and Process Engineering, The Sirindhorn International Thai–German Graduate School of Engineering (TGGS), King Mongkut's University of Technology North Bangkok, Bangsue, Bangkok 10800, Thailand; E-mail: vibhacbabu@gmail.com

Alka Yadav
Department of Biotechnology, SGB Amravati University, Amravati, Maharashtra, India;
E-mail: nanoalka@gmail.com

Abbreviations

ADR	adriamycin
ALP	alkaline phosphatase
AFM	atomic force microscopy
AGM	agar/gellan gum/MMT
AP	active packaging
ASTM	American Society for Testing Materials
BC	bacterial cellulose
BEO	basil leaf essential oil
BOPP	biaxially oriented polypropylene
BRS	boiled rice starch
BSA	bovine serum albumin
CA	contact angle
CH	chitosan
CMC	carboxymethyl cellulose
CNF	carbon nanofiber
DMA	dynamic mechanical analysis
DPA	dopamine
DSC	differential scanning calorimetry
EA	ethylene vinyl acetate
EB	elongation at break
EC	European Commission
EDX	energy dispersive X-ray analysis
EFSA	European Food Safety Agency
ENMs	engineered nanomaterials
EPA	Environmental Protection Agency
FDA	Food and Drug Administration
FEP	fluorinated ethylene-propylene copolymers
FG	Farsi gum
FTIR	Fourier transform infrared spectroscopy
GA	gum Arabic
GIT	gastrointestinal tract
GRAS	Generally Recognized as Safe
GS	glycylsarcosine

HA	hyaluronic acid
HDPE	high density polyethylene
His	histones
HNT	halloysite nanotubes
HNT/PE	halloysite nanotube-spolyethylene
ICP-MS	inductively coupled plasma mass spectroscopy
IOSP	intelligent or smart packaging
ISO	International Standard Organization
κ-CG	κ-carrageenan
LD	longitudinal direction
LDH	layered double hydroxides
LDPE	low density polyethylene
LLDPE	linear low-density polyethylene
LSPR	localized surface plasmon resonance
MAP	modified atmosphere packaging
MDR	multidrug resistance
MFC	microfibrillated cellulose
MMT	montmorillonite
MTX	mitoxantrone
NADH	nicotinamide adenine dinucleotide
NFC	nanofibrillated cellulose
NGO	non-governmental organization
n-HA	nano-hydroxyapatite
NIR	near-infrared
NMR	nuclear magnetic resonance
NPs	nanoparticles
NRC	NanoRiskCat
OECD	Organization for Economic Corporation and Development
OM	optical microscopy
OP	oxygen permeability
OPP	oriented polypropylene
OTR	oxygen transmission rate
PAG	PVA/AgNO3/ginger extract
PAI	photoacoustic imaging
PBS	poly(butylene succinate)
PCL	polycaprolactone
PE	polyethylene

PepT-1	peptide transporter-1
PET	poly(ethylene terephthalate)
PHAs	polyhydroxyalkanoates
PLCL	poly(L-lactide-co-ε-caprolactone)
PLGA	poly lactic-co-glycolic acid
PLLA	poly (l-lactic acid)
PP	polypropylene
PS	polystyrene
PTT	photothermal therapy
PVA	polyvinyl alcohol
PVC	polyvinyl chloride
PVDC	polyvinylidene chloride
RES	reticulo-endothelial system
RFID	radiofrequency identification
RGD	arginyl glycylaspartic acid
ROS	reactive oxygen species
SDEY	salted duck egg yolk
SEDs	sacrificial electron donors
SEM	scanning electron microscopy
SONs	structured oil nanoparticles
SPR	surface plasmon resonance
TD	transverse direction
TEM	transmission electron microscopy
TGA	thermogravimetric analyzer
TS	tensile strength
TTIs	time temperature indicator
USFDA	United State Food and Drug Administration
UV	ultraviolet
WAXD	wide angle X-ray diffraction
WHO	World Health Organization
WVP	water vapor permeability
WVTR	water vapor transmission rate
XRD	X-ray diffraction technique
YM	Young's Modulus

Foreword

Nanotechnology is the way nature builds up strong and smart structures in a hierarchic approach, and bioinspiration is the new frontier in materials science. Over the past 20 years, nanotechnologies have been making their mark in the materials sector, improving performance and functionality in the latter, paving the way for new applications. In particular, in food packaging, nanotechnologies are used both to improve mechanical and barrier performance and to facilitate functional, active, and intelligent packaging. This book serves as a tool for scientists and technologists to explore recent innovations in the field, with insights also in biopolymers, health, and safety issues.

—Mario Malinconico
Research Director,
International Science Council ISC CNR Representative
IPCB-CNR, Via Campi Flegrei, 34-80078 Pozzuoli (Na) Italy

Introduction

This book presents and discusses many important aspects and innovations introduced by nanotechnology in the form of polymer nanocomposites in the food packaging industry. This book introduces readers to polymers, composites, nanocomposites, polymer nanocomposites, and other various nanofillers used in them. It also provides a detailed discussion on the important advantages of using polymer nanocomposite as food packaging over conventional packages. In addition, the important preparation and characterization techniques of polymer nanocomposites are discussed in great detail. This book also covers the major food packaging applications of polymer nanocomposites such as active packaging, intelligent/smart packaging, biodegradable packaging. The applications of nanotechnology in food preservation are also discussed, which includes the intervention of nanoencapsulation technology and the use of edible nanocoatings/nanolaminates for food-quality enhancement and preservation of organoleptic properties of food. Last but not the least, a detailed information on the evaluation of the possible health and safety hazards of nanotechnology application in food industry is also provided.

By offering a detailed account on the major features and novel trends of polymer nanocomposites, this book is expected to primarily aid readers who are interested and engaged in the research and development of polymer nanocomposites. Moreover, the illustrations and tables included in the text can broaden readers' insight and help them to develop a more informed understanding of the need of using polymer nanocomposites for efficient food packaging application. This book will serve as an excellent source of general information for the beginners stepping in this field as well as an essential reference for researchers, scientists, and industrialists.

Preface

Packaging plays a crucial role in protecting all kinds of foods from farm to the consumer level. But none of the conventional packaging technology available is ideal in its functions. One of the rapidly advancing areas in this field are nanotechnology-based applications in food packaging. Nanoscale involvements in materials science has led to the development of polymer nanocomposites which result in stronger, lighter weight, cheaper, highly functional, antimicrobial, and safer food packaging materials that amend, the traditional concept of this process.

In this book, the latest trends of nanotechnology-based packaging in the food industry are addressed. It is a summary of the most relevant and recent advances made with nanostructured materials that have significant impact on the food packaging industry. This book is divided into three major sections: (1) General Aspects of Nanocomposite Food Packaging Materials, (2) Types of Nanocomposite Food Packagings, and (3) Role of Nanotechnology in Food Preservation.

Chapter 1 of this book provides a general introduction and is an overview of different types of nanocomposite food packaging and the various nanomaterials used for their development. Each type of application is presented with suitable examples. Chapter 2 discusses the commonly and traditionally used food packaging materials as well as the advantages of using nanocomposite-based food packaging materials over these conventional systems. Chapter 3 presents the important properties of polymer nanocomposite as packaging materials. It also covers the major preparative methods and the varied quantitative and qualitative analytical techniques used for characterization of nanocomposites with appropriate examples. Chapter 4 provides detailed information on the functions and the applications of active nanocomposite packaging. This chapter describes in detail the nanofillers used to design active antimicrobial and antioxidant packaging systems. Chapter 5 reviews the functions and applications of smart/intelligent nanocomposite packaging and includes a note on the various sensors, indicators, bar codes, and radiofrequency indicators used for their development.

Chapter 6 explains the functions and different applications of biopolymer-based nanocomposites. This chapter mainly focuses on the different natural and synthetic biopolymers available and their suitability for fabricating bionanocomposite food packagings. Chapter 7 discusses the role of nanotechnology in preserving and maintaining the odor and flavor of packed food. This chapter majorly explains the method of nanoencapsulation technology and its applications in preserving the organoleptic properties of packed food. Chapter 8 emphasizes nanotechnology-based edible packagings in the form of coatings and thin films. This chapter covers information regarding the general aspects, functions, and methods of preparing and applying nanocoatings. It also details the various biomaterials utilized for the fabrication of edible nanocoatings with appropriate examples.

Finally, Chapter 9 provides an assessment of the possible health and safety issues associated with the involvement of nanotechnology in food applications. It covers the factors influencing nanoparticle toxicity, the route of exposure of humans to nanoparticles, and the various tests available for risk evaluation of nanoparticles in foods.

This book is a collective effort of several postgraduate students, researchers, scientists, retired academicians, and postdoctoral fellows who are well-known, experienced, and respected in different fields of polymer nanotechnology and food nanotechnology research. Each chapter of this book is written in simple and clear language and is also accompanied with tables and figures for easy understanding. This book will be highly appealing to graduate and postgraduate students and is expected to benefit professors/teachers, scientists, researchers as well as specialists working in industries related to advanced food science, food packaging, food safety, and shelf life.

We express our sincere appreciation to all the contributing authors for their interest and effort in helping us with their contributions and suggestions for the completion of this book. We are also extremely thankful to the team members of Apple Academic Press for giving us this opportunity and their effort in publication of this book. Finally, we are also thankful to our family members for their valuable encouragement, patience, support, and love provided throughout the period of this book project.

—Shiji Mathew
E. K. Radhakrishnan

PART I

General Aspects of Nanocomposite Food Packaging Materials

CHAPTER 1

An Overview of Nanotechnology-Based Innovations in Food Packaging

GEMECHU BERHANU KERORSA[1] and MAHENDRA PAL[2*]

[1]*College of Agriculture and Veterinary Medicine, Dambi Dollo University, Ethiopia*

[2]*Narayan Consultancy on Veterinary Public Health and Microbiology, Anand, Gujarat, India*

[*]*Corresponding author. E-mail: palmahendra2@gmail.com*

ABSTRACT

Packaging of the food items is the most important and inevitable level of the food supply chain because the higher packaging leads to longer sturdiness, shelf life, and renovation of mechanical, physical, chemical, and physicochemical properties of food products. Nanotechnology-based food packaging is a new and novel technique of packaging method, which relies upon nanomaterials, has grown to be one of the fastest-growing elements in nanotechnology, and denotes critical opportunity and advancements. An essential nano-based food packaging material needs to have gas and moisture permeability integrated with strength and biodegradability. Different components of nanotechnology-based techniques used in food packaging include nanocomposites, bio-nanocomposite films, nanoemulsions, nanostructures, nanolaminates, nanoclays, nanopolymers, and nanocoatings. There are different types of nanotechnology-based food packaging including active food packaging, smart/intelligent food packaging, improved packaging, bio-based packaging, modified atmospheric packaging, and oxygen scavenging packaging. Nanotechnology in food packaging substances confers many benefits over conventional packaging methods because they improve food quality and safety, inform the consumers whether the diets are spoiled or not, renovate tears in packaging, and prolong the duration of storage of the food. Therefore, nanotechnology-based innovations are playing key roles in the modern food packaging industry.

1.1 INTRODUCTION

Food is indispensable for the continuation of the life process in all living things (Pal et al., 2019). Food products are vulnerable to attack by spoilage and pathogenic microorganisms during processing and distribution. Food processing operations, such as peeling, slicing, or washing, if no longer well conducted, can be the cause of public health issues, mainly if these ingredients are handled and disbursed under inappropriate conditions (Jaiswal et al., 2019). Food safety is a developing pattern for all food items to prevent them from physical, chemical, natural/biological, and radiation pollution, which may occur throughout production, handling, storing, processing, transportation, and distribution (Keshwani et al., 2015).

Consumers need microbiologically safe, new and fresh-like, wholesome, sturdy, fitting products made by utilizing environmentally versatile advancements (Pal, 2017). Microbial defilement prompts the development of pathogenic microorganisms and poor diet concomitant with foods and foodstuffs (Jaiswal et al., 2019). Packaging is a crucial connection between the food manufacturers and the clients which unless fulfilled efficaciously, may end up in less product satisfaction and losing the client's confidence (Pal et al., 2019). The roles of food packaging include prolonging the shelf existence of prepared meal substances through avoiding unfavorable adjustments because of microbial spoilage, chemical contaminants, temperature exchange, oxygen, moisture, moderate, external pressure, and retaining the safety and quality of food products starting from the time of production to the time of consumption (Emanuel and Sandhu, 2019; Jaiswal et al., 2019).

Nanotechnology, which is referred to as the science of very small substances, is studied to have the most powerful effect on food production and packaging. A significantly low degree of a nanoparticle is needed and thought to be enough to change the properties of packaging substances without many modifications of their density, transparency and packaging, and processing characteristics because of their huge issue ratios. Nanotechnology elucidation emphasizes diet fortification by hindering pathogenic microbial growth, delaying oxidation, comfort, and improving tamper visibility (Keshwani et al., 2015; Kuswandi, 2016). Nanotechnology provides a wide range of possibilities for the advance and application of structures, materials, or devices with new properties in numerous areas like agriculture, food, and medication (Singh et al., 2017). Nanotechnology in the food sector considers sanitation by checking the development of pathogenic microorganisms, improving alters permeability, and reducing oxidation time and expediency (Davarcioglu, 2017).

Modern rising developments in meals packaging have given an increase to food processing industries. Trends on meals packaging materials have decreased the wastages collectively by improving the product splendidly, extending its duration, and safeguarding product safety (Emanuel and Sandhu, 2019; Mihindukulasuriya and Lim, 2014). Improvements in nanotechnology have converted a number of medical and commercial enterprise areas collectively with the food enterprise (Singh et al., 2017). Novel food packaging technologies, which include aseptic packaging, energetic packaging, practical packaging, nano-packaging, and bioactive packaging deliberately concomitant with the meals merchandise, have shown to be the best technological research areas (Pal et al., 2019). There are different categories of nanosensors, additives, nanoparticles, and signs integrated with nanofood packaging strategies (Montazer and Harifi, 2017; Yotova et al., 2013). Therefore, the aim of this chapter is to present an overview of nanotechnology-based innovations that are being used in the food packaging industry.

1.2 FUNCTIONS OF FOOD PACKAGING

Food items are packaged and cleanly moved to shield and safeguard them from any inadequate alteration in quality, before arriving at the end-customer (Sharma et al., 2017). Packaging frameworks are those products that are orchestrated with any material to ensure, protect, contain, control, transport appropriately, and to distinguish each progression along with its supply network, from raw materials to consumers or end-users (Emanuel and Sandhu, 2019). Representing the primary customary materials for food packaging, metal, glass, and cardboard has been broadly replaced in numerous applications by plastics (Davarcioglu, 2017). These materials offer numerous points of interest, such as being lightweight contributes to decrease in transportation costs, easiness of handling and processing, physical and chemical properties, among them mechanical, tribological, and thermal properties, just as the penetrability to gases or dampness, and shifting in an expansive range contingent upon the specific polymer (Duncan, 2011; Rossi et al., 2017).

The primary functions of any packaging materials include to hold, protect, distribute, transport and to inform the details of the products to the customers (Kuswandi, 2016). Food packaging has a crucial function in giving safety and maintaining the quality of food. Food packaging with new features called active packaging has been established as a result of customer demand for protection and greater natural products with a longer durability and shelf life, higher cost-benefits, and accessibility (Rossi et al., 2017).

Various food products, for example, meat, poultry, fish, vegetables, fruits, and natural products, are exceptionally transient and delicate if not appropriately packaged and dealt with. Protein denaturation, lipid oxidation, and development of microorganisms because of auspicious pH and high-water activity (Dave and Ghaly, 2011) lessen the item life span as a result of spoilage of these food products (Alparslan and Baygar, 2017). The packaging must be done cautiously for these food products, and initial quality must be held for a longer subsequent time thus increasing the item life lengt h and reducing the food wastage (Emanuel and Sandhu, 2019).

1.3 NANOTECHNOLOGICAL INNOVATIONS IN FOOD PACKAGING

Nanotechnology is referred to as the science of very small substances, and it has a powerful influence in food enterprise inclusive of food production and packaging (Keshwani et al., 2015; Pradhan et al., 2015). The word "nano" is originated from the Greek language, which means "dwarf." The concept of nanotechnology was hosted by Richard Feynman in 1959, and the word "nanotechnology" was later claimed in 1974 by Norio Taniguchi (Sharma et al., 2017). Nanotechnology, which makes use of materials on an atomic, molecular, or supramolecular measure, is conjoining with biotechnology in a steady and malleable packing business organization (Montazer and Harifi, 2017). Nanotechnology is surprisingly interdisciplinary and particularly encompasses fabrication, awareness, manipulation, and characterization of systems, devices, or substances at the nanometer scale, approximately 1–100 nm in length, with constitutive factors that have at least one dimension (Kuswandi, 2016; Mihindukulasuriya and Lim, 2014). It creates and uses materials that have novel properties. It is determined that these materials have precise places in contrast to their macroscale counterparts because of the excessive surface-to-volume ratio and other novel physicochemical properties like color, solubility, power, diffusivity, toxicity, magnetic, optical, and thermodynamic (Sekhon, 2010; Singh et al., 2017).

Food packets gain extra tensile strength with the technique of nano-reinforcement. Nano-reinforcement particularly incorporates nanoclays, cellulose, and graphene. Food packets can also be blended with different agents like antimicrobial agents, oxygen scavenging retailers, and enzyme immobilization systems to form nanocomposite active packaging (Othman, 2014). Nanocomposite smart packaging includes sensors, such as gas detectors, time–temperature integrators, and different nanosensors (Rhim et al., 2013).

Nanocomposite active packaging integrates many beneficial systems (antimicrobial, enzyme immobilization systems, and oxygen scavenging) in conjunction with the food packets. Within the last two decades, polymer clay nanocomposites, a new class of clay stuffed polymers, have been advanced (Agriopoulou, 2016; Dasgupta and Ranjan, 2018). Food nanopackaging continues to be a rather unexplored area of nanoscience and food science (Kuswandi, 2016). A new packaging material that has received considerable attention is nanocellulose. Nanocellulose, nanofibrils, and nanocrystals have been included as a strengthening segment in nanocomposites. Additionally, nanocellulose is used as a base substance that is more desirable with another nanomaterial together with photocatalysts (El-Wakil et al., 2015). In other words, it may be the carrier of different antimicrobial agents with the prepared release (Eleftheriadou et al., 2018; Sundaram et al., 2016).

Nanotechnology has added many outstanding and essential modifications for better lifestyles. The usage of nanotechnology in packaging gives many opportunities to assist sustainability and cost-effective measures, enhance health, wealth, products, and quality of life (Davarcioglu, 2017). Nanotechnology affords material with more potent packaging barriers, which preserve food quality at the time of transportation, prolong fruits and vegetable freshness at the time of storage, and preserve gist or chicken from different pathogenic microorganisms (Bumbudsanpharoke and Ko, 2015; Singh et al., 2017). The application of nanotechnology in polymers involves the layout, production, processing, and application of polymer substances filled with nanoparticles and devices of nano-range (Momin and Joshi, 2015; Sharma et al., 2017). Nanotechnology additionally deals with the advanced manner of manufacturing, processing, and transportation of foods using sensors for the determination of pathogens and contaminants, devices to uphold widespread environmental information of an express product and tracking of individual transportations, structures that deliver the combination of sensing, localization, informing, and far-flung control of ingredients by smart and intelligent systems, and which can be capable of enhancing the performance and safety of the processing and conveyance of food items, carrying, protection, and shipping of useful food ingredients to their precise place of activity provided through encapsulation and shipping systems (Davarcioglu, 2017; Paul and Robeson, 2008). While the particle size of the material is decreased into nano-length, the resulting material reveals physical and chemical properties that are considerably distinct from the properties of macroscale materials composed of the identical substance (Danie et al., 2013; Kuswandi, 2016).

The advent of polymer nanotechnology in food packaging stepped forward the fundamental features of traditional packaging systems together with containment (ease of transportation and managing), convenience (being consumer-friendly), protection and maintenance (avoids leakage or break-up and protects against microbial contaminants, providing longer shelf life), marketing, and communication about the actual time information about the quality of enclosed foodstuffs, additionally the nutritional ingredients and preparatory recommendations (Sharma et al., 2017). Moreover, nanotechnology-dependent packaging shows users about the quality of diet yields, and it has considerably addressed the food quality, safety, and stability issues. Moreover, nanotechnology has been reconnoitered for meticulous discharge of additives/antimicrobials, prolonging the product storage duration inside the package (Silvestre et al., 2011; Vanderroost et al., 2014).

Nanotechnology has enabled the manufacturing of packaging that is smart and active with radical perfunctory and thermal properties to safeguard enhanced fortification of foods (Khalaj et al., 2016). Nanotechnology-based food packaging enables food packages to be embedded with nanoparticles that alert consumers while a product is not safe to devour. Sensors can warn before the food is going rotten or can tell us the exact dietary fame contained within the contents. In reality, nanotechnology is going to change the production of the whole packaging enterprise (Perez-Esteve et al., 2013; Sekhon, 2010). The mixing of nanoclays in biopolymers improved their mechanical properties, permitting their importance as an unconventional eco-friendly and biodegradable nutrients packaging (Ghanbarzadeh et al., 2013). In addition to the enrichment of the mechanical behavior, packaging may be a gas obstacle to escalate product duration with hampering oxygen entrance or preventing CO_2 escape in carbonated beverages using titanium nitride nanoparticles (Reščcek et al., 2016). Nano-enabled elucidations also endorsed the addition of abundant nanoparticles and bioactive molecules to avert food degradation and oxidation. Cellulose and selenium nanoparticles can be included in food packaging to retard or inhibit the reactive oxygen species (ROS) that can degrade food quality (Vera et al., 2016). Other essential oils can also be incorporated in nanofibers (Wen et al., 2016) to prolong the lifespan of fresh food products (Eleftheriadou et al., 2018).

1.4 IMPORTANCE OF NANO INNOVATIVE FOOD PACKAGING

Nanotechnology in packaging has also several importance (Fig. 1.1). Some of them include antibacterial in which the exploitation of silver nanoparticles in the form of the antibacterial element within the packaging of food products is developing (Chatterjee et al., 2014; Shafiq et al., 2020). It is usually lined on plastic packets to prevent food from going off and also incorporated into food storage containers. This technology is used even in the interior of fridges to reduce fungal growth. The other one is protective coatings wherein several different nanocoatings have been used for protection of the food freshness and the taste (Azeredo et al., 2016), and additionally through obstructing the rays of the sunlight (Davarcioglu, 2017; Sozer and Kokini, 2009).

Nanotechnology based food packaging materials not only aids in increasing the sturdiness of food, improve the food safety and eminence to the users, but also safeguards the food so that reaches them in good quality (Ravichandran, 2010; Singh et al., 2017). Nanotechnology importance in the diet business may be exploited to distinguish bacteria in packaging or create more compelling flavors and color quality, and safety by aggregating the barrier properties. Nanotechnology holds great promise to provide advantages not only within food products but also around food and its products (Keshwani et al., 2015; Sekhon, 2010).

Food industries are commonly attempting to find cheaper and new approaches to create and preserve diet. New movements in food packaging are associated with nano-reinforcement, within the field of nanocomposites active packaging and in the field of smart packaging

(Thiruvengadam et al., 2018). Polymer nanocomposites normally have much better polymer–filler interactions than the usual composites (Pandey et al., 2013). Gaps of packaging substances are filled with the help of nano-reinforcement methods and the very last substances have improved viability and tensile power (Agriopoulou, 2016; Luduena et al., 2007).

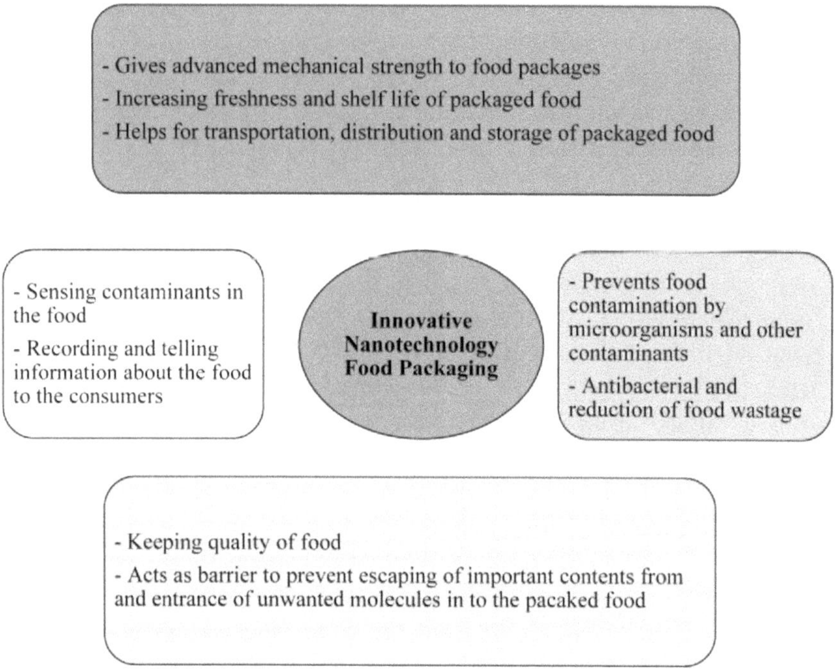

FIGURE 1.1 Importance of innovative nanotechnology-based food packaging.

1.5 REQUIREMENTS OF NANO INNOVATIVE FOOD PACKAGES

Nanotechnology allows researchers to modify the structure of packaging materials at the molecular level (Biswal et al., 2012). The basic classes of nanotechnology applications and functionalities present in the improvement of food packaging consist of the development of plastic material barriers, the integration of active ingredients that can deliver functional features beyond those of traditional active packaging, and the sensing and signaling of appropriate information (Chaudhry and Castle, 2011; Sekhon, 2010). Innovative packaging materials are rapidly growing to come across the consumer need for higher quality food with accessibility, safety, and sustainability, fortification, preservation, inhibition, food waste reduction, marketing of food, information about food, traceability, and convenience (Jaiswal et al., 2019; Ravichandran, 2010).

An essential packaging material needs to have gas and moisture permeability integrated with strength and biodegradability (Couch et al., 2016; Siracusa et al., 2008). Nano-based smart and active food packaging confer many benefits over conventional packaging methods from offering higher packaging material with superior barrier behavior, mechanical power, and antimicrobial films to nanosensing for alerting clients about the safety condition of the meals and pathogen detection (Singh et al., 2017; Sundaram et al., 2016). Because of the nondegradability of longstanding packaging material, there may be a need for the usage of natural polymer primarily based packaging substances in food packaging and wrapping that not only maintains the high-quality and protection but also extends the shelf life of food products (Keshwani et al., 2015).

As a way to meet effective food packaging requirements, nanomaterial advanced polymers assist to intensify the advantages related to an existing polymer, with improved safety and besides addressing environmental issues (Dasgupta and Ranjan, 2018). The advanced packaging material gives the decrease of any serious interface between packaging and food matrices, the effect over user's health, and drop in the quantity of waste material, developed biodegradability and protection to gases, and reduced CO_2 emissions (Sharma et al., 2017). Vital packaging devices should have moisture and gas permeability joined with biodegradability and strength (Couch et al., 2016).

1.6 NANO-BASED COMPONENTS USED IN FOOD PACKAGING

There are different components of nanotechnology innovation-based techniques used in food packaging.

1.6.1 NANOCOMPOSITES

Nanocomposites are multiphase materials resulting from the combination of two or more elements, containing a continuous phase (matrix) and an irregular nano-dimensional level with at least one nano-sized size (with less than 100 nm) (Bratovcic et al., 2015; Farhoodi, 2016; Montazer and Harifi, 2017). Polymer nanocomposites are polymer substances made in addition to nanoparticles to boost the properties of polymers (Yotova et al., 2013). Nanocomposites are mostly exploited in the area of food packaging as they are eco-friendly and biodegradable. Nanocomposites exhibit extraordinarily versatile chemical functionality and are therefore used for the growth of high complication properties (Farhoodi, 2016; Pandey et al., 2013). Nanocomposites generally tend to reduce the packaging of undesirable materials related to treated food products and assist to attain the preservation necessities of fresh foods by extending their shelf life (Azeredo et al., 2009; Paul and Robeson, 2008; Singh et al., 2017). One of the advantages of incorporating clay nanocomposites in food packaging films is that it provides extended shelf life and makes it highly heat resistant (Keshwani et al., 2015; Sekhon, 2010). Clay nanocomposite is used to make gas barriers that reduce the outflow of carbon dioxide from the bottles of carbonated drinks (Rossi et al., 2014; Yotova et al., 2013).

Nanocomposites of zinc oxide and titanium oxide work as a sensor system and detect volatile organic compounds (El-Wakil et al., 2015; Jones et al., 2008). Such sensor systems are found to be highly beneficial to the consumers; by providing accurate data on the quality of the food, and also for the producers; enabling fast distribution and validation of the quality of food products (Chellaram et al., 2014; Dasgupta and Ranjan, 2018). The development of bio-nanocomposite substances for food packaging is significant not only to reduce the environmental problem but also to enhance the functions of the food packaging materials (Othman, 2014; Sondi and Salopek-Sondi, 2004; Thiruvengadam et al., 2018). Incorporation of nanoparticles in composites can render it heat and cold stable, provide chemical functionalities, excellent barrier properties and improve the duration of the stored product. (Acosta, 2009). They inhibit microorganism contamination inside packaged food, best barrier behavior against carbon dioxide escape from carbonated products. Nanoclays are used effectually to impart these gas barrier behavior (Emanuel and Sandhu, 2019).

1.6.2 BIO-NANOCOMPOSITE FILMS

Materials, of which degradation is possible through the enzymatic activity of microorganisms, for instance, bacteria, fungi, yeasts, and the end products of the process of deterioration, such as CO_2, H_2O, and biomass under aerobic circumstances, and hydrocarbons, biomass, and methane under anaerobic situations, are expressed by the term of biodegradable substances (Davarcioglu, 2017; Mills et al., 2006). In the course of biodegradation, typically, the long polymer molecules are condensed to make molecules with shorter lengths, and that they undertake oxidation that is expressed within the attachment of oxygen groups to the molecules of polymers (Zhao et al., 2008).

For the general advantage, biodegradable plastics ought to offer benefits for the structures of waste control, in addition to the cost and performance. Biodegradable food packaging becomes progressively vital in the food sector, wherein higher interest is paid to development in overall performance, for example, handiness and portioning (Assis et al., 2017; Malathi et al., 2014). Similarly, for better recognition of sustainability, the realization of which is usually possible at various levels is ensured. At the levels of raw substances, the employment of reused substances and the usage of sustainable resources constitute two techniques to diminishing CO_2 discharge and the reliance on fossil reserves (Davarcioglu, 2017). Bio-nanocomposite gelatin-based film has also been developed as biodegradable food packaging (Del Nobile et al., 2009; Tuan Zainazor et al., 2020).

The biodegradability of packaging material may be improved through integrating inorganic elements into the biopolymeric medium and can be measured with surfactants that might be applied for the alteration of the layered silicate (Ghanbarzadeh et al., 2013). Using inorganic factors also makes it feasible for food packaging to have multiple functionalities that can aid in the improvement of the techniques to deliver fragile micronutrients within edible drugs (Sorrentino et al., 2007). Cassava-based biodegradable packaging films containing lycopene nanocapsules improved the tensile strength and elongation of the film. It additionally provided a more barrier to light transmission as compared to the control and the industrial polyethylene films (Reščk et al., 2016). When used for the storage of sunflower oil, it avoided its oxidation and improved shelf life. The organized films have a fast biodegradability for days hence displaying the capability of lycopene nanocapsules that are helpful for active biodegradable packaging (Assis et al., 2017; Emanuel and Sandhu, 2019).

1.6.3 NANOENCAPSULATION

Nanoencapsulation is defined as a technology to pack substances in minuscule using methods consisting of the nanocomposite, nanoemulsification, and nanostructuration and offers end product capability that consists of the managed release of the meal product (Farhoodi, 2016; Wen et al., 2016). The protection of bioactive compounds including vitamins, proteins, lipids, and antioxidants, as well as carbohydrates, can be accomplished using this method for the production of functional foods with improved functionality and stability (Flanagan and Singh, 2006; Moraru et al., 2003; Sekhon, 2010). Nanoencapsulation is the incorporation of substances in small vesicles or walled material with nano sizes (Surassmo et al., 2009). Nanoencapsulation within the form of micelles, liposomes, or biopolymer-based carrier structures has been used to develop delivery systems for components and supplements for use in food and beverage products. Nanoencapsulation is the addition of microencapsulation, which has been utilized by the food enterprise for food additives and components for decades (Baeumner, 2004; Burdo, 2005; Ghanbarzadch ct al., 2013; Othman, 2014).

These nanomaterials provide numerous benefits together with retaining the ingredients and additives during processing and storage, protecting unpleasant tastes and flavors, controlling the release of components, better dispersion of water-insoluble food ingredients and additives, and progressed uptake of the encapsulated nutrients and dietary supplements (Momin and Joshi, 2015; Momin et al., 2013). Nanoencapsulation is carried out with the assist of nanocapsules (Surassmo et al., 2009). They offer numerous advantages along with ease of handling, more suitable stability, protection toward oxidation, retention of risky ingredients, taste making, moisture introduced on managed release, pH caused controlled launch, consecutive transport of more than one lively ingredient, exchange in flavor behavior, long-lasting organoleptic acceptance, and increased efficacy and bioavailability (Pradhan et al., 2015). The nanoencapsulation of different entities including phenols also can protect against degradation, especially of fatty ingredients (Liu et al., 2017).

1.6.4 NANOLAMINATES

Nanolaminates consist of two or more layers of material with nanometer measurements that can be chemically or physically attached to the other (1–100 nm per layer, normally 5 nm). These can be used to encapsulate numerous hydrophilic, amphiphilic, or lipophilic elements, active well-designed agents that include anti-browning agents, antimicrobials, enzymes, flavors, antioxidants, and colors with greater moisture and gas barrier properties (Momin et al., 2013). They are evolved in food-grade components including proteins, polysaccharides, and lipids. Those functional materials can enhance the storage life and quality of coated foods consisting of meats, cheese, fruits and vegetables, confectionery, bakery, and fast food items (Momin and Joshi, 2015).

1.6.5 NANOEMULSIONS

Nanoemulsion is part of nanotechnology techniques to an existing device that assists the food enterprise. Nanoemulsions due to small droplet size possess particular rheological and textural properties to foodstuffs which is suitable in the food industry. Making nanoemulsions in food products can facilitate much less fat use without compromising creaminess (Mason et al., 2006). Consumers can have a healthier choice from this technology. Low-fat nanostructured mayonnaise spreads and ice creams are a few such examples (Chaudhry and Castle, 2011). Fat may be decreased from 16% to 1% without compromising taste, texture, and quality through making nanoemulsions in ice cream. A 2.5% fat ice cream is commercially available worldwide from a recognized top rate ice cream brand that claims to have no flavor defects because of the low fats content; but, no nanotechnology privilege is made by the manufactured goods (Momin and Joshi, 2015).

Nanoemulsions have been evolved for use in the decontamination of the food packaging system and the packaging of food products. A standard example is a nanomicelle-based product claimed to contain natural glycerin. It gets rid of pesticide residues from fruits and vegetables as well as the oil/dirt from cutlery. Nanoemulsions have lately acquired a variety of attention from the food industry due to their high clarity (Farhoodi, 2016). These enable the addition of nanoemulsified bioactive and flavors to a beverage without a change in product appearance. Nanoemulsions are effective against a variety of food pathogens such as Gram-negative bacteria. They may be used for surface decontamination of food processing plants and reduction of floor contamination of chook skin. The growth of *Salmonella typhimurium* colonies can be removed by using treatment with nanoemulsion (Sekhon, 2010). Nanoemulsions exhibit numerous advantages over traditional emulsions because of the size of the small droplets as they possess high optical precision, the best physical fidelity against gravitational partition and droplet gathering, and the better bioavailability of encapsulated substances that may make them comfortable for food systems. Nanoencapsulation is the most substantial promising technology having the opportunity to catch bioactive chemicals (Thiruvengadam et al., 2018).

1.6.6 EDIBLE NANOCOATING

Coating in food may be defined as a thin film of edible substance sited between diet components to give obstacle to bulk allocation. These coatings ought to be used as lipid, moisture, and gas barriers. Coverings are formed and applied instantly on the food product either through the addition of a liquid film-forming solution or by using molten compounds (Alparslan and Baygar, 2017). Additives of edible coatings may be divided into two classes including water-soluble polysaccharides (hydrocolloids) and lipids. Appropriate polysaccharides encompass cellulose derivatives, alginates, pectins, starches, chitosan, and different polysaccharides. Different lipid compounds including animal and vegetable fats are used to produce consumable films and coatings (Montazer and Harifi, 2017; Vital et al., 2016). Suitable lipids encompass waxes, acylglycerols, and fatty acids. Lipid films have splendid moisture barrier properties or as coating agents for including gloss to confectionery products. Waxes are typically used for coating fruits and vegetables to retard respiration and lessen moisture loss (Emanuel and Sandhu, 2019; Kuswandi, 2016; Montazer and Harifi, 2017).

Nanotechnology is allowing the development of nanoscale edible coatings as thin as 5 nm (Vital et al., 2016). Edible nanocoatings may be used on meats, cheese, fruits and vegetables, confectionery, bakery items, and fast food and could offer a barrier to moisture and gas exchange, act as material to transfer tastes, colors, enzymes, antioxidants, and antibrowning agents, and may also increase the duration of synthetic foods, even after the packaging is opened (Azeredo, 2009). The properties of mango pure edible films may be substantially advanced through cellulose nanofibers reinforcement. Consumable antibacterial nanocoating can be applied directly to bakery objects (Naoto et al., 2009; Sekhon, 2010).

Nanocoatings are used for antimicrobial and self-cleaning food contact surfaces. The waxy coating is employed significantly for certain foods like cheeses and apples (Alparslan and Baygar, 2017; Azeredo, 2009). The abovementioned films and coatings can function as barriers against moisture, lipids and gases (Kuzma, 2010). However, it is noteworthy that the properties of the nanomaterials employed in layers determine the characteristics of the edible coatings in question (Weiss et al., 2006). Polysaccharides, proteins, and lipids are utilized in layers. It is possible to perceive various competencies primarily based on the kind of biopolymeric nanoparticles involved in the coating (Cha and Chinnan, 2004). Lipid-based totally layers function as moisture barriers. But, they do not have an impact on preventing gases, and that they have inadequate mechanical strength (Davarcioglu, 2017; Kuswandi, 2016). Components and functions of nano-based food packages used in food packaging are enlisted in Table 1.1.

TABLE 1.1 Components and Functions of Nano-Based Food Packages Used in Food Packaging.

Components nano-based food packaging	Their functions	References
Nanocomposites	- Preservation necessities of fresh foods by extending their shelf life	Paul and Robeson (2008)
	- Sensor system and detect volatile organic compounds	El-Wakil et al. (2015)
Bio-nanocomposites	- Increase performance and decrease food wastage	Malathi et al. (2014)
Nanoencapsulation	- Retaining the ingredients and additives during processing and storage	Momin et al. (2013)
	- Protecting unpleasant tastes and flavors - Controlling the release of components	Momin and Joshi (2015)
Nanolaminates	- Encapsulate numerous hydrophilic, amphiphilic, or lipophilic substances - Enhance the storage life and quality of coated foods	Weiss et al. (2006) Momin and Joshi (2015)
Nanoemulsions	- Facilitate much less fat use without compromising creaminess	Mason et al. (2006)
	- Decontamination of the food packaging system	Yuan et al. (2008)
	- Active against a variety of food pathogens	Sekhon (2010)
Edible nanocoatings	- Barriers moisture, lipid, and gas	Alparslan and Baygar (2017)
	- Increase the shelf life of synthetic foods, even after the packaging is opened	Azeredo et al. (2009)
	- Antimicrobial and self-cleaning food contact surfaces	Alparslan and Baygar (2017)

1.7 ENGINEERED NANOMATERIALS USED IN FOOD PACKAGING

A variety of engineered nanomaterials have been added to food packaging as useful additives such as silver nanoparticle (AgNP), nanoclay, nano-zinc oxide (nano- ZnO), nano-titanium dioxide (nano-TiO_2), and titanium nitride nanoparticle (nano-TiN) (Adeyeye and Fayemi, 2018; Tager, 2014). Because of differences in chemical structure and characteristics, every nanomaterial introduces distinct properties to the host material, which leads to extraordinary practical packaging applications (Marambio-Jones and Hoek, 2010; Rubilar et al., 2014). AgNPs are metallic silver atom clusters that are engineered usually for antimicrobial and sterilization functions (Brody, 2007). Due to the fact that AgNP has a larger surface area per mass than micro-scale silver particles or bulk silver material, the ability to release silver ions is also greater than that of bulk silver (Marambio-Jones and Hoek, 2010). Nanoclays and layered silicates are going on fine-grained minerals (Majeed et al., 2013).

The dispersal of sheet-established nanoclay into the polymer matrix creates the enhanced barrier properties of homogeneous polymer because of the increase of the tortuous pathway against penetrating molecules (Duncan, 2011). Metal oxide nanoparticles including zinc oxide and titanium dioxide are regularly used as photocatalysis agents to degrade natural molecules and microorganisms (Sun et al., 2009). The photocatalytic reaction of nano-ZnO and nano-TiO_2 attributes to the production of ROS, ensuing in the oxidation of cytoplasm of bacterial cells and resulting in deceased cell (Bodaghi et al., 2013). ZnO is especially more efficient and appealing over silver due to its less toxicity and cost-effectiveness (Duncan, 2011; Paul and Robeson, 2008). Nano-TiN, an approved food contact material by the European Food Safety Authority, is normally synthesized by heating TiO_2 particles in a nitrogen-containing gas at high temperatures. Nano-TiN is abundantly used for mechanical processing and strength aid predominantly for polyethylene terephthalate (Bumbud-sanpharoke and Ko, 2015; Chaudhry and Castle, 2011). Clay nanoparticles and nanocrystals enhance the water vapor barrier properties of dairy and food packaging materials that are improved by incorporating nanoclays and nanocrystals (Keshwani et al., 2015).

Polymers used for the fabrication of nanocomposites consist of polyurethane and polyamide (Appendini and Hotchkiss, 2002). The engineered nanoparticle polymer composites (also termed nanocomposites) are commonly reinforced with up to 5% (w/w) of nanoparticles, and this can deliver a radical enhancement in the performance and behavior of the polymer. Engineered nanosensors are discovered by Kraft along with Rutgers University within packages to modify the color to inform the user if the diet is beginning to spoil or has been contaminated by different pathogens using digital "noses" and "tongues" to "smell" or "taste" diets (Scrinis and Lyons, 2007; Sozer and Kokini, 2009). Nestle´, British airlines and Monoprix supermarkets have been using chemical nanosensors that can effortlessly detect color change (Momin and Joshi, 2015; Pehanich, 2006; Yao et al., 2016).

1.8 TYPES OF NANOTECHNOLOGY-BASED FOOD PACKAGING

These common nano-based food packagings have their own basic features in the food industry (Table 1.2).

1.8.1 ACTIVE FOOD PACKAGING

The packaging substances are polymeric compounds either combined or grafted and have distinct functionalities for permeation of gases or vapors (Arvanitoyannis and Bosnea, 2001). Current trends require making packaging active by incorporating sachets in the package or via the addition of an active agent that makes it environmentally pleasant with much less chemical utilization, which includes preservatives or components (El-Wakil et al., 2015). However, their essential benefit is for the stable food products and now not for the liquid and has other limitations too (Emanuel and Sandhu, 2019; Yam et al., 2005). In contrast to conventional food packaging, active packaging is a deliberately designed packaging system that incorporates components that might release (antimicrobial or antioxidant agents) or take in (water vapor or oxygen) substance from or into the packaged diet or the diet situation (Nile et al., 2020; Sharma et al., 2017).

Active packaging is described as active substances in contact with diet, with the capacity to alter the content of the diet or the situation around it (Restuccia et al., 2010; Rossi et al., 2017). Active packaging shows active purposes away from being a passive hurdle comprising and guarding the diet product (Gomez-Estaca et al., 2014). It deals with extending the shelf life of foods and improving their safety and quality through releasing or eradicating affluence from or into the packaged diet or the headspace adjoining the diet. Polymer nanocomposites applied to active foods packaging have observed exciting potential applications in particular within the development of antimicrobial films and oxygen scavengers (Cerisuelo et al., 2019; Khalaj et al., 2016).

TABLE 1.2 Common Types of Nanotechnology-Based Food Packaging and Their Basic Features.

Common types nanotechnology-based food packaging	Basic features	References
Active food packaging	- Antimicrobial, inhibits the growth of bacteria, fungi, and other food pathogens	Sundaram et al. (2016)
	- Antioxidant, prevents oxidative destruction of packaged food	Vital et al. (2016)
Smart/Intelligent food packaging	- Sensing and telling information about the packaged food to the consumers	Emanuel and Sandhu (2019)
	- Detects microbial and biochemical changes occurring in the food	Davarcioglu (2017)
Improved packaging	- Integrated into the polymer matrix to enhance the barrier properties of the packages	Kuswandi (2016)
Modified atmospheric packaging	- Conversion of the packaging environment by flushing in an appropriate combination of various gases	McMillin (2008) Costa et al. (2011)
	- Enhance shelf life and inhibit pathogens growth	
Oxygen scavenging packaging	- Maintain a low oxygen concentration to prevent metmyoglobin concentration	Emanuel and Sandhu (2019)
Biobased packaging	- Control moisture and/or gas exchange to enhance safety and preserve the food	Siracusa et al. (2008)

Active packaging technology can be divided into two categories. Ethylene, oxygen, carbon dioxide, excessive water, or different undesirable compounds can be eliminated by using absorbers or scroungers, which comprise the initial classification. The next class is the controlled releasing systems, which actively adds or emits compounds at suitable proportions to the packaged diets or into the headspace of the covering including antimicrobials, carbon dioxide, antioxidants, and preservatives to supply a continuous substitute of active compounds, hinder bacterial multiplication, and prolong product duration (Agriopoulou, 2016; Duncan, 2011).

An active food packaging affords a twofold advantage, acts as a barrier system for food spoilage microorganisms, and at the same time, it beneficially interacts with the food, as an instance, casting off the factors that cause food spoilage (including water vapor and oxygen), releasing useful compounds like antioxidants and antimicrobials. Such interactions bring about stability in the food and food packaging and food stability (Dasgupta and Ranjan, 2018). In active packaging structures, the safety function of a package is improved through incorporating into it active compounds consisting of antimicrobial compounds, preservatives, oxygen absorbers, water vapor absorbers, and ethylene removers (Mihindukulasuriya and Lim, 2014). Zinc oxide quantum dots have been applied as a powder, bound in a polystyrene film (ZnO-PVP) or suspended in a polyvinylpyrrolidone gel (ZnO-PVP), which is used as antimicrobial packaging for eradication of *Salmonella enteritidis, Listeria monocytogenes,* and *Escherichia coli* O157:H7 (Silvestre et al., 2011; Sun et al., 2009).

1.8.1.1 ANTIMICROBIAL PACKAGING

Antimicrobial packaging is definitely a form of active packaging that contacts the meal's product or the headspace internal to inhibit or retard the microbial increase that may be present on meal surfaces. Many nanoparticles that consist of copper, silver, chitosan, and metal oxide nanoparticles similar to zinc oxide or titanium oxide are known to have antibacterial property (Singh et al., 2017). Antimicrobial packaging is one of the most imperative active food packaging techniques. Antimicrobial packaging substances extend shelf life and promote safety by reducing a load of microorganisms in packaged food (Tan et al., 2013). Besides, antimicrobial packaging materials can be self-sterilizing or sanitizing. Such antimicrobial packaging materials notably lessen the capacity for the recontamination of processed food products and simplify the remedy of substances to remove product contamination (Jaiswal et al., 2019).

In general, silver can release silver ions from the silver particle (Marambio-Jones and Hoek, 2010) and bind to bacterial cell wall membrane, leading to the inactivation of the enzymes (Rossi et al., 2017). Nano-silver antimicrobial packaging applications are observed to be a unique method toward the preservation of food and its shelf life extension. The structural analysis explains the intercalation of silver and titanium dioxide nanoparticles of silver inside the 20–70 nm range, which is found to be within the nanoparticles bulk polymer that shows its antimicrobial activity (Keshwani et al., 2015). Antimicrobial packaging extends the lag-time growth of microbial and decreases the growth of microorganisms, increasing the shelf-life of the meal, and keeping the quality and safety of the meal. The antimicrobial packaging film can be conveyed into direct contact with the diet surface to vaporize the antimicrobial substance or distribute it to the diet. Numerous antimicrobial substances including carbon dioxide, silver ions, chlorine dioxide, bacteriocins, antibiotics, organic acids, ethanol, critical oils, and spices inhibit the growth of microorganisms in packaged foods (Malhotra et al., 2015); however, antimicrobial packaging is commercially restricted apart from silver nanomaterials (Jaiswal et al., 2019).

There are numerous methods in that antimicrobial packaging may be used. The diet byproducts are sealed off by using pads, sachets, or sheets comprising volatile antimicrobial substances into packages. These may be in the custom of moisture and oxygen absorbers or producing ethanol vapor. Moisture and oxygen absorbers avoid microbial growth by making anaerobic situations and accordingly reducing the water activity. They are typically used for packaging determination in pasta, bakery, and meat production to impede water condensation and oxidation (Appendini and Hotchkiss, 2002; Mills et al., 2006). Essential oil coating may be applied to the food products as a thin film having antimicrobial actions for the existence of phenolic compounds. They deliberately produce antimicrobial substances on to the diet surface or transfer to the surface and prolong the product duration, and prevent the occurrence of foodborne infection as a result of pathogenic microorganisms (Emanuel and Sandhu, 2019).

Copper nanoparticles have been proven to inhibit the growth of *Staphylococcus aureus*, *L. monocytogenes, E. coli,* and *Saccharomyces cerevisiae* on a polymer combination after 4 h exposure (Cioffi et al., 2005). Copper nanoparticles antibacterial effect against *E. coli* and *Bacillus subtilis* in polyurethane nanofibers containing copper nanoparticles. Copper nanoparticles lead to multiple toxic effects that include production of reactive oxygen species, DNA degradation, protein oxidation, and lipid peroxidation, and that may be accountable for its antimicrobial action (Chatterjee et al., 2014). Zinc nanocrystals are used as an antifungal and antimicrobial substances when integrated with the plastic matrix. Distinct nanoparticles oxide(s) consisting of titanium dioxide (TiO_2), zinc oxide (ZnO), silicon oxide (SiO_2), and magnesium oxide (MgO) also shows utility in food packaging because of their capacity to act as ultraviolet blockers and photocatalytic disinfecting agents. Among all, TiO_2 particles have been most promising (Farhoodi, 2016).

The antimicrobial activity of TiO_2 nanoparticles is photocatalyzed, and those antimicrobial substances are active merely in the occurrence of UV light. TiO_2 nanoparticles displayed excellent action against *Vibrio parahaemolyticus, S. choleraesuis,* and *L. monocytogenes* in ultraviolet illumination but not within the dark (Sharma et al., 2017). Silver salts have been historically used to treat burns and infections; lately, silver ions integrated into inorganic materials together with zeolites, zirconium phosphate, and glass were employed as effective antimicrobial agents. In this regard, the effectiveness of silver salts impregnated on nanofillers including nanoclays is being intensively studied for the improvement of antimicrobial nanocomposite films (Cerisuelo et al., 2019). The antimicrobial properties of metal oxide-based nanomaterials have attracted technological interest; also, silver nanoparticles have extra performance, decrease toxicity, and greater value effectiveness (Rhim et al., 2013; Rossi et al., 2017; Yadav et al., 2016).

1.8.1.2 ANTIOXIDANT PACKAGING

Reactive unfastened radicals like superoxide, peroxyl, hydroxyl, and alkoxyl and non-radicals inclusive of hypochlorous and hydrogen peroxide cause the oxidative destruction of lipids, nucleic acids, and proteins thereby ensuing in alteration of product color and texture, toxic aldehyde development, off-flavors development, and reduction of dietary importance of the product. This is the main source of the decline in the duration of diets, mainly, chicken, meat, and seafood (Arora and Padua, 2010; Vital et al., 2016). This can be conquered by the usage of an antioxidant packaging method that may be in the form of an autonomous sachet or pad packed beside the diet product or by fusion of the active element in the packaging substance that may either rivet the unwanted compound from the diet or release the antioxidant into the diet packages (Vera et al., 2016). The sachet or pad includes a compound or combinations of compounds that both retain or release a specific vapor or gas whose presence or absence is useful for the food. The use of phenolic compounds like tannic acid, caffeic acid, and green tea extracts improves the barrier properties of biodegradable films formed from gelatin and turmeric (Zhao et al., 2008). They divulged higher tensile power, little water vapor absorptivity, and water solubility of films for the formation of cross-linking. These films when used for the packing of fresh powdered pork prolonged their duration because of the antioxidant action of the extra phenolics by hindering the lipid oxidation of the food samples (Emanuel and Sandhu, 2019).

1.8.2 SMART/INTELLIGENT FOOD PACKAGING

Smart packaging deals with packaging that comprises an internal or external indicator to give data about constituents of the quality of the food and/or the history of the package (Dobrucka and Cierpiszewski, 2014; Emanuel and Sandhu, 2019). Smart packaging responds to environmental conditions or repairs it or alert a customer to contamination and/or the presence of pathogens. Chemical Giant Bayer (Leverkusen, Germany) produces a transparent plastic film (Durethan) containing nanoparticles of clay. The nanoparticles are dispersed throughout the plastic and can block oxygen, carbon dioxide, and moisture from reaching fresh meats or different foods. The nanoclay additionally makes the plastic lighter, more potent, and greater heat resistant (Sekhon, 2010).

Smart packaging permits tracking and tracing a product to control and study outside the package to notify its producer, store, or customer on the product's situation at any duration. It is used to prolong shelf life, express information on quality, display freshness, and enhance product and customer safety (Schaefer and Cheung, 2018). An intelligent package consists of devices that do not alter the product; however, it can monitor the situation of the product, package, or packaging environment (Rossi et al., 2017). Smart/intelligent packaging detects microbial and biochemical changes occurring in the food, including recognizing certain microorganisms that are found in diet products, in addition to unique gases emerging from food deterioration. Furthermore, positive smart packaging is used as a means of tracking food security or avoiding fake food products. Nanobiosensors are used to reveal the quality of foodstuffs, and nano-bioswitches are used to release preservatives (Davarcioglu, 2017; Scrinis and Lyons, 2007).

The hallmark of intelligent packaging systems is to enhance the communication aspect of a package, such as to dynamically reflect the actual quality of the food in actual time, as opposed to the static "best before" and "use by" dating approach. This type of advanced packaging is beneficial to increase the efficacy of information transfer during the product distribution chain via innovative communication methods such as intelligent tags, radiofrequency identification tags, time–temperature indicators, oxygen and carbon dioxide sensors, and freshness indicators (Mihindukulasuriya and Lim, 2014). Intelligent packaging systems are gaining significance these days with food protection becoming an important function linked to the product (Emanuel and Sandhu, 2019). The growths of polymer nanomaterials for smart diet packaging comprehend putrefaction indicators, product identification, and traceability (Kuswandi, 2016).

Intelligent or smart packaging systems enhance the communication aspect of a package and such kind of advanced packaging may detect any behavior of the packaged diet and observe specific mechanisms to record and bring relevant information regarding the situation of the food including its protection and digestibility. Such a device makes use of exclusive modern communication methods such as nano-sensors, time–temperature indicators, oxygen sensors, and freshness indicators (Ahvenainen, 2003). The inclusion of nanosensors in food packaging systems facilitates in detecting the spoilage-related changes, pathogens, and chemical contaminants, and accordingly providing the precise situation of food products freshness (Liao et al., 2005). Nanosensors are nanotechnology-enabled sensors that are characterized through many variations. Typically, nanosensors can be implemented as markers or coverings to augment an intelligent purpose to diet packaging in relation of making guaranteed the veracity of the package by detection of trickles (foodstuffs packed in the inert atmosphere or vacuum), warning signs of time–temperature variations (freeze–thaw–refreeze) or microbial protection (the deterioration of foodstuffs) (Dong et al., 2011; Kerry et al., 2006; Mannino and Scampicchio, 2007). Gas sensors are used for revealing the gaseous analyte inside the package. Optical oxygen sensors work on the edict of luminescence appeasing or absorbance fluctuations caused by contact with the analyte; while photochemical sensors are used to plaid the eminence of yields by sensing thrill analytes, including carbon dioxide, hydrogen sulfide, and volatile amines (Marsh and Bugusu, 2007). Multi-detection test, Food Expert ID, has been developed by bio-Merieux for a nano-surveillance response to food scares. Nanoscale radiofrequency identification labels are advanced to trail basins or different foods items (Fang et al., 2017) and are being utilized in retailing chains (Asadi and Mousavi, 2006; Metak et al., 2015). The nanotech business enterprise SiNutria has also developed nano-based tracking technologies, which include an ingestible BioSilicon, which can be placed in foods for monitoring functions and pathogen detection but can also be consumed by users (Scrinis and Lyons, 2007).

1.8.3 IMPROVED PACKAGING

In improved packaging development, nanomaterials are integrated into the polymer matrix to increase the gas blockade behavior, also humidity and temperature confrontation of the packaging material. A variety of nanoparticle reinforced polymers, additionally termed as nanocomposites have been advanced, which generally incorporate up to 5% (w/w) nanoparticles through clay nanoparticle composites with better blockade behavior for the production of palatable oils, carafes for beer, carbonated food and drink, and films (Chaudhry and Castle, 2011; Kuswandi, 2016). The usage of nanocomposite having contact with food has been approved by the United States Food and Drug Administration (USFDA) (Duncan, 2011; Pal, 2017; Silvestre et al., 2011).

Nanoparticle-reinforced substances termed nanocomposites are polymers reinforced with small portions (up to 5% by weight) of nanosized particles that have been developed. These nanocomposites have high aspect ratios and can enhance the properties and overall performance of the polymer. Polymer composites with nanoclay are among the first nanocomposites available in the market as improved materials for food packaging (Schaefer and Cheung, 2018). Nanoclay with a natural nanoscaled layer structure restricts the permeation of gases. Nanoclay–polymer composites have been made from a thermoplastic polymer strengthened with nanoparticles of clay that include polyamides, polyethylene terephthalate, nylons, polyolefins, polyurethane, polystyrene, epoxy resins, ethylene-vinyl acetate copolymer, and polyimides. Commercially, some nanoclay–polymer composites are available. Known applications of nanoclay in multilayer film packaging consist of bottles for beer, edible oils, and carbonated beverages and films (Scrinis and Lyons, 2007).

1.8.4 MODIFIED ATMOSPHERIC PACKAGING

Modified atmospheric packaging (MAP) involves the conversion of the packaging environment either by flushing in an appropriate combination of various gases or by changing the permeability of the package to attain a certain particular composition of gases to acquire an enhanced shelf life, inhibit mold growth, and numerous other benefits (McMillin, 2008). MAP may be classified as an active or passive MAP. Active MAP entails tinting the package through a stable arrangement of gases to preclude the diet product from revelation to gases that may decrease their duration or consequence in microbial development. This conformation is then sustained by the assistance of blockade packaging. However, the passive packaging uses the respiration rate of the product in addition to the permeability of the package to attain the favored steady gas composition (Costa et al., 2011; Emanuel and Sandhu, 2019; Pal et al., 2019).

1.8.5 OXYGEN SCAVENGING PACKAGING

Oxygen is accountable for the deterioration of different diets whichever indirectly or directly. For example, direct oxidation returns consequences in the frying of fruits and spoilage of vegetable oils. Diet spoilage by indirect action of oxygen consists of food spoilage through aerobic microorganisms. Therefore, the incorporation of oxygen scavengers into food packages can maintain a very low oxygen level, which is beneficial for several applications since it will enhance the shelf-life of the food (Malhotra et al., 2015). Oxygen scavenger films have been successfully advanced by including titanium dioxide nanoparticles (TiO_2) in different polymers (Asadi and Mousavi, 2006). So, they are used for packaging a large diversity of oxygen-sensitive foodstuffs. Care has mostly been engrossed in the photocatalytic action of nanocrystalline titania beneath ultraviolet radiation. TiO_2 acts by a photocatalytic mechanism, and its principal drawback are the requirements of UVA light (Kuswandi, 2016; Mills et al., 2006).

Oxygen participates in several sorts of food deterioration. Low levels of oxygen may be maintained by incorporating oxygen scavenging systems into food packaging. Low levels of oxygen prevent direct oxidation reactions that result in browning reactions and rancidity, and food deterioration by indirect action of oxygen occurs by the action of aerobic microorganisms. Xiao et al. (2004) developed oxygen scavenger films by adding TiO_2 nanoparticles to exclusive polymers. Many oxygen-sensitive food products can be preserved by the use of nanocomposite substances as packaging films. The main drawback of using TiO_2 is that it requires ultraviolet-activated light for its photocatalytic action (Mills et al., 2006). Commonly used polymer polythene is known for moisture and barrier applications. However, it indicates poor oxygen barrier properties and is unsuitable to package oxygen-sensitive food products. To conquer this drawback, iron-containing kaolinite has been integrated into the high-density polythene films to create an oxygen barrier (Dasgupta and Ranjan, 2018).

Oxygen scavengers in conjunction with modified environment packaging have been used in the meat industry to maintain the color of fresh diets and to improve the optical satisfactoriness of meats (Tewari et al., 2002). The oxygen scavengers in the package are used to maintain a low oxygen concentration (0.1%) to prevent metmyoglobin concentration (Emanuel and Sandhu, 2019). These days, both producers and consumers require an entire absence of O_2 in oxygen-free food packaging. Therefore, there exists a big demand for nontoxic oxygen-free packaging systems and irreversible O_2 sensors with packaging accomplished under nitrogen or vacuum. One such instance of an O_2 sensor is an ultraviolet-activated colorimetric O_2 indicator developed (Dasgupta and Ranjan, 2018). An ideal oxygen scavenger relies on the O_2 trapped within the food, O_2 level in the product, and equipment packaging absorptivity. In tallying to being able to rivet a huge amount of oxygen, the scavenger has to be harmless to the human body and be economical (Cerisuelo et al., 2019). Oxygen scavenging packaging using enzymes among polyethylene films has also been developed (Lopez-Rubio et al., 2006).

1.9 INDICATORS USED IN ACTIVE FOOD PACKAGING MATERIALS

1.9.1 *CARBON DIOXIDE ABSORBERS AND RELEASERS*

Even though carbon dioxide is used in modified atmospheric packaging, the accumulation of carbon dioxide in the food can harm the quality of the food by-product. This is a cause of essential concern in the case of carbon dioxide-producing foods including fermented foods and fresh produce. CO_2 levels above a limitation can cause physiological damage to the produce. Calcium hydroxide is the most common CO_2 scavenger used in food packaging systems as it reacts with CO_2 to produce calcium carbonate and release water as a by-product. Magnesium hydroxide, sodium chloride, magnesium oxide, and sodium carbonate are some of the additional substances that absorb CO_2 based on a chemical reaction (Emanuel and Sandhu, 2019; Pal et al., 2019).

1.9.2 MOISTURE REGULATORS

Packaged food products have interaction with their package surroundings and can gain or lose wetness hence distressing the equilibrium relative humidity of the product. Both the reduction of moisture and moisture acceptance undesirably upset the product duration and its appropriateness (Pal et al., 2019). Building up of moisture in minimally processed food can lead to pathogen growth as well as dietary and sensory loses to the food. Moisture regulation machinery controls the temperature oscillations in the package. It desorbs or absorbs the moisture to soothe the whole moisture content of the package to the wanted levels (Emanuel and Sandhu, 2019).

1.9.3 DETECTION OF SPOILAGE AND PATHOGENIC MICROORGANISMS

Antibodies combined with nanomaterials, such as quantum dots, have been developed to identify bacteria. The rationale for using quantum dots is large because of their high fluorescence efficiency, stability against photobleaching, long decay lifetime, and higher sensitivity (Mihindukulasuriya and Lim, 2014; Pal et al., 2019).

1.9.4 TIME, TEMPERATURE, AND HUMIDITY INDICATORS

Temperature abuses encountered by using food during distribution are one of the environmental factors that lead to a reduction in product shelf-life. For many products, raising the temperature will increase the rate of deterioration reaction, unwanted phase change, and increase in water activity, all of which will lead to quality issues (Pal et al., 2019). For others, a decrease in temperature below a critical level can cause product chill injury or undesirable phase change that causes structural damages. By and large, the impact of temperature is accumulative, depending on the extent to which the product is exposed to the abuse. Time-temperature indicators (TTIs) are useful to track thermal history throughout diet handling, storage, and delivery (Mihindukulasuriya and Lim, 2014).

1.9.5 FRESHNESS INDICATORS

Freshness indicators provide actual-time information to the manufacturer, retailer, and/or consumers on the actual product quality at the time of storage and distribution. Freshness indicators depend upon the detection of marker spoilage compounds or microbial metabolites produced as a food product spoiled, including volatile sulfides and amines. A transition metal (silver and/or cupper) coating of 1–10 nm thick on plastic film or paper is used in packaging systems. Upon reacting with sulfide volatiles produced as fresh meats undergo spoilage, the thin coating turns into distinctive dark color (Mihindukulasuriya and Lim, 2014).

The actual time data of products that include actual product safety and quality during storage and distribution can be provided by freshness indicators to the manufacturer, retailer, and/or consumers. Food products may be tested to check whether a food is spoiled or not by detecting the metabolites released by the spoilage microorganisms like volatile sulfides and amine and these metabolites or compounds affords in food products while food products are spoiled (Shafiq et al., 2020). The different dark color appears while this thin coating reacts with sulfide volatiles components found in spoiled food products. Peptide receptor-based transportable bioelectronic nose is created to test the freshness of food by detecting trimethylamine, which presents only when the raw seafood is spoiled (Dasgupta and Ranjan, 2018; Jh et al., 2016).

1.10 DIFFERENT NANOSENSORS USED FOR FOOD PACKAGING

The use of nanosensors in food packages helps in the determination of the nutrient content of food along with its quality (Keshwani et al., 2015). Nanosensors integrated with nanofibrils of perylene-based fluorophores can verify spoilage by detecting gaseous amines in fish and meat (Dasgupta and Ranjan, 2018). Nanosensors together with polymers are used to screen food pathogens and chemical substances during storage and transit procedures in smart packaging. Additionally, smart packaging confirms the integrity of the food package and the authenticity of the food product (Pathakoti et al., 2017; Thiruvengadam et al., 2018). Nanosensors assist in detecting any type of alteration in the color of the diet and it additionally helps in the revealing of any wastes formed due to spoilage. The sensors are generally sensitive towards gases including hydrogen, hydrogen sulfide, nitrogen oxides, sulfur dioxide, and ammonia (Pradhan et al., 2015).

Packaging containing nanosensors is used to present information of enzymes produced inside the breakdown of diet substances making them perilous for human intake. These packages can also be used to permit air and other enzymes out but not in, therefore increasing shelf life, as well as the reduction of man-made preservatives in our foods. Some other imperative impending application of nanoparticles in diet packaging is the breakage of ripening gas, including ethylene (Hu and Fu, 2003; Sekhon, 2010). Packaging equipped with nanosensors is likewise intended to trail either the external or internal situations of diet products, containers, and pallets, all over the supply series. For example, such packaging may display humidity or temperature over time after which offer relevant information on these situations, for instance by changing color. Nanosensors in plastic packaging can detect gases given off by food when it spoils and the packaging itself modifies color to alert the consumer. The so-called nanosensors are capable of responding to environmental changes (temperature or humidity in storage rooms, levels of oxygen acquaintance), microbial spoilage, or deprivation products (Bouwmeester et al., 2009; Kuswandi, 2016).

The incorporation of nanosensors in food packaging plays a critical role. Nanosensors identify the spoilage of diet by detecting the alteration in the color of the product or by detecting the gas produced because of spoilage of the product including hydrogen, hydrogen sulfide, nitrogen oxides, sulfur dioxide, and ammonia. The sensors are prepared from a data processing unit beside a detecting portion which may either sense the gas formation, color change, or chemical substance changing it into electronic signals. Nanosensors are used in the recognition of pesticides in fresh vegetables and fruits, cancer-causing agents in food materials, adulteration in food and beverages, pathogens such as Staphylococcal enterotoxin B and cholera toxin (Rai et al., 2012), detection of volatile organic compounds and temperature abuse indication. Nanosensors used for food packaging may be steel (platinum, palladium, and gold) dependent nanosensors used for recognition of aflatoxin B1 in the milk, detection of color change or gas formation in the diet because of spoilage or the encounter of any alter in heat, light, humidity or gas into electric indicators. Single-walled carbon nanotubes and DNA have been used in the detection of pesticides on fruits and vegetable surfaces (Sozer and Kokini, 2009). Carbon black and polyaniline nanosensors have been used for detecting carcinogens present in the food substances, detection of microorganisms that infest the food, and for detection of foodborne pathogens. Surface plasmon-coupled emission biosensor (with Gold) has also been used to detect pathogenic organisms in food (Emanuel and Sandhu, 2019).

Metal-based nanosensors(palladium, platinum, and gold) are used for detection of any type of change in the color of the food, Detection of any gases being produced because of spoilage (Kang et al., 2007), Detection of any alternate in light, heat, humidity, gas, and chemicals into electrical indicators (Fernández et al., 2008; Mao et al., 2006), detection of toxins which include aflatoxin B1 in milk (Meetoo, 2011; Pradhan et al., 2015); carbon black and polyaniline are used detection cancer-causing agents found in the food substances (Vidhyalakshmi et al., 2009), foodborne pathogens (Yuan et al., 2008) and microorganisms that generally contaminate the food; array biosensors, electronic noses, nano-test strips, and nanocantilevers are used for detection of changes in color on coming in contact with any sign of contamination in the food material (Biswal et al., 2012); nanobiosensors are used for detection of the viruses and the bacteria , and nano-smart dust is used for detection of any sort of environmental contamination (Coles and Frewer, 2013).

1.11 NANOPARTICLES USED FOR FOOD PACKAGING AND PRESERVATION

Through the addition of definite nanoparticles into packaging equipment and bottles, packages of diet can be made lighter, with stronger mechanical and thermal properties, in addition to fire-resistant (Keshwani et al., 2015). The incorporation of nanomaterials into food packaging gives three different advantages to food packaging: Barrier resistance, Incorporation of active components to offer functional overall performance, and Sensing of relevant information. Organic or inorganic nanoparticles might not be used in food systems directly but may be utilized in coating or packaging substances to get the antimicrobial effect without sacrificing the quality of the packaged food (Jaiswal et al., 2019).

Silicon dioxide is used in minimizing the leakage of moisture (Coma, 2008), anticaking, and desiccating agent (Horner et al., 2006), and absorbing the water molecules in food, showing hygroscopic importance and titanium dioxide acts as a food colorant (Jones et al., 2008), and photocatalytic sterilizing agent (Zhao et al., 2008), and also used as a food whitener for food harvests such as milk, and other dairy merchandises (Acosta, 2009); Zinc oxide decrease the movement of oxygen in the packaging basins; silver nanoparticles act as an antibacterial agent and protect the food from microbial infestation (Arshak et al., 2007), extend the shelf life of the vegetables and fruits by gripping and disintegrating ethylene; inorganic nanoceramic is used in catering oil for deep-frying diet, and polymeric nanoparticles are identified to be proficient conveyance schemes and are bactericidal (Bouwmeester et al., 2009).

1.12 PUBLIC ACCEPTANCE OF NANOTECHNOLOGY INNOVATIONS FOOD PACKAGING

Consumer's attitudes to, and acceptance of, emerging technologies and their packages are crucial determinants of their successful implementation and commercialization (Giles et al., 2015). Nanotechnology is more recognized in diet packaging in preference to be incorporated into food products. Trust and confidence in agri-food nanotechnology desire to be fostered, to increase consumer acceptance (Asadi and Mousavi, 2006). Presenting evidence to users on the remunerations of nanotechnology, and making sure an up-to-date community could aid to decrease customer anxiety and might encourage food nanotechnology purchases. But future research is needed to recognize what consumers perceive as beneficial, in addition to how they understand by perils. Espousing theoretically buttressed methods to understanding customer insights and approaches simplify relative investigation across many groups of clients, different diet nanotechnology packages, and permit evaluation of trends in consumer priorities and concerns with time (Giles et al., 2015).

Although the usage of nanomaterials is growing in food packaging applications, the situation of toxicity affects consumer perceptions and acceptance. Nanoparticles can travel from packaging into the diet. Even though the observed amount of nanomaterial migration is lower than the migration limit in legislation, the regulation was written only for general materials, and the table does not cover all the forms of nanomaterials, which exist in the market. Until their safety has been fully established, the public may still have questions as to their possible fitness impact. Numerous media outlets and non-governmental organizations (NGOs) have brought up this issue via their communication channels (Bumbudsanpharoke and Ko, 2015).

1.13 LIMITATIONS OF NANOTECHNOLOGY IN FOOD PACKAGING

The essential disadvantage related to food packaging substances is their absorbent property. None of the packaging substances provides thorough resistance to the water vapors, atmospheric gases, and diet and packaging substances. Among the accessible alternatives include organic polymer substances viz. polypropylene, polyethylene, polyethylene terephthalate, polystyrene, and polyvinyl chloride, stay the main preference material within the meals packaging enterprise, favored because of their lesser price, easy processing, and lightweight. Nevertheless, the crucial problem respites in their essential absorptivity to gases and different small molecules (Sharma et al., 2017). Non-sustainable production, lack of recyclability, inadequate mechanical, and barrier properties are some of the continued challenges faced by the food and packaging industries. Although metal and glass are good barrier substances to prevent unwanted mass transport in food packaging, plastics are still popular because of their lightweight, formability, value effectiveness, and versatile characteristics. As a result, the packaging industry consumes greater than 40% of the plastics through half of it for diet packaging (Rhim et al., 2013). Conversely, the widely held packaging substances are petroleum-based and not maintainable from a feedstock supply perspective. Furthermore, petroleum-based plastics are not biodegradable (Arvanitoyannis and Bosnea, 2001; Mihindukulasuriya and Lim, 2014).

Weak barrier properties to water vapor and gases are important points in food packaging. For instance, fresh goods that are alive including vegetables, fruits, and meats need to be packaged in O_2 permeable substances with an optimal transmission rate, while processed products do not need such large transfer. Matching barrier presentations with explicit products to prolong their shelf-life with the limited product thermoplastics accessible is challenging. Even though polymer blends and multilayered composite structures have been developed to increase the functional characteristics of thermoplastics, problems remain unresolved such as high cost and difficulty in recycling these materials (Mihindukulasuriya and Lim, 2014).

Another problem encountered by many food producers is to attain an adequate shelf-life for their products while sustaining the optimal quality and safety of the products. This is large for producers in developing countries where there is a lack of sufficient food distribution and preservation infrastructures (Xiao et al., 2004). The growth of microorganisms due to contamination and temperature abuse, decrease of nutritional qualities because of oxidation, loss of organoleptic and dietary qualities due to interaction with deleterious extrinsic factors including light, oxygen, and water are some of the known food quality and safety issues that still need to be overcome. Previously, intensive study and development activities in the academic and industry have been focused on manipulating nanotechnology to address many of the challenges (Mihindukulasuriya and Lim, 2014).

Consumers most of the times pay more for better quality foods and conveniently packed foods. The rapid improvement of nanotechnology affects inevitable human acquaintance to nanomaterials. While there are numerous studies on the improvement of nanomaterials in food packaging, few kinds of research occur on potential toxicity instigated to human health. By means of it disquiets the carbon nanotubes, there is some information suggesting that carbon nanotubes may additionally have a cytotoxic influence on human cells, at least during interaction through skin or lungs (Prasad et al., 2014; Warheit et al., 2004). When occuring in the diet packaging substance, the nanotubes may ultimately migrate into food. Consequently, it is obligatory to recognize any potential health effects of ingested carbon nanotubes. Besides, the migration of materials from packaging material to food is a perilous concern and can upset the wellbeing of the diet, triggering unlimited concern to the customer (Agriopoulou, 2016). The incorporation of nanomaterials into foods affords an entire novel collection of perils for the community, personnel in the diet industry, and farmers because chemically they are greater reactive than larger particles with greater contact to our bodies than the greater constituent part. They have enhanced toxicity because of greater bioavailability and even compromise our resistance response and can have pathological results in the long term (Chaudhry and Castle, 2011).

1.14 CONCLUSION

Food packaging is one of the very important areas in the food industry. Nanotechnology-based food packaging is a new area of packaging method which relies upon nanomaterials, has grown to be one of the fastest-growing elements in nanotechnology, and denotes a critical opportunity and advancements over the traditional food packaging system. Recently, packaged foods should fulfill the consumer's demands. There are different components of nanotechnology-based food packages including nanocomposites, nanoencapsulation, decomposable polymer films, nanoemulsions, nanolaminates, and eatable nanocoatings, in which each of them has its own basic functions in food packaging. Additionally, A variety of engineered nanomaterials have been added to food packaging as useful additives such as silver nanoparticles, nano-zinc oxide, titanium nitride, and nano-titanium dioxide nanoparticles may be available in diet packaging. Nanofood packaging can be categorized as active, smart/intelligent, improved, modified, oxygen scavenger, and biobased packaging. An essential packaging material needs to have gas and moisture permeability integrated with strength and biodegradability. Nano-based smart and lively food packaging confer many benefits over conventional packaging methods from offering higher packaging material with superior mechanical strength, barrier properties, and antimicrobial films to nanosensing for pathogen detection and alerting clients about the safety condition of the meals. Nanoparticles and materials used in nanofood packaging can detect and sense carbon dioxide, temperature change, humidity, time, pathogenic microorganism contamination, and freshness of the packaged food. Although it has many advantages in the food sectors, there are some limitations of nano-based food packaging, which should be improved for the future.

KEYWORDS

- **food packaging**
- **food quality**
- **nanotechnology**
- **nanoparticles**

REFERENCES

Acosta, E. Bioavailability of Nanoparticles in Nutrient and Nutraceutical Delivery. *Curr. Opin. Coll. Interf. Sci.* **2009,** *14* (1), 3–15.

Adeyeye, S. A. O.; Fayemi, O. E. Nanotechnology and Food Processing: Between Innovations and Consumer Safety. *J. Culinary Sci. Technol.* **2018,** *7* (5), 435–452.

Agriopoulou, S. Nanotechnology in Food Packaging. *EC Nutr.* **2016,** *5* (2), 1137–1141.

Ahvenainen, R. *Novel Food Packaging Techniques*; Wood Head Publishing: Cambridge, UK, 2003; pp 536–541.

Alparslan, Y.; Baygar, T. Effect of Chitosan Film Coating Combined with Orange Peel Essential Oil on the Shelf Life of Deepwater Pink Shrimp. *Food Bioprocess Technol.* **2017,** *10* (5), 842–853.

Appendini, P.; Hotchkiss, J. H. Review of Antimicrobial Food Packaging. *Innov. Food Sci. Emerg. Technol.* **2002,** *3* (2), 113–126.

Arora, A.; Padua, G. W. Review: Nanocomposites in Food Packaging. *J. Food Sci.* **2010,** *75* (1), R43–R49.

Arshak, K.; Adley, C.; Moore, E.; Cunniffe, C.; Campion, M.; Harris, J. Characterization of Polymer Nanocomposite Sensors for Quantification of Bacterial Cultures. *Sens. Actuat. B: Chem.* **2007,** *126* (1), 226–231.

Arvanitoyannis, I. S.; Bosnea, L. A. Recycling of Polymeric Materials Used for Food Packaging: Current Status and Perspectives. *Food Rev. Int.* **2001,** *17* (3), 291–346.

Asadi, G.; Mousavi, M. Application of Nanotechnology in Food Packaging. In *13th World Congress of Food Science and Technology*, 2006 p 739.

Assis, R. Q.; Lopes, S. M.; Costa, T. M. H.; Flôres, S. H.; de Oliveira Rios, A. Active Biodegradable Cassava Starch Films Incorporated Lycopene Nanocapsules. *Indust. Crops Prod.* **2017,** *109*, 818–827.

Azeredo, H. C.; Mattoso, L. H.; Wood, D. F.; Williams, T. G.; Avena-Bustillos, R. D.; Mc Hugh, T. H. Nanocomposite Edible Films from Mango Puree Reinforced with Cellulose Nanofibers. *J. Food Sci.* **2009,** *74* (5), N31–N35.

Azeredo, H. M. C. Nanocomposites for Food Packaging Applications. *Food Res. Int.* **2009,** *42*, 1240–1253.

Azeredo, H. M. C.; Rosa, M. F.; Mattoso, L. H. C. Nanocellulose in Bio-Based Food Packaging Applications. *Indust. Crops Prod.* **2016,** *97*, 664–671.

Baeumner, A. Nanosensors Identify Pathogens in Food. *Food Technol.* **2004,** *58* (8), 51–55.

Biswal, S. K.; Nayak, A. K.; Parida, U. K.; Nayak, P. L. Applications of Nanotechnology in Agriculture and Food Sciences. *Int. J. Sci. Innov. Discov.* **2012,** *2* (1), 21–36.

Bodaghi, H.; Mostofi, Y.; Oromiehie, A.; Zamani, Z.; Ghanbarzadeh, B.; Costa, C. et al. Evaluation of the Photocatalytic Antimicrobial Effects of a TiO_2 Nanocomposite Food Packaging Film by In Vitro and In Vivo Tests. *LWT-Food Sci. Technol.* **2013,** *50* (2), 702–706.

Bouwmeester, H,; Dekkers, S.; Noordam, M. Y.; Hagens, W. I.; Bulder, A. S., de Heer, C. et al. Review of Health Safety Aspects of Nanotechnologies in Food Production. *Regulat. Toxicol. Pharmacol.* **2009,** *53* (1), 52–62.

Bratovcic, A.; Odobaši´c, A.; ´C ati´c, S.; Šestan, I. Application of Polymer Nanocomposite Materials in Food Packaging. *Croatian J. Food Sci. Technol.* **2015,** *7* (2), 86–94.

Brody, A. L. Case Studies on Nanotechnologies for Food Packaging. *Food Technol.* **2007,** *07*, 102–107.

Bumbudsanpharoke, N.; Ko, S. Nano-Food Packaging : An Overview of Market, Migration Research, and Safety Regulations. *J. Food Sci.* **2015,** *80* (5), 910–923.

Burdo, O. G. Nanoscale Effects in Food-Production Technologies. *J. Eng. Phys. Thermophys.* **2005,** *78* (1), 90–96.

Cerisuelo, J. P.; Hernandez-Munoz, P.; Domínguez, I.; Gracia, L.; Catala R.; Gavara, P. R. Nanotechnology in Food Packaging. *Nanomater. Food App.* **2019,** 205–232.

Cha, D. S.; Chinnan, M. S. Biopolymer-Based Antimicrobial Packaging: A Review. *Crit. Rev. Food Sci. Nutr.* **2004,** *44* (4), 223–237.

Chatterjee, A. K.; Chakraborty, R.; Basu, T. Mechanism of Antibacterial Activity of Copper Nanoparticles. *Nanotechnology* **2014,** *25*, 135101.

Chaudhry, Q.; Castle, L. Food Applications of Nanotechnologies: An Overview of Opportunities and Challenges for Developing Countries. *Trends Food Sci. Technol.* **2011,** *22* (11), 595–603.

Chellaram, C.; Murugaboopathi, G.; John, A. A.; Sivakumar, R.; Ganesan, S.; Krithika, S.; Priya, G. Significance of Nanotechnology in Food Industry. *APCBEE Procedia* **2014,** *8*, 109–113.

Cioffi, N.; Torsi, L.; Ditaranto, N.; Tantillo, G.; Ghibelli, L.; Sabbatini, L. et al. Copper Nanoparticle/Polymer Composites with Antifungal and Bacteriostatic Properties. *Chem. Mater.* **2005,** *17*, 5255–5262.

Coles, D.; Frewer, L. J. Nanotechnology Applied to European Food Production—A Review of Ethical and Regulatory Issues. *Trends Food Sci. Technol.* **2013,** *34* (1), 32–43.

Coma, V. Bioactive Packaging Technologies for Extended Shelf Life of Meat-Based Products. *Meat Sci.* **2008,** *78* (2), 90–103.

Costa, C.; Lucera, A.; Conte, A. Mastromatteo, M.; Speranza, B.; Antonacci, A.; Del Nobile, M. A. Effects of Passive and Active Modified Atmosphere Packaging Conditions on Ready-to-Eat Table Grape. *J. Food Eng.* **2011,** *102* (2), 115–121.

Couch, L. M.; Wien, M.; Brown, J. L.; Davidson, P. Food Nanotechnology: Proposed Uses, Safety Concerns and Regulations. *Agro Food Indust. Hi Technol.* **2016,** *27*, 36–39.

Danie, K. J.; Shivendu, R.; Nandita, D.; Proud, S. Nanotechnology for Tissue Engineering: Need, Techniques and Applications. *J. Pharm. Res.* **2013,** *6*, 200–204.

Dasgupta, N.; Ranjan, S. Nanotechnology in Food Packaging. *Environ. Chem. Sustain. World* **2018,** 129–150.

Davarcioglu, B. Nanotechnology Applications in Food Packaging Industry. *Nanotechnology* **2017,** 87–113.

Dave, D.; Ghaly, A. E. Meat Spoilage Mechanisms and Preservation Techniques: A Critical Review. *Am. J. Agric. Biol. Sci.* **2011,** *6* (4), 486–510.

Del Nobile, M. A.; Conte, A.; Buonocore, G. G.; Incoronato, A. L.; Massaro, A.; Panza, O. Active Packaging by Extrusion Processing of Recyclable and Biodegradable Polymers. *J. Food Eng.* **2009,** *93* (1), 1–6.

Dobrucka, R.; Cierpiszewski, R. Active and Intelligent Packaging Food—Research and Development—A Review. *Polish J. Food Nutr. Sci.* **2014,** *64* (1), 7–15.

Dong, S.; Chen, X., Gu, L.; Zhou, X.; Xu, H.; Wang, H.; et al. Facile Preparation of Mesoporous Titanium Nitride Microspheres for Electrochemical Energy Storage. *ACS Appl. Mater. Interf.* **2011,** *3* (1), 93–98.

Duncan, T. V. Applications of Nanotechnology in Food Packaging and Food Safety: Barrier Materials, Antimicrobials and Sensors. *J. Colloid Interf. Sci.* **2011,** *363* (1), 1–24.

Eleftheriadou, M.; Pyrgiotakis, G.; Demokritou, P. Nanotechnology to the Rescue: Using Nano-Enabled Approaches in Microbiological Food Safety and Quality. *Curr. Opin. Biotechnol.* **2018,** *44*, 87–93.

El-Wakil, N. A.; Hassan, E. A.; Abou-Zeid, R. E.; Dufresne, A. Development of Wheat Gluten Nanocellulose Titanium Dioxide Nanocomposites for Active Food Packaging. *Carbohydrate Polym.* **2015,** *124*, 337–346.

Emanuel, N.; Sandhu, H. K. Food Packaging Development: Recent Perspective of Food Packaging Development: Recent Perspective. *J. Thin Films Coat. Sci. Technol. App.* **2019,** *6* (3), 13–29.

Fang, Z.; Zhao, Y.; Warner, R. D.; Johnson, S. K. 2017. Active and Intelligent Packaging in Meat Industry. *Trends Food Sci. Technol.* **2017,** *61*, 60–71.

Farhoodi, M. Nanocomposite Materials for Food Packaging Applications: Characterization and Safety Evaluation. *Food Eng. Rev.* **2016,** *8* (1).

Fernández, A.; Cava, D.; Ocio, M. J.; Lagarón, J. M. Perspectives for Biocatalysts in Food Packaging. *Trends Food Sci. Technol.* **2008,** *19* (4), 198–206.

Flanagan, J.; Singh, H. Microemulsions: A Potential Delivery System for Bioactives in Food. *Crit. Rev. Food Sci. Nutr.* **2006,** *46* (3), 221–237.

Ghanbarzadeh, B.; Oleyaei, S. A.; Almasi, H. Nanostructured Materials Utilized in Biopolymer-Based Plastics for Food Packaging Applications. *Crit. Rev. Food Sci. Nutr.* **2013,** *55* (12), 1699–1723.

Giles, E. L.; Kuznesof, S.; Clark, B.; Hubbard, C.; Frewer, L. J. Consumer Acceptance of and Willingness to Pay for Food Nanotechnology: A Systematic Review. *J. Nanopart. Res.* **2015,** *17* (12), 1–26.

Gomez-Estaca, J.; Lopez-de-Dicastillo, C.; Hernandez-Munoz, P. Catalá, R.; Gavara, R. Advances in Antioxidant Active Food Packaging. *Trends Food Sci. Technol.* **2014,** *35* (1), 42–51.

Horner, S. R., Mace, C. R., Rothberg, L. J.; Miller, B. L. A Proteomic Biosensor for Enteropathogenic *E. coli*. *Biosens. Bioelectron.* **2006,** *21* (8), 1659–1663.

Hu, A. W.; Fu, Z. H. Nano Technology and Its Application in Packaging and Packaging Machinery. *Packag. Eng.* **2003,** *24*, 22–24.

Jaiswal, L.; Shankar, S.; Rhim, J. Applications of Nanotechnology in Food Microbiology. In *Nanotechnology*, 1st ed., Vol. 46, 2019.

Jh, L.; Park, J.; Jh, A.; Jin, H. J.; Hong, S.; Park, T. H. A peptide receptor—Based Bioelectronic Nose for the Realtime Determination of Seafood Quality. *Biosens. Bioelectron.* **2016,** *39* (1), 244–249.

Jones, N.; Ray, B.; Ranjit, K. T.; Manna, A. C. Antibacterial Activity of ZnO Nanoparticle Suspensions on a Broad Spectrum of Microorganisms". *FEMS Microbiol. Lett.* **2008,** *279* (1), 71–76.

Kang, S.; Pinault, M.; Pfefferle, L. D.; Elimelech, M. Single Walled Carbon Nanotubes Exhibit Strong Antimicrobial Activity. *Langmuir* **2007,** *23* (17), 8670–8673.

Kerry, J. P.; O'Grady, M. N.; Hogan, S. A. Past, Current and Potential Utilization of Active and Intelligent Packaging Systems for Meat and Muscle-Based Products: A Review. *Meat Sci.* **2006,** *74* (1), 113–130.

Keshwani, A.; Malhotra, B.; Kharkwal, H. Advancements of Nanotechnology in Food Packaging. *World J. Pharm. Pharm. Sci.* **2015,** *4* (4), 1054–1057.
Khalaj, M. J.; Ahmadi, H.; Lesankhosh, R.; Khalaj, G. Study of Physical and Mechanical Properties of Polypropylene Nanocomposites for Food Packaging Application: Nano-Clay Modified with Iron Nanoparticles. *Trends Food Sci. Technol.* **2016,** *51*, 41–48.
Kuswandi, B. Nanotechnology in Food Packaging Nanotechnology in Food Packaging. *Sustain. Agric. Rev.* **2016,** *20*, 152–184.
Kuswandi, B.; Wicaksono, Y.; Abdullah, A.; Heng, L. Y.; Ahmad, M. Smart Packaging: Sensors for Monitoring of Food Quality and Safety. *Sens. Instrument. Food Qual. Safe.* **2011,** *5* (4), 137–146.
Kuzma, J. Nanotechnology in Animal Production-Upstream Assessment of Applications. *Livestock Sci.* **2010,** *130* (3), 14–24.
Liao, F.; Chen, C.; Subramanian, V. Organic TFT as Gas Sensors for Electronic Nose Applications. *Sens. Actuat. B Chem.* **2005,** *107* (2), 849–855.
Liu, F.; Avena-Bustillos, R. J.; Chiou, B. S.; Li, Y.; Ma, Y.; Williams, T. G. et al. Controlled Release of Tea Polyphenol from Gelatin Films Incorporated with Different Ratios of Free/Nanoencapsulated Tea Polyphenols into Fatty Food Simulants. *Food Hydrocoll.* **2017,** *62*, 212–221.
Lopez-Rubio, A.; Gavara, R.; Lagaron, J. M. Bioactive Packaging: Turning Foods into Healthier Foods through Biomaterials. *Trends Food Sci. Technol.* **2006,** *17* (10), 567–575.
Luduena, L. N. Alvarez, V. A.; Vazquez, A. Processing and Microstructure of PCL/Clay Nanocomposites. *Mater. Sci. Eng.* **2007,** *460*, 121–129.
Majeed, K.; Jawaid, M.; Hassan, A.; Abu Bakar, A.; Khalil, H. P. S. A.; Salema, A. A.; Inuwa, I. Potential Materials for Food Packaging from Nanoclay/Natural Fibres Filled Hybrid Composites. *Mater. Design* **2013,** *46*, 391–410.
Malathi, A. N.; Santhosh, K. S.; Udaykumar, N. Recent Trends of Biodegradable Polymer: Biodegradable Films for Food Packaging and Application of Nanotechnology in Biodegradable Food Packaging. *Curr. Trends Technol. Sci.* **2014,** *3* (2), 73–79.
Malhotra, B.; Keshwani, A.; Kharkwal, H. Antimicrobial Food Packaging: Potential and Pitfalls. *Front. Microbiol.* **2015,** *6*, 611.
Mannino, S.; Scampicchio, M. Nanotechnology and Food Quality Control. *Vet. Res. Commun.* **2007,** *31* (1), 149–151.
Mao, X.; Huang, J. Leung, M. F.; Du, Z.; Ma, L.; Huang, Z. Novel Core-Shell Nanoparticles and Their Application in High-Capacity Immobilization of Enzymes. *Appl. Biochem. Biotechnol.* **2006,** *135* (3), 229–246.
Marambio-Jones, C.; Hoek, E. M. V. A Review of the Antibacterial Effects of Silver Nanomaterials and Potential Implications for Human Health and the Environment. *J. Nanopart. Res.* **2010,** *12* (5), 1531–1551.
Marsh, K.; Bugusu, B. Food packaging—Roles, Materials, and Environmental Issues. *J. Food Sci.* **2007,** *72* (3), R39–R55.
Mason, T. G.; Wilking, J. N.; Meleson, K.; Chang, C. B.; Graves, S. M. Nanoemulsions: Formation, Structure, and Physical Properties. *J. Phys. Condensed Matter* **2006,** *18* (41), R635–R666.
McMillin, K. W. Where Is MAP Going? A Review and Future Potential of Modified Atmosphere Packaging for Meat. *Meat Sci.* **2008,** *80* (1), 43–65.

Meetoo, D. D. Nanotechnology and the Food Sector: From the Farm to the Table. *Emirates J. Food Agric.* **2011,** *23* (5), 387–407.
Metak, A. M.; Nabhani, F.; Connolly, S. N. Migration of Engineered Nanoparticles from Packaging into Food Products. *LWT-Food Sci. Technol.* **2015,** *64*, 781–787.
Mihindukulasuriya, S. D. F.; Lim, L. Nanotechnology Development in Food Packaging : A Review. *Trends Food Sci. Technol.* **2014,** *40*, 149–167.
Mills, A.; Doyle, G.; Peiro, A. M.; Durrant, J. Demonstration of a Novel, Flexible, Photocatalytic Oxygen-Scavenging Polymer Film. *J. Photochem. Photobiol. A Chem.* **2006,** *177*, 328–331.
Momin, J. K.; Chitra, J.; Prajapati, J. B. Potential of Nanotechnology in Functional Foods. *Emirates J. Food Agric. (EJFA)* **2013,** *25* (1), 10–19.
Momin, J. K.; Joshi, B. H. Nanotechnology in Foods. *Nanotechnol. Food Agric.* **2015,** 1–22.
Montazer, M.; Harifi, T. New Approaches and Future Aspects of Antibacterial Food Packaging: From Nanoparticles Coating to Nanofibers and Nanocomposites, with Foresight to Address the Regulatory Uncertainty". In *Food Package*; Grumezescu, A. M., Ed.; Academic Press, 2017; pp 533–559.
Moraru, C. I.; Panchapakesan, C. P.; Huang, Q.; Takhistov, P.; Liu, S.; Kokini, J. L. Nanotechnology: A New Frontier in Food Science. *Food Technol.* **2003,** *57* (12), 24–29.
Naoto, S.; Hiroshi, O.; Mitsutoshi, N. Micro- and Nanotechnology for Food Processing. (Food Safety Series). *Resour.: Eng. Technol. Sustain. World* **2009,** *16* (6), 19.
Nile, S. H.; Baskar, V.; Selvaraj, D.; Nile, J. X.; Kai, G. Nanotechnologies in Food Science: Applications, Recent Trends, and Future Perspectives. *Nano-Micro Lett.* **2020,** *12*, 45.
Othman, S. H. Bio-Nanocomposite Materials for Food Packaging Applications: Types of Biopolymer and Nano-Sized Filler. *Agric. Agric. Sci. Procedia* **2014,** *2*, 296–303.
Pal, M. Nanotechnology: A New Approach in Food Packaging. *J. Food Microbiol. Safet. Hygiene* **2017,** *2*, 121.
Pal, M.; Devrani, M.; Hadush, A. Recent Developments in Food Packaging Technologies. *Bev. Food World* **2019,** *46* (1), 21–25.
Pandey, S.; Zaidib, M. G. H.; Gururani, S. K. Recent Developments in Clay-Polymer Nanocomposites. *Sci. J. Rev.* **2013,** *2* (11), 296–328.
Pathakoti, K.; Manubolu, M.; Hwang, H. Nanostructures: Current Uses and Future Applications in Food Science. *J. Food Drug Analy.* **2017,** *25* (2), 245–253.
Paul, D. R.; Robeson, L. M. Polymer Nanotechnology: Nanocomposites. *Polymer* **2008,** *49*, 3187–3204. doi: 10.1016/j.polymer.2008.04.017.
Pehanich, M. Small Gains in Processing and Packaging. *Food Process* **2006,** *11*, 46–48.
Perez-Esteve, E.; Bernardos, A.; Martínez-Máñez, R.; Barat, J. M. Nanotechnology in the Development of Novel Functional Foods or Their Package. An Overview Based in Patent Analysis. *Recent Patents Food, Nutr. Agric.* **2013,** *5* (1), 35–43.
Pradhan, N.; Singh, S.; Ojha, N.; Shrivastava, A.; Barla, A.; Rai, V.; Bose, S. Facets of Nanotechnology as Seen in Food Processing, Packaging, and Preservation Industry. *Biomed Res. Int.* **2015,** *365672*, 1–17.
Prasad, R.; Kumar, V.; Prasad, K. S. Nanotechnology in Sustainable Agriculture: Present Concerns and Future Aspects. *Afr. J. Biotechnol.* **2014,** *13*, 705–713.
Rai, V.; Acharya, S.; Dey, N. Implications of Nanobiosensors in Agriculture. *J. Biomater. Nanobiotechnol.* **2012,** *3* (2), 315–324.

Ravichandran, R. Nanotechnology Applications in Food and Food Processing: Innovative Green Approaches, Opportunities and Uncertainties for Global Market. *Int. J. Green Nanotechnol. Phys. Chem.* **2010,** *1* (2), 72–96.
Reščèk, A.; Ščetar, M.; Hrnjak-Murgić, Z.; Dimitrov, N.; Galić, K. Polyethylene/Polycaprolactone Nanocomposite Films for Food Packaging Modified with Magnetite and Casein: Oxygen Barrier, Mechanical and Thermal Properties. *Polym.-Plastics Technol. Eng.* **2016,** *55* (14), 1450–1459.
Restuccia, D.; Spizzirri, U. G.; Parisi, O. I.; Cirillo, G.; Curcio, M.; Iemma, F. et al. New EU Regulation Aspects and Global Market of Active and Intelligent Packaging for Food Industry Applications. *Food Control* **2010,** *21* (11), 1425–1435.
Rhim, J. W.; Park, H. M.; Ha, C. S. Bio-Nanocomposites for Food Packaging Applications. *Progress Polym. Sci.* **2013,** *38*, 1629–1652.
Rossi, M.; Cubadda, F., Dini, L.; Terranova, M. L.; Aureli, F.; Sorbo, A. et al. Scientific Basis of Nanotechnology, Implications for the Food Sector and Future Trends. *Trends Food Sci. Technol.* **2014,** *40*, 127–148.
Rossi, M.; Passeri, D.; Sinibaldi, A.; Angjellari, M.; Tamburri, E.; Sorbo, A. et al. Nanotechnology for Food Packaging and Food Quality Assessment. *Adv. Food Nutr. Res.* **2017,** *82*, 150–193.
Rubilar, O.; Diez, M. C.; Tortella, G. R.; Briceno, G.; Marcato, P. D.; Duran, N. New Strategies and Challenges for Nanobiotechnology in Agriculture. *J. Biobased Mater. Bioenergy* **2014,** *8* (1), 1–12.
Schaefer, D.; Cheung, W. M. Smart Packaging: Opportunities and Challenges. *Procedia CIRP* **2018,** *72*, 1022–1027.
Scrinis, G.; Lyons, K. The Emerging Nano-Corporate Paradigm: Nanotechnology and the Transformation of Nature, Food and Agri-Food Systems. *Int. J. Sociol. Food Agric.* **2007,** *15* (2), 22–44.
Sekhon, B. S. Food-Nanotechnology-An-Overview. *Nanotechnol. Sci. App.* **2010,** *3*, 1–15.
Shafiq, M.; Anjum, S.; Hano, C.; Anjum, I.; Abbasi, B. H. An Overview of the Applications of Nanomaterials and Nanodevices in the Food Industry. *Foods* **2020,** *9*, 148.
Sharma, C.; Dhiman, R.; Rokana, N.; Panwar, H. Nanotechnology : An Untapped Resource for Food Packaging. *Front. Microbiol.* **2017,** *8*, 1735.
Silvestre, C.; Duraccio, D.; Cimmino, S. Food Packaging Based on Polymer Nanomaterials. *Progress Polym. Sci.* **2011,** *36*, 1766–1782.
Singh, T.; Shukla, S.; Kumar, P.; Wahla, V.; Bajpai, V. K. Application of Nanotechnology in Food Science: Perception and Overview. *Front. Microbiol.* **2017,** *8*, 1501.
Siracusa, V.; Rocculi, P.; Romani, S.; Dalla, R. M. Biodegradable Polymers for Food Packaging: A Review. *Trends Food Sci. Technol.* **2008,** *19* (2), 634–643.
Sondi, I.; Salopek-Sondi, B. Silver Nanoparticles as Antimicrobial Agent: A Case Study on *E. coli* as a Model for Gram-Negative Bacteria. *J. Colloid Interf. Sci.* **2004,** *275* (1), 177–182.
Sorrentino, A.; Gorrasi, G.; Vittoria, V. Potential Perspectives of Bio-Nanocomposites for Food Packaging Applications. *Trends Food Sci. Technol.* **2007,** *18* (2), 84–95.
Sozer, N.; Kokini, J. L. Nanotechnology and Its Applications in the Food Sector. *Trends Biotechnol.* **2009,** *27* (2), 82–89.
Sun, D.; Jin, T.; Su, J. Y.; Zhang, H.; Sue, H. J. Antimicrobial efficacy of Zinc Oxide Quantum Dots against *Listeria monocytogenes, Salmonella enteritidis* and *Escherichia coli* O157:H7. *J. Food Sci.* **2009,** *74* (1), M46–M52.

Sundaram, J.; Pant, J.; Goudie, M. J.; Mani, S.; Handa, H. Antimicrobial and Physicochemical Characterization of Biodegradable, Nitric Oxide-Releasing Nanocellulose–Chitosan Packaging Membranes. *J. Agric. Food Chem.* **2016,** *64* (25), 5260–5266.

Surassmo, S.; Min, S. G.; Bejrapha, P.; Choi, M. J. Effects of Surfactants on the Physical Properties of Capsicum-Oleoresin-Loaded Nanocapsules Formulated through the Emulsion-Diffusion Method. *Food Res. Int.* **2009,** *43* (1), 8–17.

Tager, J. Nanomaterials in Food Packaging: FSANZ Fails Consumers Gain. *Chain Reaction* **2014,** *122*, 16–17.

Tan, H.; Ma, R.; Lin, C.; Liu, Z.; Tang, T. Quaternized Chitosan as an Antimicrobial Agent: Antimicrobial Activity, Mechanism of Action and Biomedical Applications in Orthopedics. *Int. J. Mol. Sci.* **2013,** *14* (1), 1854–1869.

Tewari, G.; Jayas, D. S.; Jeremiah, L. E.; Holley, R. A. Absorption Kinetics of Oxygen Scavengers. *Int. J. Food Sci. Technol.* **2002,** *37* (2), 209–217.

Thiruvengadam, M.; Rajakumar, G.; Min, I. Nanotechnology: Current Uses and Future Applications in the Food Industry. *Biotechnology* **2018,** *8* (3), 74.

Tuan Zainazor, T. C.; Fisal, A.; Goh, E. G.; CheSulaiman, N. F.; Sarbon, N. M. Emerging of Bio-Nanocomposite Gelatine-Based Film as Bio-Degradable Food Packaging: A Review. *Food Res.* **2020,** 1–13.

Vanderroost, M.; Ragaert, P.; Devlieghere, F.; Meulenaer, B. D. Intelligent Food Packaging: The Next Generation. *Trends Food Sci. Technol.* **2014,** *39* (1), 47–62.

Vera, P.; Echegoyen, Y.; Canellas, E.; Nerín, C.; Palomo, M.; Madrid, Y.; Cámara, C. Nano Selenium as Antioxidant Agent in a Multilayer Food Packaging Material. *Analyt. Bioanalyt. Chem.* **2016,** *408* (24), 6659–6670.

Vidhyalakshmi, R.; Bhakyaraj, R.; Subhasree, R. S. Encapsulation 'the Future of Probiotics': A Review. *Adv. Biol. Res.* **2009,** *3* (4), 96–103.

Vital, A. C. P.; Guerrero, A.; Monteschio, J.; de, O.; Valero, M. V.; Carvalho, C. B.; de Abreu Filho, B. A. Effect of Edible and Active Coating (with Rosemary and Oregano Essential Oils) on Beef Characteristics and Consumer Acceptability. *PLoS One* **2016,** *11* (8), 1–15.

Warheit, D. B. Laurence, B. R.; Reed, K. L.; Roach, D. H.; Reynolds, G. A.; Webb, T. R. "Comparative Pulmonary Toxicity Assessment of Single-Wall Carbon Nanotubes in Rats". *Toxicol. Sci.* **2004,** *77* (1), 117–125.

Weiss, J.; Takhistov, P.; Mcclements, D. J. Functional Materials in Food Nanotechnology. *J. Food Sci.* **2006,** *71* (9), R107–R116.

Wen, P.; Zhu, D. H.; Wu, H.; Zong, M. H.; Jing, Y. R.; Han, S. Y. Encapsulation of Cinnamon Essential Oil in Electrospun Nanofibrous Film for Active Food Packaging. *Food Control* **2016,** *59*, 366–376.

Xiao, L.; Green, A. N. M.; Haque, S. A.; Mills, A.; Durrant, J. R. Light-Driven Oxygen Scavenging by Titania/Polymer Nanocomposite Films. *J. Photochem. Photobiol. A Chem.* **2004,** *162*, 253–259.

Yadav, H. M.; Kim, J. S.; Pawar, S. H. Developments in Photocatalytic Antibacterial Activity of Nano TiO_2: A Review. *Korean J. Chem. Eng.* **2016,** *33* (7), 1989–1998.

Yam, K. L.; Takhistov, P. T.; Miltz, J. Intelligent Packaging: Concepts and Applications. *J. Food Sci.* **2005,** *70* (1), 1–10.

Yao, Z. C.; Chang, M. W.; Ahmad, Z.; Li, J. S. Encapsulation of Rose Hip Seed Oil into Fibrous Zein Films for Ambient and on Demand Food Preservation via Coaxial Electrospinning. *J. Food Eng.* **2016,** *191*, 115–123.

Yotova, L.; Yaneva, S.; Marinkova, D. Biomimetic Nanosensors for Determination of Toxic Compounds in Food and Agricultural Products (Review). *J. Chem. Technol. Metall.* **2013,** *48* (3), 215–227.

Yuan, Y.; Gao, Y.; Zhao, J.; Mao, L. Characterization and Stability Evaluation of β-Carotene Nanoemulsions Prepared by High Pressure Homogenization under Various Emulsifying Conditions. *Food Res. Int.*, **2008,** *41* (1), 61–68.

Zhao, R.; Torley, P.; Halley, P. J. Emerging Biodegradable Materials: Starch- and Protein-Based Bio-Nanocomposites. *J. Mater. Sci.* **2008,** *43* (9), 3058–3071.

CHAPTER 2

Benefits of Nanocomposite Food Packaging Over Conventional Packaging

M. GIRILAL* and SONY GEORGE

Department of Food Technology, Saintgits College of Engineering, Kottayam, Kerala, India

**Corresponding author. E-mail: girilal@gmail.com*

ABSTRACT

A suitable packaging is crucial for protecting food materials from its surroundings until it reaches the hands of the customer. In the case of the food items, an ideal packaging material is inevitable to protect the food materials from various extrinsic as well as intrinsic factors, which cause its spoilage and at the same time it should be cost effective as well as should not make any reaction with the food material changing its original flavor or appearance. In most of the conventional packaging materials, some of these requirements need to be sacrificed which leads to the invention of numerous novel packaging materials. Nanocomposites are one such promising packaging material with unique potential which could be tailor made for a specific food product by manipulating the matrix and filler materials. The specific physical, chemical, and biological properties and their improved properties when scaled down and the high surface to volume ratio are making them stand out. The nano-size makes them unpredictable and thus the safety of the nanocomposite packaging material needs to be experimented properly before commercial usage.

2.1 INTRODUCTION

In many agriculture-based countries, the foremost concern is to keep the food products as much safe and fresh as possible till it reaches the consumer's hand. Proper aseptic packaging is the best method to increase the shelf life of the food products. From decades lots of research and innovations is being going on in the field of packaging materials as well as packaging technologies.

Proper packaging will protect and contain the goods or products, which starts from processing, manufacturing through handling and storage until it reaches the hands of the final consumer. The Packaging Institute International defines Packaging as any material acting as an enclosure to products in a wrapped box, pouch, cup, box, tray, can, bag, bottle, tube, or any other container form to perform one or more functions like protection and/or preservation, containment, communication, and utility or performance. Packaging is thus a co-ordinated system designed for the efficient delivery of high quality and safe food throughout the supply chain. It can be considered a combination of art, science, and technology used for the storage, transportation, and consumption of food. A pre-requisite to any food product, it offers protection to the food one buys from processing till consumption by maintaining the basic attributes of food such as temperature, color, texture, flavor. In addition to this, attractive packaging also draws customers to buy the product and can hence act as a communication channel.

Packaging thus be any product of any nature and material which may serve the purpose of protection, containment, delivery, handling, and may convey the details of the product for the awareness of the consumer from the raw materials to the finished product. This definition takes into consideration the four quintessential aspects associated with packaging which includes packaging materials, its functions, type of products, and materials that constitute a package and actors involved in the packaging process.

2.2 PRINCIPLES OF PACKAGING

Packaging should ensure four basic primary functions which include providing protection, communication, convenience, and containment followed by the secondary functions of traceability and tamper indication (Fellows, 2000). Food packaging should ensure the attributes which add convenience and that allow the consumer to utilize the product much easier. It may appeal to the consumers to buy the product and should convey the details about nutritional content, contained ingredients, method of usage, and all other relevant information. Packaging should be designed to minimize the environmental and social impacts of its disposal and should comply with recommendations by the Food and Drug Authority India. Types of Packaging materials may depend on the variety of foods being packaged and it varies over a wide range such as primary packaging, secondary packaging, and tertiary packaging, which involved several layers of packaging material employed for a common aim (Sun, 2006; Cheruvu et al., 2008).

In short, the prime functions of packaging include

i. To contain the product
ii. To protect the product

2.3 ADVANTAGES OF FOOD PACKAGING

The most recommended function of packaging is to protect the food from spoilage due to physical, chemical, or biological damage. It ensures that the product stays intact and is not affected by heat, vibrations, humidity, which in turn influences the products shelf life and provides a barrier between the food and the environment thereby controlling the rate of transfer of heat, movement of microorganisms or insect's, moisture and gases and light transmission. Packaging plays a crucial role since it provides aesthetics thereby attracting customers. A package is assumed as the face of a product and is the only exposure of the product that consumers experience prior to purchase. They also contain information and instructions such as its contents, date of manufacturing, method of preparation along with the other messages it conveys and hence acts as a silent salesman.

Ease of access, handling, disposal, resealable, product visibility and ability to do microwave may add convenience and greatly influence package innovation and sales. Innovations like anti-theft devices, RFID tags have been applied on some packets (Gupta and Dudeja, 2017). Packaging may also help in grouping, categorizing, and appropriate storage food products. Increasing trend toward grazing and the consumer demand for convenient and safe products along with change in market conditions led to the exploration of innovative ideas replacing the earlier materials used. Packaging now plays a tremendous role in the food chain, from farm to table, as it helps rapid distribution, which ends in the removal of local food surpluses, and which gives consumers more choices in the selection of foods and thus helps to reduce malnutrition. It aids in the prevention of post-harvest losses, thus allowing the access to larger markets and help producers to enhance their incomes. Therefore, in today's competitive market, packaging should satisfy all legislative, marketing, and functional requirements effectively and economically (Janjarasskul and Krochta, 2010).

2.4 FUNCTIONS, PROPERTIES, AND APPLICATIONS OF PACKAGING

The functions, properties, and applications of packaging are meant to ensure the safety and quality of the food products during transportation and further in storage facilities. It includes maintaining the viability of food products by preventing access of chemical contaminants, microorganisms, moisture oxygen, and light. Qualities of an ideal packaging material are mentioned below.

2.4.1 BARRIER TO ENVIRONMENT

The mass transfer phenomena contribute much to the deterioration of most packaged foods. Hence, the most typical aspect of a packaging is to prevent this. Chances of permeation of volatile components from the environment to the food arises and hence the effects of temperature and moisture content on biopolymer materials require that the permeabilities of the edible film reflect the conditions of intended use which should be calculated under specific conditions.

2.4.2 MOISTURE PROPERTY

Edible films, coatings, and biodegradable packaging produced from biological materials offer numerous advantages such as prevention of exchange of moisture between the food product and the environment since it has been indicated that changes in relative humidity lead to the microbial deterioration, textural changes, and chemical and enzymatic deterioration of the product. Studies associated with the water vapor permeability (WVP) values of plastic and various edible packaging films have concluded that hydro-colloid films exhibit higher WVP in contrast to edible wax films. An increase in WVP of hydrophilic films was observed at higher relative humidity and plasticizer concentration owing to their high substantial polarity thereby restricting their use to short time or for low moisture food. On the contrary, lipids and other hydrophobic molecules, because of their low polarity, low water affinity, and dense network matrix, are employed to enhance the moisture barrier properties of other hydrocolloid films.

2.4.3 OXYGEN BARRIER

Oxidation of lipids and food ingredients, discoloration of myoglobin in fresh meat or enzymatic browning of fresh-cut produce brings forth deterioration of food. Packaging material can be exploited having low oxygen permeability (OP) in place of O_2 barrier to avoid deterioration and maintain the quality and palatability of the food (Janjarasskul and Krochta, 2010).

2.4.4 AROMA BARRIER

The migration of volatile organic compounds from the food into the environment and vice versa can be prevented by the application of good packaging films. This ensures that the characteristic aroma of the material is not lost. This can be done by optimizing the barrier efficiency of packaging so that a migrating compound will have low affinity to the film materials and low diffusivity through the polymer matrix.

2.4.5 CARRIER OF ANTIMICROBIALS, ANTIOXIDANTS, AND OTHER ADDITIVES

Packaging can act as a carrier of antimicrobials and antioxidants and release them in a controlled manner. This particular property has attracted many to run a research in this field and formulate desired food additives into them and act as carriers. They can also be used for the microencapsulation of flavoring agents.

2.5 CONVENTIONAL FOOD PACKAGING MATERIALS

There are many packaging materials which we are using from centuries. But in those days, the major function was covering the unprocessed raw materials or cooked food which was meant for immediate consumption. During the time of scarcity, they could not pack and store the food for future use. Later on, many trial and error methods with different packaging materials were tried as well as experimented. Some of the conventional packaging materials and different food commonly packed are mentioned in Table 2.1. Some of the packaging materials are discussed below.

2.5.1 LEAVES, VEGETABLE FIBERS, AND TEXTILES

Natural materials such as leaves have been used to wrap food. Traditionally, cheese such as guava cheese and other products were wrapped inside banana or plantain leaves and provide less mechanical support for the product and lack rigidity. Since the textile containers have poor moisture and gas barrier properties, they are mainly used for the transportation of bulk foods like flour, grain, salt, and sugar. Plant fiber sacks have the advantage of more flexible, highly durable, lightweight, resistant to tearing, good durability, and maybe chemically treated to prevent them rotting. Woven jute sacks are often chemically treated to reduce rotting and flammability. They are non-slip and have a high tear resistance, have low extensibility and good durability compared to synthetic fiber sacks, and they are bio-degradable but are proven to be difficult to clean when contaminated with the decaying organism. They can also cause damage to the commodity if stacked tightly due to the high pressure.

2.5.2 WOOD

Timber crates or boxes are commonly used for the transportation of fruits and vegetables. They impart rigidity to the product when stacked for long-distance transport while also exhibiting a good weight to strength ratio. Nowadays, it's been widely used as a protective outer package for the delicate bottles of wine, tea, beer because of its high impact resistance and stackability. Wood packaging has been replaced with plastic containers since it has the disadvantage of damaging the product inside due to its sharp edges and splinters.

TABLE 2.1 Common Conventional Packaging Materials Used for Different Foods Products.

Packaging Product type	Glass		Metal				Paper/paper hard board			Plastic		
	Jar	**Bottle**	**can**	**Tin**	**Foil**	**Tube**	**Wrap**	**Carton**	**Bottle**	**Cellulose**	**DPE/ HDPE**	**OPP Film**
Dairy products		x						x				
Fresh fruit/vegetables							x			x		
Fresh meat/fish							x					
Dried foods	x						x			x	x	x
Preserves	x		X									
Baked goods	x			X				x		x		x
Beverages		x	X									
Frozen foods		x					x		x			
Pastes and purée's	x		X			x						
Snack food										x	x	x
Sterilized foods	x	x	X									
Cooking oils		x							x			
Syrups/honey	x	x							x			

2.5.3 LEATHER

Camel, pig, or kid goat hides can be used to manufacture leather cases and pouches that have been traditionally used as flexible, non-breakable containers, lightweight for water, milk and wine manioc flour, and solidified sugar. Now the usage of leather as a packaging material has been ceased for most commercial food applications because of legal and aesthetic factors.

2.5.4 EARTHENWARE

Properly manufactured and glazed pots from clay can act as well-sealed barriers to oxygen, light, and heat, thereby making them suitable for the storage of commodities such as wine and oils. They are also being used in rural areas for the storage of curd, yogurt, etc.

2.5.5 ALUMINUM

Aluminum is among the most popular materials, found in bauxite ore along with oxygen in the form of alumina, used for packaging various kinds of food products. It is recyclable in nature, environment friendly, and most importantly easy and low cost on manufacturing. The natural aluminum oxide coating present provides effective barrier properties against oxygen, temperature, moisture, and chemical attack, thereby preventing it from any form of corrosion. The light but strong material imparts high physical, chemical, and mechanical properties such as barrier effect, dead-fold properties, and suitability for food contacts enable a wide range of applications in the food industry. Aluminum cans are used mainly for packaging canned goods, aerated drinks, meat, baked beans, condensed milk sweets, and similar items.

2.5.6 GLASS

Glass containers are made by heating silica, sodium carbonate, limestone, and alumina at very high temperatures and moulded into the required shape. Although very brittle and heavy, glass containers possess very good barrier properties and are inert as well. They are odorless, impermeable to both gas and liquid and recyclable. They are heavily used in the product packaging industry and are still preferred for the storage of different items like jellies, jams, beer, pickles, juices, and cooking oils. Some of the disadvantages of glass include its high brittleness and weight resulting in increased transportation weight, vulnerability to thermal shocks, potential contamination due to splinters and the limitation to store products prone to light-catalyzed reactions (Kenneth and Betty, 2007).

2.5.7 PAPER AND CARDBOARD

Paper and cardboards are versatile materials used to package food made with wood pulp and additives to give it the required properties. Sulfite papers are produced by treating wood pulp with peroxide or hypochlorite and are lighter and weaker compared to sulfate papers which are strong and hence is used in case of single or multi-walled paper sacks for flour, sugar, etc. Greaseproof papers used to wrap candy bars and other oily foods, are made by a process called beating where the cellulose fibers undergo longer hydration period than normal causing the fibers to break up becoming gelatinous. Glassine is a greaseproof paper having a glossy and highly smooth appearance. When used as primary packaging, paper is treated, coated, and laminated with materials such as waxes or resins for moisture barrier properties.

With more companies emphasizing on renewable and environment-friendly packaging materials like cardboard and paper have become a preferred choice for them. Cardboard tubes, small tubs, and cans are made and items are packed inside the same. Although dry and lightweight items like nuts, candies, salt, spices, tea, etc., can be packed and transported easily in them, heavy and wet items cannot be packed inside these cardboard boxes. Cardboard cartons are recyclable and display high shock resistive capacity as well.

2.5.8 FLEXIBLE PLASTIC FILMS

Plastics are now become the common packaging material for food items. Some items that are commonly stored and packed in them include snack foods, bread, cheese condiments, frozen goods, cooking oil, etc. The permeability of a given plastic material to oxygen, water vapor, aromas, and carbon dioxide mostly depends on the polymer composition and structure. They are often pigmented to eliminate the light-catalyzed reactions in the food. They offer the advantages of being cheap, easy to handle, and good barrier properties among others. But plastic additives and any residual monomers have the potential for migrating into foods and they can absorb aroma and flavor in the food into the plastic material, thereby deteriorating its quality (Mahalik and Nambiar, 2010).

Some of the flexible plastic films include:

2.5.8.1 CELLULOSE

Plain cellulose, a glossy transparent film that is tasteless, odorless, biodegradable, tough, and puncture resistant, and has high recommendable dead-folding properties that make it suitable for twist-wrapping (e.g., sugar confectionery).

2.5.8.2 POLYETHYLENE (OR POLYTHENE)

Polyethylene is the most simple, versatile, and inexpensive plastic synthesized by the polymerization of ethylene. They are majorly classified into low-density polyethylene (LDPE), high-density polyethylene (HDPE), and linear LDPE (LLDPE) based on the density, crystallinity and chain branching. LDPE is usually used as a film for packaging fresh produce because of its heat sealability, inert, and odor-free properties. It can act as a good moisture barrier but is moderately permeable to oxygen and is a poor odor barrier. HDPE molecules can fold and pack into a highly crystalline structure making it brittle. It is a better barrier to gases and moisture comparatively. It has higher melting point, greater tensile strength, and better chemical resistance. They are blow-molded to make plastic bottles for substances such as water or milk, food containers (Paine and Paine, 1992).

2.5.8.3 POLYPROPYLENE

Polypropylene is a linear, crystalline polymer having the lowest density among plastics and has higher tensile strength and hardness. It offers moderate barrier to moisture, gases, and odor, which is not have any significance changes based on humidity. The property like higher melting point making it suits for retorting purposes. It also serves an important application in packaging films. Oriented polypropylene (OPP) films shows improved strength, better gas barrier whereas unoriented PP has better clarity, good dimensional stability, and good heat-seal strength.

2.5.8.4 ORIENTED POLYPROPYLENE

It is clear and glossy film with good optical properties, puncture resistance, and high tensile. It is having a moderate permeability to moisture, odors, and gases. It is thermoplastic in nature and have moderate stretch, less than polyethylene. The low friction which minimizes the static build up and make them suitable for high-speed filling equipment. Biaxially OPP (BOPP) has similar characteristics to oriented polypropylene but is much stronger. They are used for jars, crisp packets bottles, biscuit wrappers jars, and boil-in-bag films among many other applications (Robertson and Marcel, 1993).

2.5.8.5 POLYETHYLENE TEREPHTHALATE (PET)

Polyethylene terephthalate (PET) is the major polyester in packaging. Its amorphous form is moulded into bottles for carbonated soft drinks, juices, and edible oils. It is a thermoplastic polyester synthesized by the condensation of terephthalic acid and ethylene glycol. It offers a glass like appearance, good gas barrier properties, light weight, and shatter resistance.

2.5.8.6 UNCOATED POLYVINYLIDENE CHLORIDE

It is an addition polymer of polyvinylidene chloride (PVDC) with its major advantage being the oxygen and moisture barrier. They have good clarity and oil resistance. It is heat sealable with its application in poultry, cured meat, cheese, coffee, tea, and confectionery.

2.5.8.7 POLYSTYRENE

Polystyrene (PS) is an amorphous with excellent clarity having a relatively low melting point (88°C). It is hard and brittle in nature. It is thermoformed or injection moulded into food containers, closures, cups, etc. It is used as protective packaging for egg cartons, food trays, and others.

2.5.8.8 ETHYLENE-VINYL ACETATE (EVA)

It is a copolymer of vinyl acetate and ethylene. It has high mechanical strength and flexibility at low temperatures. Ethylene-vinyl acetate (EVA) is as flexible as PVC without even plasticizers and it has greater resilience and greater flexibility than LDPE and PVC. The weight percent of vinyl acetate may vary from 10% to 40%, and the rest is contributed by ethylene. EVA copolymer which contains less than 5% vinyl acetate, is usually used for deep-freeze applications and with 6–10% are used in bag-inbox applications and milk pouches, and above 10% vinyl acetate the material is used as a hot-melt adhesive.

2.6 NOVEL PACKAGING MATERIALS

Although many conventional food packaging methods exists, there was always research for better packaging technology and packaging materials. Nanotechnology-based inventions became the center of attention during the recent decades not only in the field of food industries, but also in health, pharmacy, and medicine industries. In this scenario, nanoparticles-based packaging technology has attracted majority interest due to its unique optical, thermal, and antimicrobial properties. Many types of metal nanoparticles as well as polymeric nanoparticles are significantly contributing to the basic sciences and have been tested for its vast applications. Exploiting the potential applications of metal nanoparticles, various researches have been going on in the field of food nanotechnology for the use of protective coating and suitable packaging material (Fayaz et al., 2009).

The new functional property attained by the packaging materials after coupling or incorporation with nanomaterials is the major attraction to the research community. Based on the coupling or incorporation of nanomaterials, the packaging materials are of different types. The major one is in which the desired nanoparticles/nanomaterials are mixed with the polymer matrix. This in turn changes the properties of the matrix like resistance to temperature, humidity. and also increase the gas barrier properties. Another one is active packaging in which the nanomaterials/

nanoparticles react with the surroundings and make food safer to consume. In most of the cases, the antimicrobial properties of the nanomaterials will keep away or eradicate all bacterial, fungal, and viral attacks on the food material. Another type is the intelligent or otherwise known as smart packaging which captures the attention of many. The nanoparticles/ nanomaterials incorporated can be utilized as a biosensor by exploiting the physical, chemical, and biological properties of the nanomaterials. The nanomaterials can be tailored for detecting the food pathogens, by products formed inside the food due to food spoilage, expiry date of food, adulterants etc. (Charles, 2014).

2.7 NANOCOMPOSITES

Among the various novel food packaging technologies and food packaging materials, nanocomposites-based materials are found to be promising candidate. Nanocomposites open up a large possibility to specifically design the materials for food packaging exploiting the functional groups and unique properties. Nanocomposites are composite materials which have their dimensions in the nanometre range in at least one dimension. Since they are made of one or more separate components, the final product will be expected to have the unsurpassed properties of each of the ingredients of the composites. The fillers in nanocomposites like clay, nanoparticles (metal or polymeric nanoparticles), can have a crucial role in formulating a tailor-made packaging material (Pedro et al., 2009).

Nanocomposite can be considered as multiphase solid materials at least in one phase the size less than 100 nanometers (nm), or structures with nanoscale repeat distances between the different phases that make up the material. The thermal, electrical, optical, mechanical, catalytic, electrochemical properties of the nanocomposite will differ evidently from that of the bulk materials (Charles, 2013).

The significance of size reduction is that when the size of a particle is getting decreased, the surface is to volume ration is drastically increasing. At this smaller dimension, the properties of the materials will be entirely different from that of its bulk material and there comes the significance of critical size. In the case of nanoparticles which is in the scale of nanometers (1 nm = 10^{-9} m), mostly ranging in the size between 1 and 100 nm at least in one dimension. It is normally considered as the lowest size of a particle possible in which it has at least some similar properties of its bulk material. Also, the interactions at phase interfaces become largely improved which is crucial in enhancing the properties of a material. Therefore, nanocomposites can also be projected as a potential applicant in many fields including food industry, engineering design, construction, electrical, electronics and many more, exploiting the unique properties. Another major advantage of nanocomposites is, majority of them are environment friendly and biodegradable (Henriette, 2009).

2.8 BENEFITS OF NANOCOMPOSITES IN FOOD PACKAGING

2.8.1 GENERAL BENEFITS OF NANOCOMPOSITES

As discussed earlier, the differences between a conventional composite and nanocomposites are due to the difference in the basic properties of the material when size is scaled down. The high surface area to the volume ratio is drastically increased in the case of nanocomposite which make them highly reactive and at the same time, specific. The materials can be reinforced using minerals, clay, metals, fibers, nanotubes, etc. which gave the nanocomposite a unique property which the normal composite does not possess. For instance, adding metal nanoparticles like silver will prevent food from bacteria, fungus and virus attack, and other metal nanoparticles will enhance the optical properties, thermal properties, and electrical conductivity, etc. which will make the food safer by making it undergo through various thermal and non-thermal food processing even after packing. Since the mechanical or physical strength is getting enhanced by nanocomposites, the stability and stiffness will be increased and damage through handling can be minimized (Charles, 2013).

By reducing the size of the materials to a nanoscale, it is possible to manipulate the basic characteristic features which are not at all possible by using the bulk material as packaging material. The fundamental properties like optical property magnetic property, melting temperature, electrical conductivity, and elasticity can be thus manipulated which makes them perfect for various food processing technologies. In all these cases, the interesting thing is that, there is no change in the chemical composition when comparing the same composite and its nanocomposite material but the fundamental properties will be significantly different (Shiv and Jong-Whan, 2016). More benefits of using nanocomposite as packaging material are illustrated in Figure 2.1.

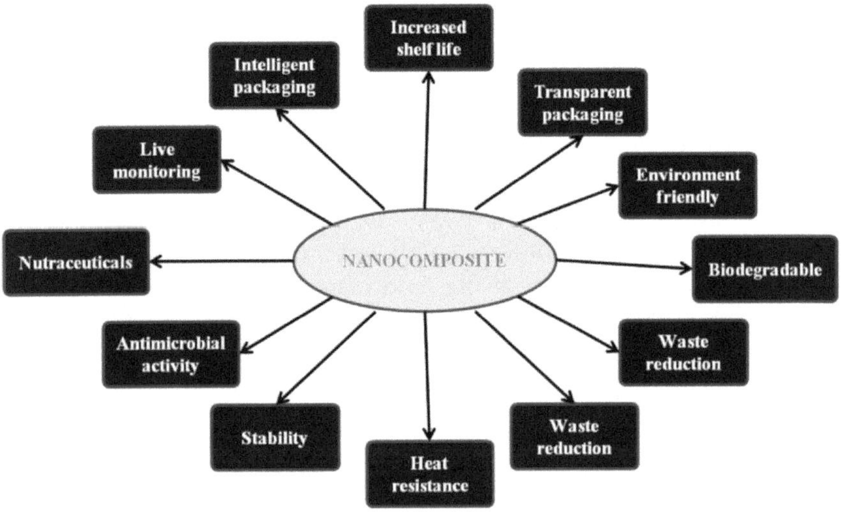

FIGURE 2.1 Key benefits of nanocomposites used as packaging materials.

While improving or manipulating the mechanical, thermal, and other barrier properties of the packaging materials, the shelf life of the food products could be increased without much wastage. In case of changing the optical property of nanocomposite, a completely transparent packaging material can be developed which help the consumer to see the food product before selection. By changing the thermal stability of the material, the heat resistance, stability, and damage of the material by exposing to high processing temperature can be minimized. While coming to the case of active packaging, the antimicrobial property of the filler material and oxygen scavenging property will make available the best food product to the consumers. Also control release and delivery of extra components into the food are possible by nanocomposites which put light to the field of nutraceuticals (Charles, 2014).

Many of the functional groups in the nanocomposite have multiple interactions with inside and outside surroundings which make them use as biosensors. Along with advanced properties like self-cleaning and self-healing properties, detection of various pathogenic microorganisms, by products during spoilage aid in making the nanocomposite a promising material which can be used in intelligent packaging technology. For providing the details of the food products, manipulations also can be done so that proper traceability can be maintained by incorporating technologies like RFID or nano barcode which can make the packaging technology more intelligent (Amra et al., 2015).

2.9 APPLICATIONS OF NANOCOMPOSITES IN FOOD PACKAGING

Application or success of any type of food packaging technology or material is by keeping food safe until consumption and at the same time retain the original quality of the food material. As far as food industry is concerned, the main advantage in using nanocomposite is increase in shelf life of the product by inhibiting spoilage by various food spoiling microorganisms. At the same time, it is possible to regulate other intrinsic as well as extrinsic factors affecting food spoilage like change in temperature, availability of oxygen, water activity, chemicals involved, exposure of various gaseous substance, etc. All the above-mentioned factors have to be maintained and regulated from the starting of packaging and need to survive storage period, transportation, and finally until consumption. By the help of the packaging material and technology used, all the contacts of food product with external factors are totally prevented or avoided by acting as a barrier. The filler materials used in nanocomposites can prevent as well protect the food product from air, water, light, heat, pressure, and microorganism (Shiv and Jong-Whan, 2016; Jong-Whan et al., 2013).

Even though the novel material prevents all physical, chemical, and biological contaminations into the food, it should be environment friendly and biodegradable too. Majority of the conventional packaging material were extremely good at preventing water or moisture, gaseous substance like oxygen, carbon dioxide but possess huge environmental problem and are not sustainable. These actually made researchers to find a packaging material composed of nanocomposites which enhance normal shelf life, keep the food safer and fresh and finally easily degradable as well as environment friendly (Charles, 2014).

2.10 TYPES OF NANOCOMPOSITE FOR PACKAGING

Nanocomposites are broadly divided into organic and inorganic irrespective of polymeric or multi-layered nature. This classification is mainly based on the nanofiller materials used to support the matrix. Majority of the polymer used now are biodegradable materials including proteins, lipids, and complex polysaccharides like chitin, chitosan, gums, lignocelluloses, collagen, gelatine, gluten, etc. There are non-biodegradable materials also like polyolefins and nylon which are not commonly supported and developed since they are not eco-friendly.

Among the organic fillers, the most common and abundantly present one is cellulose and chitin compounds. Cellulose nanofillers such as cellulose nanoparticles, cellulose nanofibrils, cellulose nanocrystals, etc. are different forms of cellulose which is one among the naturally available polymer present abundantly throughout the world inside plant cell wall, agro-wastes, wood, and other organic wastes. The crystalline forms of the cellulose can be made in nanoscale by chemical reactions and can be used as the filler with the matrix materials. The cellulose nanofillers were found to be excellent support for nanocomposite due to their unique properties like low molecular weight and at the same time possessing high strength as well as stiffness. All the cellulose-based nanofillers are eco-friendly and easily biodegradable. Other commonly available nanofillers which are present in fungi, and other crustacean group is chitin. Chitin is a complex polysaccharide and on specific treatments can be converted to good nanofillers like nanowhiskers and nanofibrils. The chitin-based materials are biocompatible and biodegradable without any threat to the environment (Shiv and Jong-Whan, 2016; Young Teck et al., 2014).

Regarding the inorganic nanofillers, the major ones are nanoclay and metallic nanoparticles. Nanoclay or clay nanoparticles are one of the cheapest and easily available inorganic nanoproduct. Silicate is the major component of the nanofiller which makes them highly reactive and a good hydrophilic nanocomposite. Metal nanoparticles can be easily formulated and manipulated in size, structure, and dimensions. The distinguished properties which make them special are the quantum properties including optical, thermal, and magnetic properties (Girilal et al., 2013). Many of the metal nanoparticles including silver have dominant antimicrobial properties which are proved to be efficient antimicrobial agent with a broad-spectrum activity (Sathiyaseelan et al., 2017). Variety of metal nanoparticles including silver, gold, iron, copper, etc. can be synthesized through physical methods, chemical methods, and biological methods.

Majority of these nanocomposites are used as packaging material for dairy products, fruit juices, food grains, bakery products, cheese, and meat products. Comparatively, the production cost of nanocomposite-based packing materials is more but in a long run it is beneficial to nature as it is more biocompatible and more green technology is involved.

2.11 NANOCOMPOSITES AS BIODEGRADABLE PACKAGING

Eco-friendliness is the base of any sustainable development and there is more acceptability for green technologies in this new era. Almost all the conventional packaging materials are petroleum-based plastic materials and its derivatives which are poor in degradability. While using these materials by food consumers, the packaging material accumulation will be more, which will be a great threat to the environment and so there is a huge demand for innovations in packaging material of natural origin and majority of the research is happening in that aspect (Ilke et al., 2014).

Nanocomposites are such a kind of innovation by being biocompatible and easily degradable when compared to other conventional packing materials. The addition or incorporation of fillers will help nanocomposites more eco-friendly. One such report says that biodegradability can be increased by the addition of natural fibers to the nanocomposites. Many of such renewable sources can be better utilized for making the biodegradable nanocomposite and many such are reported which will get degraded within a month. Since the usage of food packaging in the food industries will be much more when compared to other industries, there will be more acceptability for biodegradable nanocomposite as packaging material (Pedro et al., 2009).

Various types of biodegradable food packaging films have been developed using biopolymers. The major challenge for the synthesized nanocomposite is that, it should be mechanically strong, acts as gas as well as water vapor barrier and the same time biodegradable. Many nanofillers from renewable resources were put under study and one of such materials reported was carrageenan biopolymer with chitin and also with cellulose. A study with reinforcement with clay as well as gelatin also was reported with better barrier properties and at the same time eco-friendly (Ahmed, 2013).

2.12 NANOCOMPOSITES AS ACTIVE PACKAGING

Once packed, there are many intrinsic and extrinsic factors that play a major role in food getting spoiled. This includes the influence of presence of oxygen, carbon dioxide, water activity, food pathogens, and many others. If the packaging material interacts and counteracts with these intrinsic and extrinsic factors on a regular basis only it will be possible to extend the shelf life of the food product during the storage, transport, and distribution process. The packaging material which is an effective barrier and should absorb the unwanted by products formed or avoid the external surrounding gases. At the same time, it should stop the growth or remove the microorganisms which lead to the spoilage of food. Technology also has been introduced for supplementing additional compounds including antioxidants, vitamins, and minerals to the food while it is in contact with the food. Controlled release of the supplements to the food can be regulated by manipulating the system. The controlled release of water vapor, oxygen, and carbon dioxide also is possible by the advancement of active packaging (Shiv and Jong-Whan, 2016; Farhoodi, 2016).

The presence of gaseous products and microbial metabolite inside the packed food is a real problem in the food industry since it causes serious sensory problems like loss in nutrients, change in color, flavor, and finally provides a favorable condition for many aerobic microorganisms to grow inside. In the case of fruits and vegetable packaging, the presence or absence of oxygen will influence the production of ethylene which is crucial in their ripening processes. Complete removal of oxygen is also a challenge in the food industry since many of the food products need a considerable amount of oxygen to maintain their organoleptic characteristic features. Complete removal of oxygen also favors the growth of anaerobic microorganisms inside the food (Henriette, 2009).

High protein content in food materials pose a risk of contamination from aerobic microorganism and to pack such kind of food materials, packages releasing inert gases including carbon dioxide can be done to prevent microbial growth. This can be achieved by adding oxygen scavenging compounds inside the package. Creating a vacuum inside the container is another choice but is not applicable in the case of flexible packaging.

The packaging materials should not cause a problem to the environment after disposal too. Since food particles will retain in the packaging material, there will be further chance of invading microorganisms creating problems. This challenge also can be avoided with the help of antimicrobial substances present in the packaging material until it gets degraded. The nanocomposite material shall be specially premeditated for absorbing unwanted odor and at the same time release of various aromatic compounds in a regular time (Henriette, 2009).

In short, active nanocomposite food packaging enables the removal of undesirable microorganisms, removal or retaining of water vapor, ethylene, ethanol, enhance oxygen scavenging, and sustain release of antimicrobial compounds. By exploiting the optical and thermal capability of the packaging material food materials can be protected from light, microorganism, and even from UV light. The headspace of the container or package can accommodate gases and food supplements which can be designed for sustain release. As part of this controlled release of oxygen, carbon dioxide and water vapor, microbial metabolite and other volatile compounds are possible.

2.13 NANOCOMPOSITES AS INTELLIGENT PACKAGING

Even though many of the nanocomposite packagings are transparent, it is the right of the consumer to know better about the quality of the product they wish to buy. In the case of intelligent packaging, information about the condition of food inside the packaging will be conveyed to the consumer without opening the packaging. There comes the importance of an intelligent packaging which is quite possible with the advancement of nanocomposites. This is possible by the easily incorporation of biosensors in the nanocomposite matrix. These nanobiosensors can interact with various gaseous or other by-products formed inside the food, and the resulting reaction will be reflected as a change in color or similar which the consumer can experience by just looking at the package. The same can happen when the food package or food inside reacts with any of the extrinsic or intrinsic factor which affects the quality of the food (Zohreh et al., 2016).

The nanosensors specific for common microbial pathogens or food spoilage microorganisms can be incorporated so that it will help the food producers as well as the consumers to be aware of the risk involved. Along with the pathogens the nanocomposite packages can be designed to detect pesticides, any chemical toxins, biological toxins including aflatoxins or patulin which are very difficult to normally identify by any consumers. Any flavor-changing compounds or gaseous substances also can be set to monitor using the nanosensors. Sensors can be specified to monitor temperature also which will give an idea about how was the storage condition and how safe is the food to be consumed due to inappropriate storage temperatures. Similarly, many other sensors can be incorporated or designed by manipulating the packaging materials like monitoring leakage in the containers, temperature indicators, humidity indicators, radiation indicators, agglomeration indicators, etc. to increase the acceptability of the product in the market. Once properly implemented and validated, there will be no need to put expiry dates on intelligent nanocomposite packaging since any consumer can personally visualize and verify whether the food product is safe to consume. Even though the cost of production of nanosensors will be high, unnecessary as well as costly quality checks and physical/chemical/microbiological analysis can be avoided with the help implementing sensors (Charles, 2014).

2.14 NANOCOMPOSITES AS ANTIMICROBIAL PACKAGING

The use of protective nanocoating with antimicrobial properties on suitable packaging is an ever-interesting topic in food nanotechnology. Various nanocomposites with metal nanoparticles as the filler material have the potential for increased shelf life of many food products by its enhanced broad-spectrum antimicrobial activity (Fayaz et al., 2009). There has been a strong push toward the formulation of silver nanoparticles containing nanocomposite packaging materials for commercial use. Silver nanoparticles have been proved with its broad spectrum of antimicrobial activity against various Gram-positive and Gram-negative bacteria, yeast, mold, and viruses (Ip et al., 2006).

Nanocomposite films having antimicrobial properties could increase the shelf life by preventing the growth of spoilage as well as microorganisms. For instance, the work reported by Fayaz and co-workers, recommended the simplest way of silver nanoparticles incorporated sodium alginate films enhancing the shelf life of fruits and vegetables without any protein loss or weight loss (Fayaz et al., 2009). Nanoparticle incorporated in the film (Fig. 2.2A) retained the antimicrobial activity (Fig. 2.2B), which prevents vegetable or fruit from spoiling microorganisms and at the same time prevented protein as well as weight loss from the vegetable and fruit. Through the work, it was also proved that the sensory attributes of the vegetable and fruits also are not affected by using nanoparticles-incorporated film by performing a sensory analysis test. The prepared nanocomposite films were tested for antimicrobial activity against different Gram-positive and Gram-negative bacteria. There was no significant difference in soluble protein content during the storage period (Fig. 2.3A). Regarding the weight loss, silver nanoparticles-incorporated edible film-coated fruits and vegetables were observed to have minimum weight loss compared to film without nanoparticles (Fig. 2.3B).

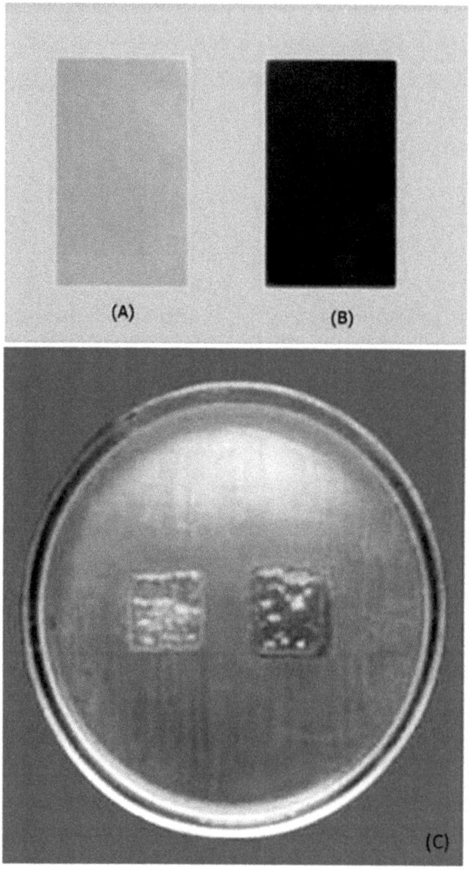

FIGURE 2.2 (A) Edible film control (without nanoparticles). (B) Edible film incorporated with silver nanoparticles. (C) Zone of inhibition test using control film and silver nanoparticle incorporated films against bacteria.

Regarding the sensory analysis test, the overall acceptability for uncoated and coated vegetables (sodium alginate film and silver nanoparticle-incorporated sodium alginate) is represented in Figure 2.4. Vegetables coated with silver nanoparticle incorporated film were found acceptable up to 10 days of storage, as judged based on the texture, color, aftertaste, and appearance compared to uncoated and film-coated vegetable. Acceptance decreased in the other control cases due to the gradual increase in microbial attack leading to the weight loss as well as protein loss. In short, antimicrobial nanocomposite film proved its sensory acceptance due to its structural integrity and enhanced barrier properties, and the antimicrobial properties contributed by silver nanoparticles.

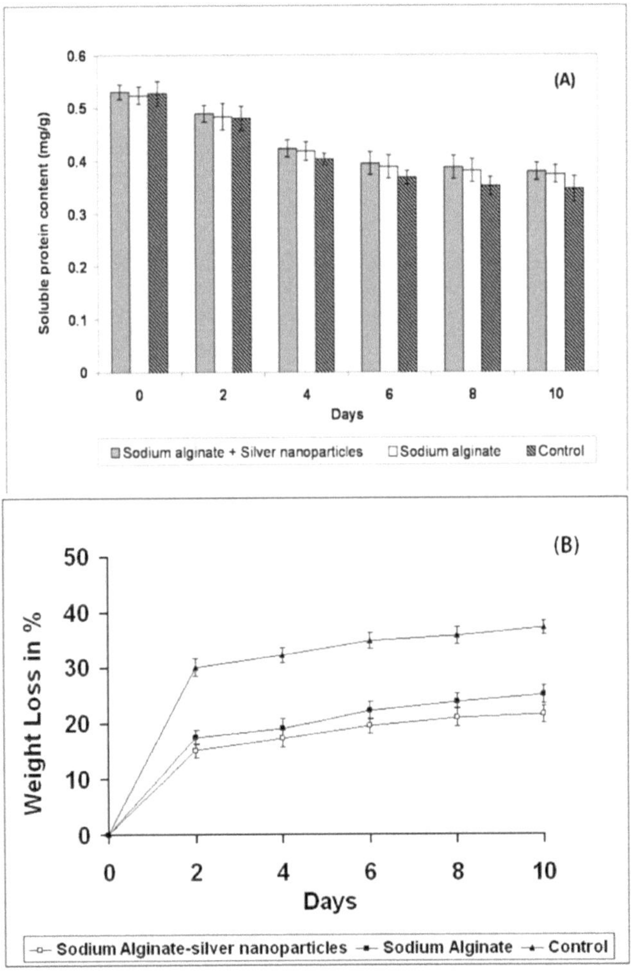

FIGURE 2.3 (A) Soluble protein content of vegetables coated with silver nanoparticle incorporated film. (B) Percentage of weight loss of vegetables coated with silver nanoparticle incorporated film.

Source: Reprinted (adapted) with permission from Fayaz et al. (2009), Copyright © 2009 American Chemical Society.

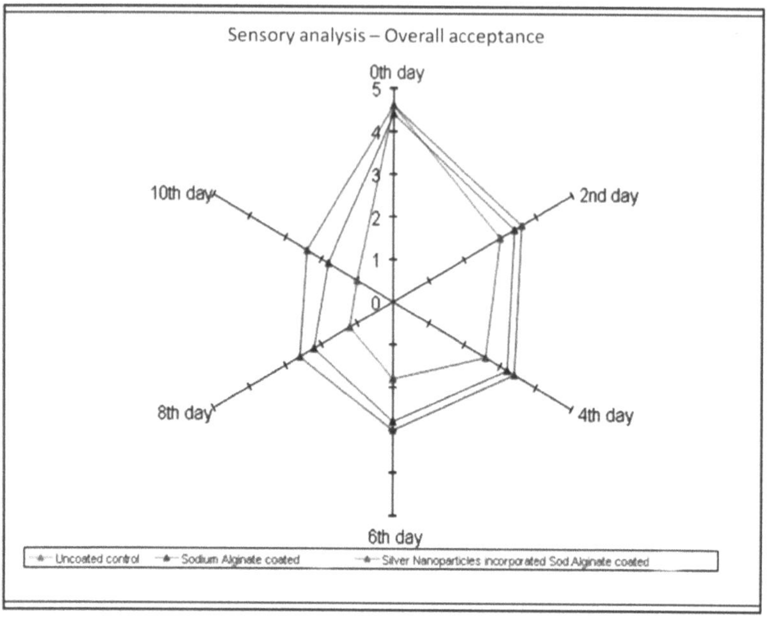

FIGURE 2.4 Sensory analysis of overall acceptance on carrots.

Source: Reprinted (adapted) with permission from Fayaz et al. (2009), Copyright © 2009 American Chemical Society.

2.15 NANOCOMPOSITES AS EDIBLE NANOCOATING

The basic functional properties of conventional edible coatings and films are subjected to the characteristic features of the film-forming materials used. At present, the primary film-forming materials used to construct the edible coatings and films are polysaccharides, proteins, and lipids. Mostly, lipid-based films act as good barriers to moisture, and at the same time less resistance to gas transfer and poor mechanical strength (Fayaz et al., 2009). In the case of biopolymer-based films, they are good as oxygen and carbon dioxide barriers, but they provide little protection against moisture (Jochenweiss and McClements, 2006). Normal cases, edible coatings or films act as barriers to moisture, lipid, and gas. Instead, they increase the textural properties of food or act as carriers of functional agents such as flavors, colors, antimicrobials, nutrients, and antioxidants.

In the case of fruits and vegetables, different microorganisms are responsible for the spoilage, thus decreasing their quality and shelf life.

Antimicrobial coating without preservative has been getting more attention in regulating and preventing food-borne microbial outbreaks. Naturally occurring enzymes such as lysozymes, lactoperoxidase, and glucose oxidase have been used as antimicrobial agents for food preservations (Davidson, 2001).

Edible films from plant, animal, or microbial origin are normally composed of complex polysaccharides which can degrade easily by microorganisms themself. There are limited numbers of chemically synthesized polymers like polylactic acid, thermoplastic starch, and poly (butylene adipate-co-terephthalate) which can be consider as edible films which have the capability of biodegradation.

2.16 HEALTH RISK IN USING NANOCOMPOSITES

The major risks related with nanocomposites are the toxicity-related problems. The wide applicability of metal nanoparticles in various industries like food, medicine, and pharmaceutical industries leads to their over utilization. As the nanoparticles behave entirely different from their corresponding bulk material, it is difficult to predict its side effects and distant effects (Cornelia, 2018). Same as in the case of polymeric nanoparticles too, that they are significantly contributing to the basic sciences and have been tested for its vast applications. The toxicity studies about polymeric nanoparticles also is not fully analyzed or studied. The toxic effects of nanoparticles are mostly because of the hazardous chemicals that have been used in the synthesis process, especially the reducing agents, which limit their usage (Girilal et al., 2015).

In many of the nanocomposites, the nanomaterials which are used as fillers are attached to the matrix properly with the help of functional groups. But the risk of these nanoparticles migrating to the food materials cannot be neglected and, in many cases, it was proved this migration happens even though in a minute quantity. This is a major issue to be considered as far as health of consumers is considered (Zohreh et al., 2016; Cornelia, 2018). A proper testing analysis is required since nanoparticles have been proved to have many toxic effects on different living systems. A living system, either plant or animals when exposed to stress factors like nanoparticles, it will respond back by synthesizing highly conserved proteins called heat shock proteins (HSP) or stress proteins. It has been proved that muscle cells, kidney cells, liver cells, etc. are badly affected when exposed to metal nanoparticles, especially chemically synthesized nanoparticles counterparts

(Fig. 2.5). Most of the membrane proteins, phospholipid layers, and even DNA damage are reported leading to cell death due to the presence of Reactive Oxygen Species (ROS) formed in the presence of stress factors like nanoparticles (Girilal et al., 2015; Girilal et al., 2018).

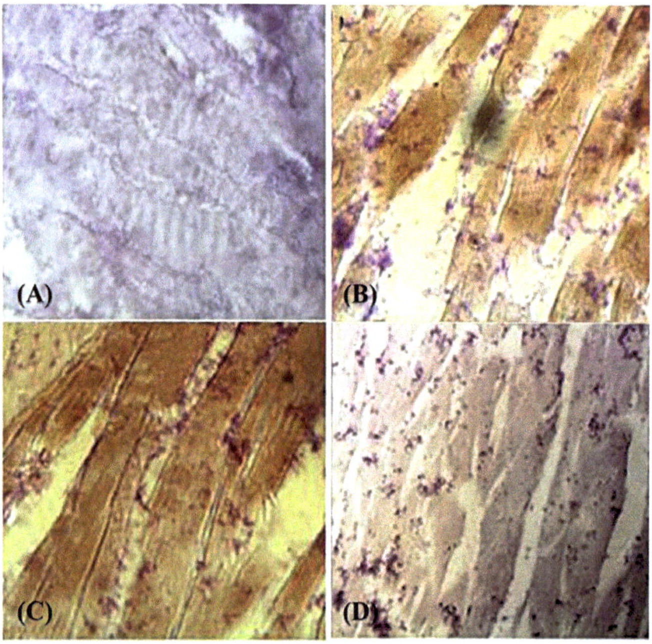

FIGURE 2.5 Immunohistochemical expression of HSP70 in fish muscle cells (A) Control fish without any nanoparticles exposure (B) Fish exposed to silver ions (C) Fish exposed to chemically synthesized nanoparticles (D) Fish exposed to biologically synthesized nanoparticles.

To conclude, it is proved from many of the studies that all since the nanoproducts extremely small and it have very less comparable properties similar to the bulk materials, proper studies need to be done before human use. Proper care and attention in the nanocomposite production as well application process need to be given due to their unpredictable toxic effects; especially the chemically synthesized nanoparticles. The immediate effect of the nanowastes is visible on water sources and animals in water, which is the main gateway for the entry of nanoparticles into higher living systems including animals, plants, and humans. So, without proper trials and experiments, it needs not to be commercialized for extensive use.

2.17 CONCLUSION

Nanocomposite opens a great chance of evolution in the field of food packaging. Since it is having a matrix and nanofillers, a variety of applications can be assigned to the nanocomposites starting from preservation, removal of microorganism, food safety, etc. Almost all nanosensors can be incorporated to monitor various extrinsic and extrinsic factors influencing food spoilage. The nanosensors give the consumers an extra confidence by monitoring themselves visually. The consumers need not even look into the expiry date since they have to go through various sensors to know the quality of the food they wish to consume. Total traceability of the food is possible for the consumers by proper implementation of nanocomposites in food packaging since it will let them know even the storage temperature. Food producers also will be welcoming the advantage of using nanocomposites because once nanocomposite efficiency is tested and validated; they can even skip the costly and tiresome quality analysis.

But in terms of consumer safety, it is important to ensure that there is no nano-products coming into the food or in other words migration of nanoparticles from nanocomposite to food particles is not happening. There are only limited number studies executed in this aspect and that too in many of the studies, only immediate side effects of the nanocomposites have been analyzed and studied. Proper evaluation and potential migration studies need to carry out to become aware of the potential health risk related to nanocomposites once it reaches inside the body.

KEYWORDS

- **food preservation**
- **food packaging**
- **packaging materials**
- **conventional packaging**
- **nanocomposite**
- **biosafety**

REFERENCES

Ahmed, M. Y. Polymer Nanocomposites as a New Trend for Packaging Applications. *Polym.-Plastics Technol. Eng.* **2013,** *52*, 635–660.

Amra, B.; Amra, O. S.; Indira, S. Application of Polymer Nanocomposite Materials in Food Packaging. *Croatian J. Food Sci. Technol.* **2015,** *7* (2), 86–94.

Charles, C. O. Nanocomposites—An Overview. *Int. J. Eng. Res. Dev.* **2013,** *8*, 17–23.

Charles, C. O. The Benefits and Applications of Nanocomposites. *Okpala Int. J. Adv. Eng. Technol.* **2014,** *5*, 12–18.

Cheruvu, P.; Kapa, S.; Mahalik, N. P. Recent Advances in Food Processing and Packaging Technology. *Int. J. Auto. Control* **2008,** *2*, 418–435.

Cornelia, V. Polymeric Nanocomposites and Nanocoatings for Food Packaging: A Review. *Materials* **2018,** *11*, 1834–1883.

Davidson, P. M. Chemical Preservatives and Natural Antimicrobial Compounds. In *Food Microbiology Fundamentals and Frontiers*; Doyle, M. P., Beuchat, L. R., Montville, T. J., Eds.; ASM Press: Washington, DC, 2001; pp 593–627.

Farhoodi, M. Nanocomposite Materials for Food Packaging Applications: Characterization and Safety Evaluation. *Food Eng. Rev.* **2016,** *8*, 35–51.

Fayaz, A. M.; Balaji, K.; Girilal, M.; Kalaichelvan, P. T.; Venkatesan, R. Mycobased Synthesis of Silver Nanoparticles and Their Incorporation into Sodium Alginate Films for Vegetable and Fruit Preservation. *J. Agric. Food Chem.* **2009,** *57*, 6246–6252.

Fellows, P. J. *Food Processing Technology: Principles and Practice*, 2nd ed.; Woodhead Publishing Limited, CRC Press, 2000.

Girilal, M.; Fayaz, A. M.; Elumalai, L. K.; Sathiyaseelan, A.; Gandhiappan, J.; Kalaichelvan, P. T. Comparative Stress Physiology Analysis of Biologically and Chemically Synthesized Silver Nanoparticles on *Solanum lycopersicum* L. *Colloid Interf. Sci. Commun.* **2018,** *24*, 1–6.

Girilal, M.; Fayaz, A. M.; Mohan, B.; Kalaichelvan, P. T. Augmentation of PCR Efficiency Using Highly Thermostable Gold Nanoparticles Synthesized from a Thermophilic Bacterium, *Geobacillus stearothermophilus. Colloids Surf. B Biointerf.* **2013,** *106*, 165–169.

Girilal, M.; Krishnakumar, V.; Poornima, P.; Fayaz, A. M.; Kalaichelvan, P. T. A Comparative Study on Biologically and Chemically Synthesized Silver Nanoparticles Induced Heat Shock Proteins on Fresh Water Fish *Oreochromis niloticus. Chemosphere* **2015,** *139*, 461–468.

Gupta, R.K.; Dudeja, P. Food Packaging. In *Food Safety in 21st Centuary*; Academic Press, 2017; pp 547–553.

Henriette, M. C. Nanocomposites for Food Packaging Applications. *Food Res. Int.* **2009,** *42*, 1240–1253.

Ilke, U. U.; Guido, C.; Eva, M.; Carlo, A. C.; Stefano, F. Nanocomposite Films and Coatings Using Inorganic Nanobuilding Blocks (NBB): Current Applications and Future Opportunities in the Food Packaging Sector. *RSC Adv*. **2014,** *4*, 29393–29428.

Ip, M.; Lui, S. L.; Poon, V. K. M.; Lung, I.; Burd, A. J. Antimicrobial Activities of Silver Dressings: An In Vitro Comparison. *Med. Microbiol.* **2006,** *55*, 59–63.

Janjarasskul, T.; Krochta, J. Edible Packaging Materials. *Annu. Rev. Food Sci. Technol.* **2010,** *1*, 415–448.

Jochenweiss, P. T.; McClements, D. J. Functional Materials in Food Nanotechnology. *J. Food Sci.* **2006,** *71*, 107–116.
Jong-Whan, R.; Hwan-Man, Park.; Chang-Sik, H. Bio-Nanocomposites for Food Packaging Applications. *Progress Polym. Sci.* **2013,** *38*, 1629–1652.
Kenneth, M.; Betty, B. Food Packaging—Roles, Materials, and Environmental Issues. *J. Food Sci.* **2007,** *72* (3), 39–55.
Mahalik, N.; Nambiar, A. Trends in Food Packaging and Manufacturing Systems and Technology. *Trends Food Sci. Technol.* **2010,** *21*, 117–128.
Paine, F.A.; Paine, H. Y. *A Handbook of Food Packaging*, 2nd ed.; Blackie Academic and Professional: London, 1992.
Pedro, H.; Cury, C.; Kestur, G. S.; Fernando, W. Nanocomposites: Synthesis, Structure, Properties and New Application Opportunities. *Mater. Res.* **2009,** *12*, 1–39.
Robertson, G.L.; Marcel, D. *Food Packaging- Principles and Practice*; New York, 1993.
Sathiyaseelan, A.; Shajahan, A.; Girilal, M.; Ramachandran, R.; Kaviyarasan, V.; Kalaichelvan, P. T. Synthesis, Characterization and Biological Applications of Mycosynthesized Silver Nanoparticles. *3 Biotech* **2017,** *7*, 333–342.
Shiv, S.; Jong-Whan, R. Polymer Nanocomposites for Food Packaging Applications. In *Functional and Physical Properties of Polymer Nanocomposites*; Aravind, D., James, N., Eds.; John Wiley & Sons, Ltd. Published, 2016; pp 29–55.
Sun, D. W. *Handbook of Frozen Food Processing and Packaging*; CRS Press, Taylor and Francis Group Eds, 2006.
Young Teck, K.; Byungjin, M.; KyungWon, K. Chapter 2–General Characteristics of Packaging Materials for Food System. In *Innovations in Food Packaging*, 2nd ed., 2014; pp 13–35.
Zohreh, H.; Zahra, H.; Morteza, M. Nanocomposites in Food Packaging Applications and Their Risk Assessment for Health. *Electron. Phys.* **2016,** *8*, 2531–2538.

CHAPTER 3

Characterization of Polymer/Clay Nanocomposites

SHIJI MATHEW and E. K. RADHAKRISHNAN*

School of Biosciences, Mahatma Gandhi University, Kottayam 686560, Kerala, India

**Corresponding author. E-mail: radhakrishnanek@mgu.ac.in*

ABSTRACT

Polymer/clay nanocomposites have established a protuberant position in the food industry as effective food preserving and food packaging materials. Polymer/clay nanocomposites are composed of two phases; a continuous phase, which is the polymer matrix into which a dispersion phase, that is, nanofillers (clay or metallic nanoparticles) is mixed to obtain the desired physicochemical properties. This chapter provides a brief description on the various properties of polymer nanocomposites and adds a note on their different methods of preparation, such as in situ polymerization, solution dispersion/solvent casting and melt intercalation method. After preparation, the developed nanocomposites must be well characterized for better understanding the structure of nanocomposites and to check the effective dispersion of the nanofillers in it. This chapter provides a brief description on the various properties of polymer nanocomposites and adds a note on their different methods of preparation. The later section gives a detailed information on the morphological and physicochemical analytical techniques used for the characterization of polymer/clay nanocomposites with suitable examples.

3.1 INTRODUCTION

One of the fastest growing fields in nanotechnology is the nanocomposite food packaging, which is a good alternative for conventional packaging. Polymer nanocomposites developed with the incorporation of nanosized fillers are found to be a solution for improving the properties of commonly available food packaging material. Nanoparticles display unique physiochemical properties apart from their macroscale counterpart by virtue of their small size, large surface area-to-volume ratio and surface activity. Hence, nanomaterials can play important role in improving the mechanical and barrier properties along with imparting their antibacterial properties to the polymers designed for packaging purposes (Sharma et al., 2017). The inclusion of nanoclays, such as smectite, halloysite, hectorite, saponite, kaolinite, mica or montmorillonite in the packaging material is found to successfully overcome the pitfalls faced by the pristine polymer.

Among the varied types of polymer nanocomposites, polymer–clay nanocomposites are the most promising ones. The delamination of a very small amount of clay is found to drastically improve the characteristics of the polymer. This technology relies on the high surface area of clay particles of 750 m^2/g and high aspect ratio of 100–500 (Arora and Padua, 2010). Polymer–clay nanocomposites first appeared in 1950 and since then, there has been a growing interest in such composites. Polymer–clay nanocomposites can be classified into three major types: (1) exfoliated (2) intercalated, and (3) flocculated or tactoid nanocomposites (Carrado, 2000). Exfoliated nanocomposites are found to exhibit best properties as the clay layers extensively delaminate and get randomly dispersed in the polymer matrix showing optimal clay–polymer interactions. In intercalated nanocomposites, an ordered multilayer sheet with alternating polymer/clay layers is formed. Here, the polymer chain gets penetrated into the interlayer region of clay and a moderate affinity is seen between both. Tactoid or flocculated nanocomposites show immiscibility of polymer and clay tactoids, indicating poor affinity (Arora and Padua, 2010). Here the interlayer space of clay gallery does not expand and hence a true composite cannot be formed.

For better understanding of the structure of polymer/clay nanocomposites, synergistic combinations of different characterization methods are to be used. Various qualitative and quantitative methods, such as microscopic, spectroscopic, and other analytical techniques are available for characterization of polymer/clay nanocomposites. The first section of this chapter discusses in detail about the important properties introduced in a polymer composite after the inclusion of nanofillers. Then, in the later section, the important methods of preparing polymer/clay nanocomposites are discussed briefly. This is followed by a comprehensive description on the various qualitative and quantitative techniques used for the characterization of polymer/clay nanocomposites with appropriate examples.

3.2 PROPERTIES OF POLYMER/CLAY NANOCOMPOSITES

The proper dispersion of nanofillers in polymer matrices can lead to the development of polymer/clay nanocomposites which can have better properties than the pristine polymer used for food packaging purpose. These nanofillers can significantly modify or improve different properties of polymers in which they are incorporated. The properties improved include high tensile strength (TS) and modulus, heat resistance, decreased gas permeability and flammability, and enhanced biodegradability of biodegradable polymers. These properties are achieved as a result of strong interfacial interaction between the nanofillers and polymer, compared with conventional fillers.

3.2.1 MECHANICAL PROPERTIES

The mechanical properties of nanocomposites, such as TS, Young's Modulus (YM) and elongation at break (EB) give information about their stiffness, brittleness, and ductility. In a recent study, tensile testing was performed to investigate the impact of blending of boiled rice starch (BRS) and starch-mediated silver nanoparticles (sAgNPs) on the mechanical properties of polyvinyl alcohol (PVA)-based bionanocomposite film (Mathew et al., 2019a). For this, YM, TS and EB of the films and the control were analyzed (Fig. 3.1).

Figure 3.1a shows that the E of nanocomposite is significantly influenced by the presence of boiled rice starch and the sAgNPs. The E of neat PVA increased from 290 to 368 MPa following the formation of PVA-mediated AgNPs in it. In the case of PVA/BRS blend film, a relative increase in the value of E to 1196 MPa was observed, which the authors stipulate to be due to the enhanced H-bond interaction between PVA and starch molecules. However, when compared with PVA/BRS blend film, the value of E is reduced to 682 MPa in the case of PVA/BRS/sAgNPs with the generation of silver nanoparticles. The authors explained this to be due to the weak interaction between the polymer chains and AgNPs which probably resulted in chain separation and reduced chain entanglement.

The TS of nanocomposite films also articulated a similar trend as that of YM (Figure 3.1b). TS of neat PVA film was observed to be doubled (from 16.8 ± 0.8 to 32.2 ± 0.6 MPa) after blending with BRS. Notably, the generation of sAgNPs in the nanocomposite blend film showed a decrease in the TS of PVA/BRS binary blends from 32.2 ± 0.61 to 26.5 ± 0.3 MPa which was explained to be due to the weak chemical interaction between polymer chains and AgNPs. At the same time, Figure 3.1c shows that the EB of control PVA decreased after blending with rice starch water. In addition, the presence of sAgNPs in PVA/BRS/sAgNPs also resulted in a decreased EB which the authors presumed to be due to restricted polymer chain movement due to the presence of sAgNPs. After all, the developed PVA/BRS/sAgNP bionanocomposite film showed highly improved TS and YM at the expense of EB when compared with the control PVA film.

The formation of polymer–clay nanocomposites has demonstrated to have marked improvement in the mechanical properties of native polymers materials even with the loading of very small quantity of clay. A recent work reported the development of highly improved PVA-based nanocomposites modified with 0.03% montmorillonite K10 (MMTK10) clay mineral and decorated with ginger-mediated silver nanoparticles for sausage packaging application (Mathew et al., 2019b). Among all the films prepared, the nanocomposite film, PAGM showed highly improved TS (29.4 ± 0.47 MPa), and increased YM (0.982 ± 1.9 GPa) (Table 3.1). The authors explain this increase in TS and E of the nanocomposite films to be due to the proper dispersion of nanofillers in the polymeric matrix which supported the regulation of the stress transfer at the interface of the filler and matrix. On the other hand, eB for neat P film was found to be reduced in the PAGM nanocomposite after blending with ginger extract, MMTK10 clay mineral, and AgNPs. This reduction in eB was elucidated to be due to the immobilization of the polymer chains by the presence of rigid nanoparticles, such as clay and AgNPs. But, ultimately, the mechanical stability of the developed nanocomposite was much improved than the neat P films, thereby confirming the role of nanoclay and silver nanoparticles in it.

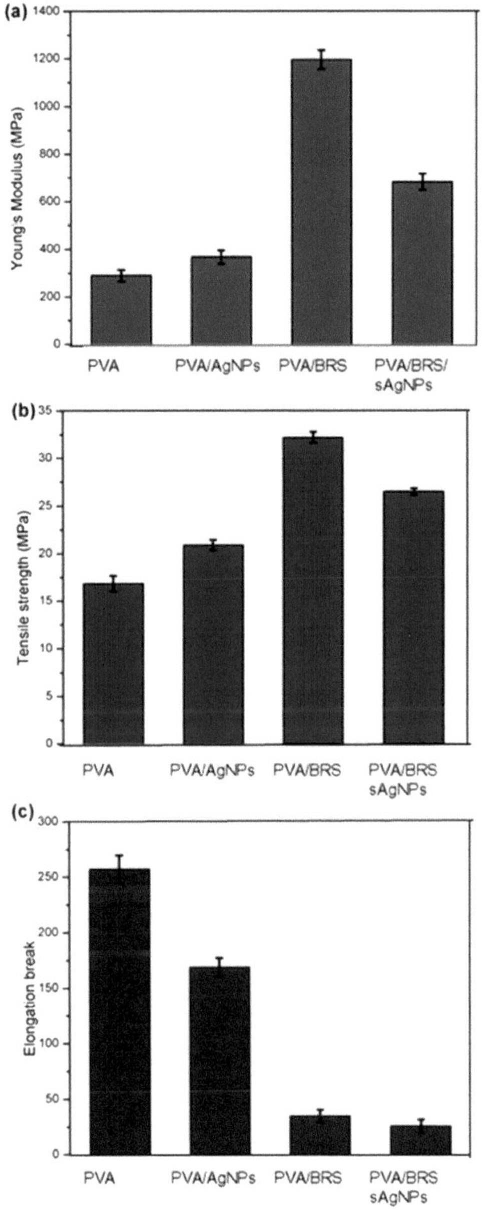

FIGURE 3.1 Mechanical properties (a) Young's Modulus. (b) Tensile strength. (c) Elongation at the break of PVA/BRS/sAgNPs nanocomposite films and its controls.

Source: Reprinted with permission from Mathew et al. (2019a). Copyright @ 2019, Elsevier B.V.

TABLE 3.1 Thickness and Mechanical Properties of the Nanocomposite Films.

Film sample	Thickness (mm)	Tensile strength (MPa)	Young's modulus (GPa)	Elongation at break (%)
P	0.08 ± 0.001	16.85 ± 0.425	0.482 ± 1.45	257 ± 1.8
PG	0.1 ± 0.06	18.5 ± 0.325	0.521 ± 2.6	60.1 ± 2.03
PM	0.12 ± 0.07	22.4 ± 0.52	0.926 ± 2.3	125.2 ± 2.5
PGM	0.21± 0.007	25.6 ± 0.28	0.882 ± 2.4	31.5 ± 1.45
PAGM	0.13 ± 0.08	29.4 ± 0.47	0.882 ± 1.9	92.8 ± 1.41

Values are given as mean ± SD (n = 3)

Source: Reprinted with permission from Mathew et al. (2019c).

3.2.2 BARRIER PROPERTIES

The polymeric materials commonly used for packaging applications possess a major disadvantage of being highly permeable to gases, water, and other molecules. The incorporation of clay nanoparticles in the polymer or blending with other polymers is a useful strategy to achieve improved gas and water barrier properties. In addition, the inclusion of nanoparticles in the polymer composites can result in decrease in the rate of transfer of gases and solvents across the matrix, thereby increasing its suitability as gas/water repellent food packaging material. The reason for increase in barrier properties of polymer–clay nanocomposites can be attributed to two important mechanisms (Müller et al., 2017).

1) By creation of tortuous pathway for the diffusion of gases and solvents and thus forcing them to follow a tortuous path surrounding the nanofillers, resulting in increased path length for diffusion
2) By changing the interfacial region of the polymer matrix, itself. The following paragraph shows the effect of nanoclays/metallic nanoparticles in improving/increasing the barrier properties of nanocomposites.

Water Barrier Properties

In a recent study, the role of MMT clay in improving the water barrier property of agar/gellan gum/MMT (AGM) nanocomposite film was checked (Lee et al., 2019). The water vapor permeability (WVP) of the control composite film without MMT clay, AG was 1.9 and with the loading of 10 wt.% MMT, the WVP of AGM-10 nanocomposite film decreased to 1.70. This clearly indicated the role of MMT in significantly improving the water barrier property of nanocomposites.

In another work, De Silva et al. developed nano–MgO reinforced chitosan nanocomposites and evaluated the role of the nanoparticles in improving the water barrier property (De Silva et al., 2017). The water vapor transmission rate (WVTR) of pure chitosan films was found to be 172.3 g/m^2 h. This was found to be significantly reduced to 127 g/m^2 h (27%) with the addition of 5 w/w% MgO nanoparticles and 117.3 g/m^2 h (32%) with the addition of 10 w/w% MgO nanoparticles. This thereby proves the role of MgO nanoparticles and similar nanomaterials in improving the water resistance of nanocomposites.

Oxygen Barrier Properties

Tas and colleagues measured the oxygen transmission rate (OTR) of Halloysite nanotube- polyethylene (HNT/PE) nanocomposite film to examine the role of HNT clay in improving the gas barrier properties of the nanocomposite film (Tas et al., 2017). The results showed that when the PE matrix was loaded with 1 wt.% halloysite nanotube, the resultant nanocomposite exhibited 22% decrease in OTR compared with neat PE films. At the same time, the greatest OTR was shown by HNT/PE nanocomposite loaded with 5 wt.% HNT. The authors explained that at higher concentration of HNT, large aggregates of clay particles were formed which thereby formed voids in the matrix, leading to faster movement of gas through it. Similarly, in another study, Dabbaghianamiri et al. reported the development of self-assembled MMT clay–polyvinyl alcohol nanocomposite as a safe and efficient gas barrier material. The developed MMT/PVA nanocomposite showed an extremely high oxygen barrier capacity with an OTR rate of zero when compared with clean PET film which had an OTR of 14 cc/m^2 day.atm (Dabbaghianamiri et al., 2020).

3.2.3 THERMAL PROPERTIES

Another disadvantage of the conventional petroleum packaging materials is that they are very less heat tolerant. When heated under the absence or presence of O_2, nonoxidative or oxidative degeneration occur respectively which can lead to the release of large number of volatile compounds into the food package. The reinforcement of metallic nanoparticles and nano-clays into the pristine polymer plays an important function in improving its thermal resistance and can effectively shield the release of volatile compounds into the food by acting as an effective insulator and even encapsulate the volatile compounds (Mathew and Radhakrishnan, 2019).

Thermogravimetric analysis (TGA) is a very useful tool for assessing the thermal properties and moreover to determine the weight change in a film sample as the function of temperature. In a recent study, TGA nanocomposite film samples was conducted to evaluate the concentration of AgNPs present in the PVA/BRS/sAgNPs bionanocomposite and also to study the influence of AgNPs on the thermal properties of the nanocomposite (Mathew et al., 2019a). Figure 3.2 shows the thermogram of the film samples wherein in the case of neat PVA film and PVA/BRS blend film, a major weight loss in the range between 300 and 639°C was detected which indicates its complete decomposition without any residual mass. The inset of Figure 3.2 demonstrates that the presence of AgNPs in PVA/AgNPs and PVA/BRS/sAgNPs films has resulted in a significant change in the thermal degradation pattern where the decomposition temperatures were found to be shifted to higher degrees. Unlike the neat PVA and PVA/BRS blend film, at 699°C, a residual mass of 1.11 and 1.05 wt.%, respectively for PVA/AgNPs and PVA/BRS/sAgNPs was seen left after the degradation, which confirms the influence of AgNPs in improving the thermal stability of these nanocomposites.

3.3 PREPARATION OF POLYMER/CLAY NANOCOMPOSITES

Various methods are followed for preparation of polymer/clay nanocomposites which include high-energy ball milling, electrospinning, centrifugal infiltration, liquid phase infiltration, solution intercalation, melt intercalation, grafting melt intercalation, rapid solidification, and 3D printing. Among these, the most commonly used methods are solvent casting, melt intercalation, and in situ intercalative polymerization which are briefly defined here.

3.3.1 SOLVENT CASTING

This method is also known as solution blending. Here, first, the polymer is dissolved in a suitable solvent. Simultaneously, the clay mineral is also dispersed in the same solvent separately. Then the clay/solvent dispersion is mixed with polymer/solvent mixture. The final polymer/clay/solvent solution is allowed to homogenize for some time and then casted on to a flat support like petri dish. During this process, the removal of the solvent occurs by evaporation, leaving behind the nanocomposite film (Gong et al., 2014).

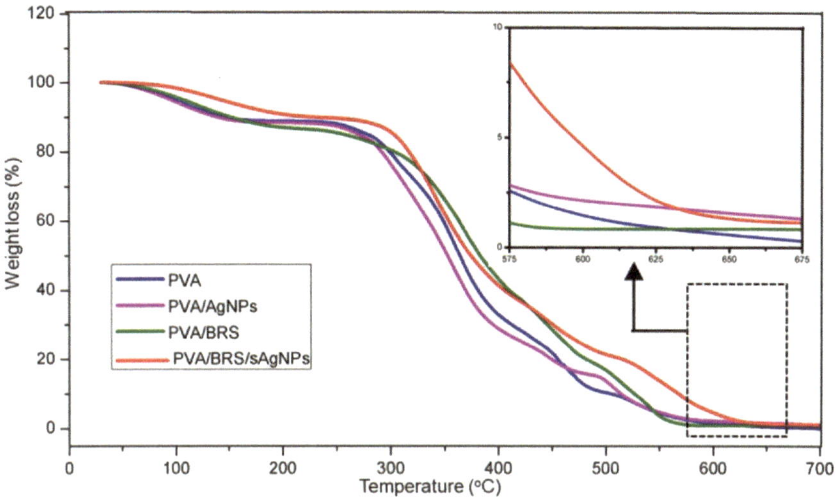

FIGURE 3.2 Thermogram of PVA/BRS/sAgNPs nanocomposite and its controls.
Source: Reprinted with permission from Mathew et al. (2019a). Copyright @ 2019, Elsevier B.V.

3.3.2 IN SITU POLYMERIZATION

This method involves the swelling of the clay mineral by adsorption in a suitable monomeric polymer solution. This method thereby increases the compatibility between the polymer and the clay mineral and also results in uniform dispersion of clay particles in the polymeric matrix. With the help of radiation, heat, or presence of catalysts, polymerization is allowed to occur. During this process, the monomeric solution migrates into the galleries of the layered silicates and thereby polymerization occurs within the intercalated sheets (Song et al., 2007). This thin layer of polymer/clay formed can be then reinforced into bulk polymeric matrix and further subjected to secondary processing techniques.

3.3.3 MELT INTERCALATION

Here, the clay mineral is directly infused into the polymeric solution. For achieving this, a temperature higher than the melting point of polymer is employed so that the clay mineral and the polymer undergo annealing process termed as melt-integration (Najafi et al., 2012). This is an environment friendly approach, as no additional solvents are needed. This method also provides better mixing of polymer and clay mineral when compared with solvent casting method. The resultant polymer/clay can be then subjected to secondary processing techniques.

3.4 TECHNIQUES FOR CHARACTERIZATION OF POLYMER/CLAY NANOCOMPOSITES

The success in the preparation of polymer/clay nanocomposites can be investigated/confirmed by several methods. Characterization data can give a clear idea about the physical and chemical properties of the nanocomposites, interaction between filler and polymer, the dispersion of nanoparticles in the polymer matrix, changes occurring in the matrix and the type of particle–polymer interface and the resulting effect of all the process parameters on the properties of nanocomposites (Farhoodi, 2016; Honarvar et al., 2016). In the case of food packaging nanocomposites, characterization is inevitable for understanding the morphological, physiochemical, mechanical, barrier, optical as well as migration properties. Nanocomposite materials intended for food packaging applications can be characterized by all the techniques used for materials science and nanotechnology (Rossi et al., 2017). For better understanding of the structure of nanocomposites, synergistic combinations of different characterization methods are to be used. Basically, the physicochemical properties and structure of nanocomposites can be characterized with the use of various microscopic, spectroscopic, and other standard analytical techniques. In this section, we discuss on how to use these techniques for characterization of nanocomposites with some suitable examples.

3.4.1 MORPHOLOGICAL CHARACTERIZATION

Morphological characterization of polymer nanocomposites can be made by several microscopic and spectroscopic techniques.

3.4.1.1 MICROSCOPICAL TECHNIQUES

The microscopical techniques include optical microscopy (OM), scanning electron microscopy (SEM), transmission electron microscopy (TEM), and atomic force microscopy (AFM) which provide the images of nanocomposites from which size, shape, microstructure, and spatial distribution of phases can be assessed. Optical microscopy is commonly used for general inspection, particularly of a highly transparent or thin nanocomposite (Oksman and Moon, 2014).

Scanning Electron Microscopy

SEM is an imaging method which provides a highly magnified view at micro range and provides a three-dimensional image of the nanocomposite by utilizing high beam of electrons. Using SEM, the details of the topography, dispersion, composition, orientation, and crystalline structure of the nanocomposites can be understood (Pavlidou and Papaspyrides, 2008). Majority of the studies have used SEM for characterizing fractured surfaces of nanocomposites and to observe the distribution of nanoparticles in the polymer matrix. Here, we are presenting one such example. SEM was used to assess the dispersion pattern and study the influence of ginger-mediated AgNPs on the morphology of so formed polyvinyl alcohol-based nanocomposites (Mathew et al., 2019b). Figure 3.3(a–d) shows the SEM micrographs of neat PVA film, PVA/AgNPs, PVA/ginger, and PVA/gAgNP films, respectively. The authors reported that the neat PVA film had a smooth, homogenous, and nonporous surface with compact morphology. And following the incorporation of ginger extract and the subsequent generation of ginger-mediated silver nanoparticles (gAgNP), the nanocomposites expressed a rough surface. Moreover, the SEM image of PVA/gAgNP nanocomposite showed uniform distribution of white spots; clearly indicating the presence of AgNPs which was biofabricated by ginger extract.

Transmission Electron Microscopy

TEM provides a qualitative data of the internal structure, spatial distribution of different phases and any defects if formed in the nanocomposite. TEM can also be used to confirm the presence, determine the size and shape and the dispersivity of nanoparticles in the nanocomposite (Fu et al., 2019). In one such attempt, TEM analysis was done to study the internal structure of PVA/$AgNO_3$/ginger extract (PAG) nanocomposite film (Mathew et al., 2019c). Figure 3.4(a–c) shows the TEM micrograph showing the presence, shape, and distribution of AgNPs throughout the composite matrix. Moreover, using TEM analysis, the particle size histogram of AgNPs were also analyzed (Fig. 3.4d).

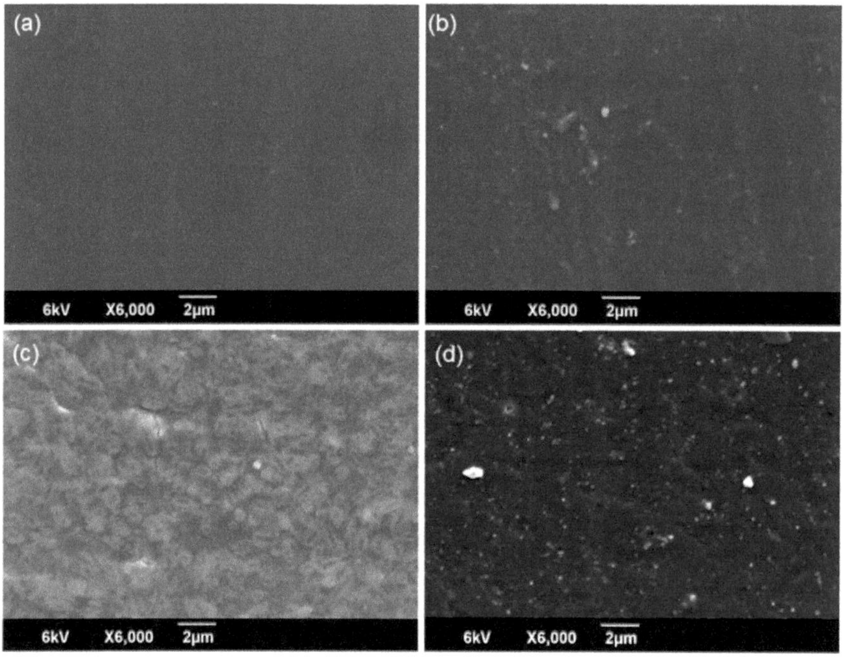

FIGURE 3.3 SEM micrographs of (a) neat PVA, (b) PVA/AgNPs, (c) PVA/ginger, and (d) PVA/gAgNPs films.

Source: Reprinted with permission from Mathew et al. (2019b). Copyright © 2019, Springer Nature.

Energy Dispersive X-ray Analysis (EDX)

This is a nondestructive analytical tool which is often combined with eectron microscopy, such as SEM and TEM to provide elementary composition of nanomaterials by X-ray mapping technique. EDX is commonly employed for analyzing the physical, chemical, and compositional properties of nanocomposites. Figure 3.4e showed the EDX spectrum of PVA/ $AgNO_3$/ginger extract nanocomposite film. The well defined peak seen at 3 KeV indicated the presence of metallic silver in the nanocomposite

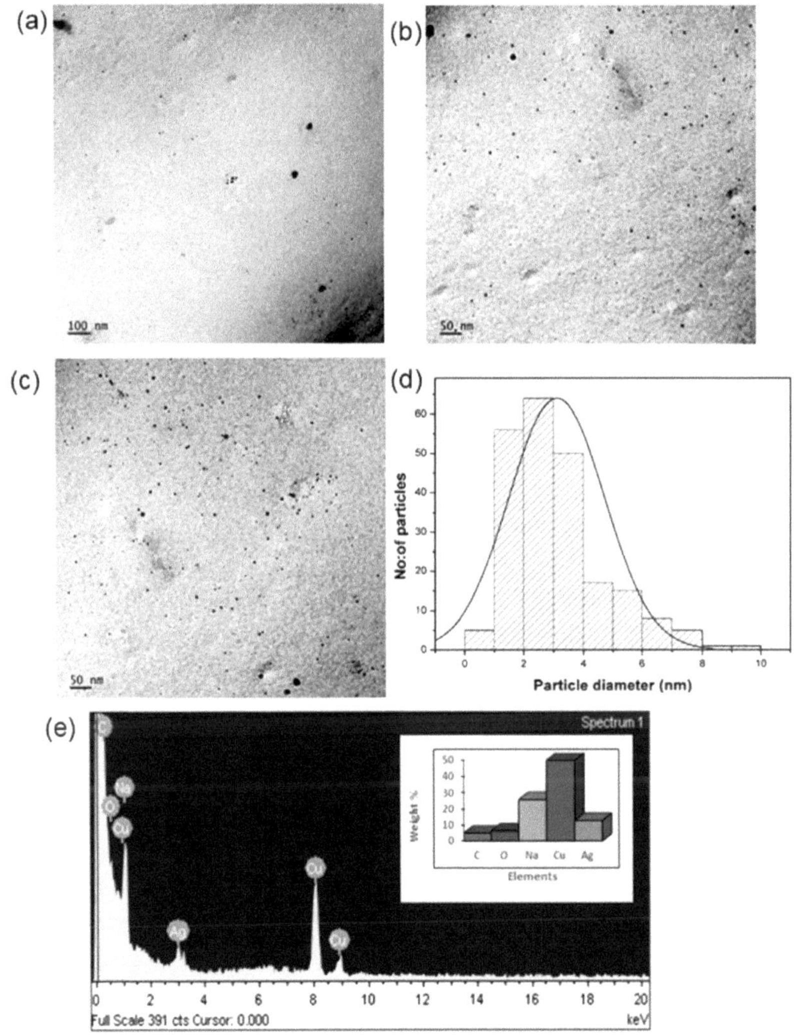

FIGURE 3.4 (a–c) TEM micrograph of PVA/$AgNO_3$/ginger extract nanocomposite film, (d) size distrbution histogram, and (e) EDX spectrum of PVA/$AgNO_3$/ginger extract nanocomposite film.

Source: Reprinted with permission from Mathew et al. (2019c).

matrix which was successfully *in situ* fabricated by ginger extract in the PVA/$AgNO_3$/ginger extract film. Other signals like C and O are also seen along with Ag, which confirm the successful formation of AgNPs in PAG nanocomposite film. In addition, the presence of Na signal may be attributed to the ginger extract added into the PVA/gAgNP film.

Atomic Force Microscopy (AFM)

AFM is an extremely useful technique for the morphological characterization of nanocomposites and is often more used than SEM because of its easiness in sample processing and lower equipment maintenance (Sousa and Scuracchio, 2014). It is a scanning probe microscopy which provides essential qualitative and quantitative information about the nanocomposite's surface chemistry and topography. Besides these, AFM can also be used to identify various molecular structures and to detect the compositional variations, mechanical mapping of nanosurfaces and to understand the homogeneity in dispersion of nanoparticles. AFM technique provides three different modes of images; contact mode, noncontact mode, and tapping mode. Figure 3.5a shows the AFM micrograph of pure pullulan which showed smooth topography with 1.2 nm average roughness. With the addition of Na^+- MMT clay, the AFM image of nanocomposite surface (Fig. 3.5b) showed more jaggy and wrinkled topography, and hence, the roughness was also found to be increased to 14.7 nm approximately (Introzzi et al., 2012).

3.4.1.2 SPECTROSCOPIC TECHNIQUES

The various spectroscopic techniques used for the characterization of nanocomposites include X-ray diffraction (XRD) technique, wide angle X-ray diffraction (WAXD), UV-visible spectroscopy, Fourier transform infrared spectroscopy (FTIR) and attenuated total reflection-Fourier transform infrared (ATR-FTIR). The most commonly applied techniques are discussed below.

X-ray Diffraction

XRD is widely used for the characterization of nanocomposites and is primarily employed for the identification of intercalated structures. The crystalline arrangement, dispersion of nanofiller, and elemental composition of the nanocomposites can be better understood by XRD analysis. Moreover, the type and shift in the pattern of diffractogram can be used to study the nanocomposite formed. However, XRD alone is not sufficient to infer the actual morphology of the nanocomposite. Figure 3.6 shows the diffractogram for PVA/gAgNP films. Here, all films showed well defined reflections at $2\theta = 19.4°$ and diffused scattering in the halo regions, which are characteristic and corresponds to (101) crystal plane of polyvinyl alcohol inferring its semi-crystalline nature. At the same time, the XRD profile of PVA/gAgNP films showed the appearance of two new reflectance at 38° and 44° which corresponded to the characteristic reflectance of silver nanoparticles belonging to the planes (111) and (200) (Mathew et al., 2019b).

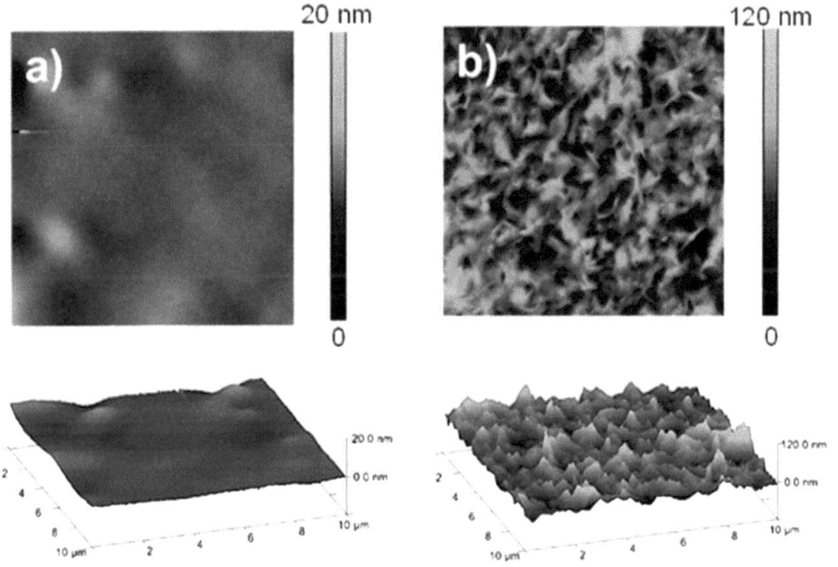

FIGURE 3.5 AFM height and 3D images (upper and lower respectively) of (a) pure pullulan and (b) pullulan nanocomposite.

Source: Reprinted with permission from Introzzi et al. (2012). Copyright © 2012, American Chemical Society.

UV-Visible Spectroscopy

UV-visible spectroscopy is a very useful technique for the qualitative and quantitative characterization of biochemical compounds as well for determining the properties of polymeric and ceramic nano-based materials (Anukiruthika et al., 2020). UV-visible spectroscopy is a cost-effective method and is extensively used as a supporting technique to identify the presence of nanoparticles in composites (Huang et al., 2015). This instrument consists of various components, such as a light source, optical table, monochromator and detector, slots for sample and reference, as illustrated in Figure 3.7. The working principle is based on the absorption of light in the UV-visible range which undergoes electronic transition and differs based on the colors perceived by the chemicals involved (Saxena et al., 2010).

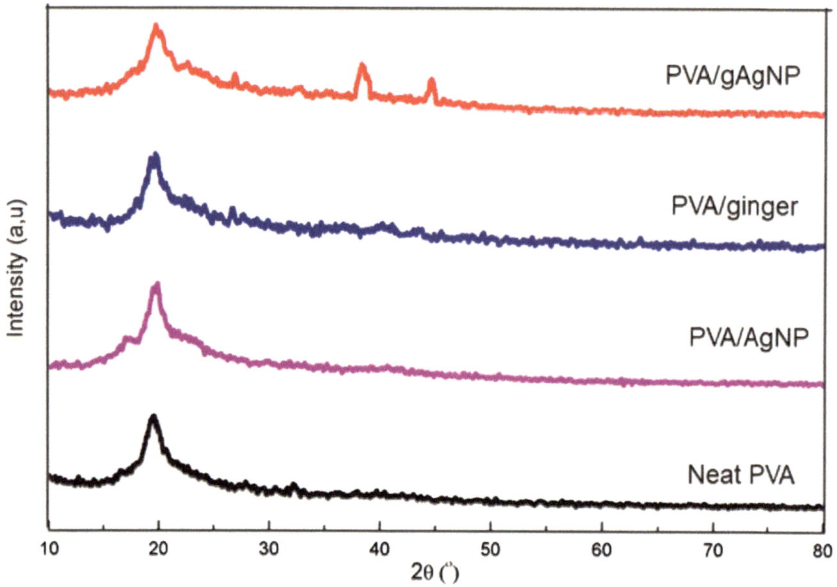

FIGURE 3.6 XRD Diffractogram of PVA/gAgNPs nanocomposite films and its controls.
Source: Reprinted with permission from Mathew et al. (2019b). Copyright © 2019, Springer Nature.

Figure 3.8 shows the UV-visible spectrum of PVA–gAgNPs nanocomposite films (Mathew et al., 2019b). The spectrum of PVA/AgNP and PVA/gAgNP films presented a peak around 420 nm due to the surface plasmon resonance of silver nanoparticles formed by reduction of Ag^{+}. At the same time, such a peak was absent in the case of neat PVA and PVA/ginger films. Thus, UV-vis spectroscopy is a reliable tool to detect the successful generation of nanoparticles in the nanocomposite matrix.

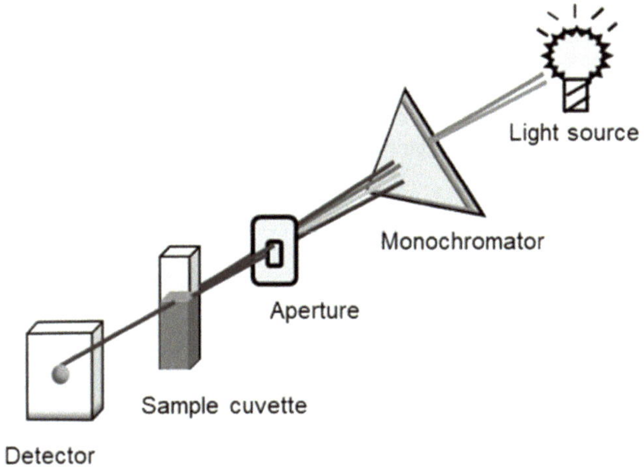

FIGURE 3.7 Schematic of working model of UV-vis spectroscopy.

Source: Adapted from http://faculty.sdmiramar.edu/fgarces/LabMatters/Instruments/UV_Vis/Cary50.htm

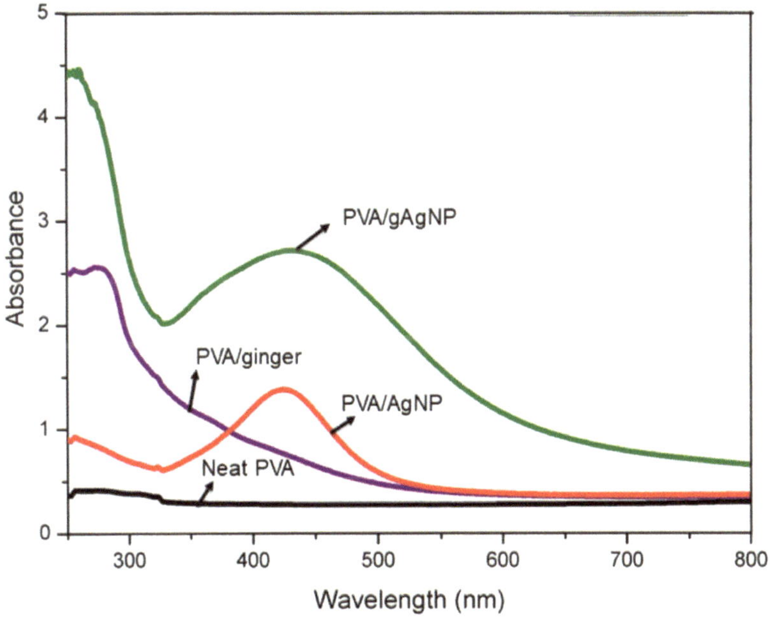

FIGURE 3.8 UV-visible spectrum of PVA/gAgNPs nanocomposite films.

Source: Reprinted with permission from Mathew et al. (2019b). Copyright © 2019, Springer Nature.

Fourier Transform Infrared Spectroscopy

FTIR plays a vital role in the identification of organic compounds and for the qualitative characterization of nanocomposites. This instrumentation involves absorption, reflection, and emission of IR light by a sample (Lin et al., 2014). FTIR is very useful to detect the functional groups involved and to understand the structure of the nanocomposite. Figure 3.9 schematically demonstrates the working principle of FTIR. The use of FTIR in understanding the structure of nanocomposite is presented below with a suitable example.

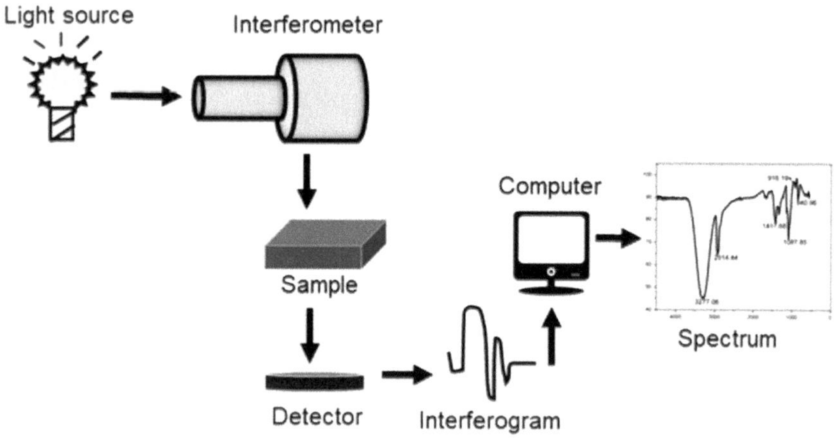

FIGURE 3.9 Working of FTIR.

Source: Adapted and modified from Mohamed et al. (2017).

Figure 3.10 represents the FTIR spectrum of PVA/gAgNPs nanocomposite film and its controls (Mathew et al., 2019b). From the spectra, it is clearly seen that the absorption spectra in the case of control films, corresponding to the –O–H stretching vibrations (3277 cm^{-1}), asymmetric stretching vibrations of C–H (2941 cm^{-1}), and also the moderate band corresponding to C–O vibrations and C–H bends seen at 1087 at 1417 cm^{-1}, respectively were found to be shifted to higher wavenumbers in the case of the nanocomposite film. This slight shift of –O–H band to a higher wave number (3292 cm^{-1}) and the shift and enhanced intensity of C–O stretching band (1091 cm^{-1}) in the case of PVA/gAgNP film is explained by the authors to be due to the interaction of ginger extract and AgNPs. In addition, a slight shift with decreased intensity of band seen at 2941 cm^{-1} in the case of PVA/gAgNP film is expected to be due to the interaction of ginger extract and PVA matrix and the associated reduction of silver ions. Moreover, the authors described the appearance of a new moderate band at 1645 cm^{-1} in the case of PVA/ginger and PVA/gAgNP film to be due to the stretching vibrations of C=C bonds of heterocyclic compounds present in the ginger extract. Therefore, FTIR can be used as an effective analytical tool to study the changes occurring in the structure of the nanocomposite.

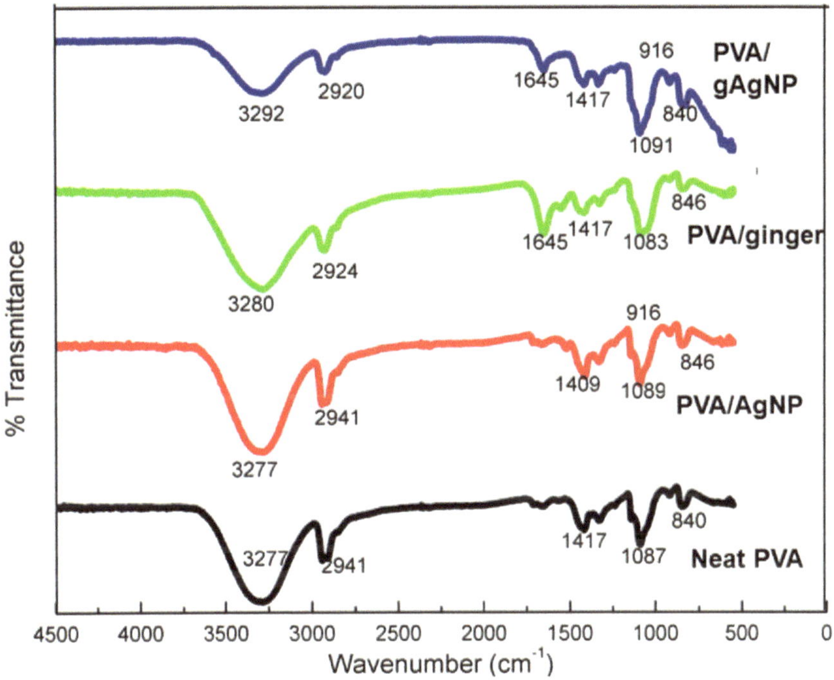

FIGURE 3.10 FTIR spectra of PVA/gAgNPs nanocomposite films and its controls.
Source: Reprinted with permission from Mathew et al. (2019b). Copyright © 2019, Springer Nature.

Nuclear Magnetic Resonance (NMR) Spectroscopy

Other spectroscopical techniques include NMR spectroscopy, infrared and Raman spectroscopy for the structural analysis of polymer nanocomposites which is discussed in the following section.

High-resolution solid state-NMR spectroscopy is a nondestructive tool which is often explored for the characterization of polymer nanocomposites. It gives idea about the nanocomposite morphology, its phase structure and surface chemistry, polymer chain dynamics, alterations in molecular mobility, interaction between components, the nature of dispersion (whether exfoliation, intercalation, or flocculated), interspatial distance between filler molecules, presence of voids, bonds, etc. (Li et al., 2013; Sadasivuni et al., 2016). With NMR spectroscopy, it is possible to analyze the polymer–filler interfaces (Bokobza, 2017) and learn about the polymer mobility and immobilized polymer chains surrounding the filler present inside the composite. Figure shows the working principle of NMR spectroscopy. Figure 3.11 illustrates the working principle of NMR spectroscopy.

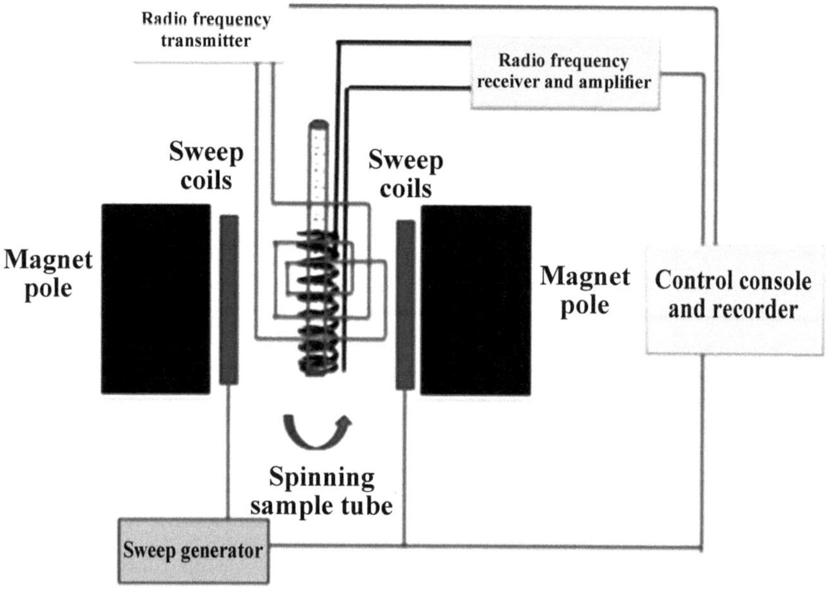

FIGURE 3.11 Working of NMR spectroscopy.

Source: Reprinted with permission from Sadasivuni et al. (2016). Copyright © 2016 Elsevier Inc.

Raman and Infrared Spectroscopy

Raman and infrared spectroscopical techniques are commonly employed for obtaining information about polymer nanocomposites based on their vibrational properties. Here, the vibrational spectra help in the identification of the vibrational bands associated with the functional groups present in the polymer and the nanofillers. In addition, using both Raman and infrared spectra, the details about the interaction of organic and inorganic phases, degree of orientation of polymer chains and the reinforcing nanofillers, the type of dispersion and functionalization of nanomaterials in the composite can be understood. Raman spectroscopy is a nondestructive technique which is an important technique used for the characterization of nanocomposites, especially those containing carbon nanomaterials. One of the major differences between infrared transmission spectroscopy and Raman spectroscopy is that the latter can be used for the analysis of thick nanocomposite samples, whereas the former can only be used to analyze thin samples (Bokobza, 2017).

Figure 3.12 represents the Raman spectra of neat PLA, TiO_2, and PLA/TiO_2 nanocomposite (Mallick et al., 2018). The Raman spectra of PLA composite film showed bands at 1787, 1059, and 2949 cm^{-1} which were ascribed to C=O, C-O stretching bands, symmetric and asymmetric stretching bands, respectively. In the case of Raman TiO_2, the spectra showed the absorption band between 200 and 1000 cm^{-1} which corresponded to the Ti-O and Ti-O-Ti bonds. The same absorption pattern was observed in the case of PLA–TiO_2 nanocomposites also.

3.4.2 OTHER CHARACTERIZATION METHODS

Besides morphological and structural characterization, the physiochemical characteristics of should be analyzed to check if any improvement has occurred in its mechanical and barrier properties of the resultant food packaging nanocomposites. For this, standardized test methods of American Society for Testing Materials (ASTM) or using International Organization for Standardization (ISO) are available. Some examples are listed below:

3.4.2.1 CHARACTERIZATION OF BARRIER PROPERTIES

To investigate whether the surface of packaging materials is hydrophilic or hydrophobic in nature, surface tension contact angle (CA) can be analyzed (Vilarinho et al., 2020). Others include WVP and WPTR. For determining gas/oxygen barrier properties, the methods include oxygen permeability (OP) and OTR.

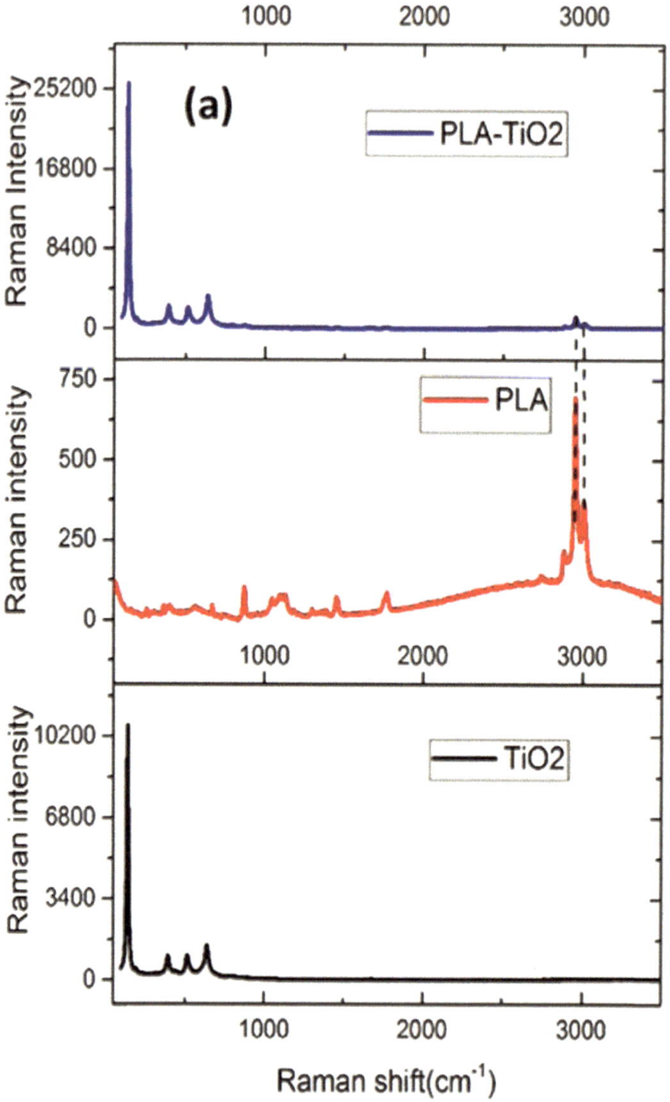

FIGURE 3.12 Raman spectra of TiO_2, PLA, PLA–TiO_2 nanocomposite.

Source: Reprinted with permission from Mallick et al. (2018). Copyrights © 2018 Elsevier Ltd and Techna Group S.r.l.

3.4.2.2 MECHANICAL CHARACTERIZATION

Mechanical characterization of nanocomposites can give an insight into the performance of the formed material in real-time applications. Mechanical characterization can be done using various theoretical and experimental analyses. The most commonly followed mechanical tests include tensile tests, longitudinal direction (LD) test, and transverse direction (TD) test, impact test, shear test, compression test, hardness test, flexure test, and dynamic mechanical analysis (DMA). As the different materials respond to these operational conditions in a different manner, no single test can determine the mechanical stability of nanocomposites (Vinyas et al., 2019).

3.4.2.3 THERMAL CHARACTERIZATION

Nanocomposite films intended for food packaging applications must undergo thermal characterization in order to study its behavior under different conditions of temperature, heat, moisture etc. The thermal properties of the nanocomposites can be analyzed using differential scanning calorimetry (DSC), moisture absorption test, thermomechanical analysis, TGA, and Melt Index Rheology Analysis (Vinyas et al., 2019).

3.4.2.4 OTHERS

Porosity percentage of nanocomposites can be determined by bulk density method. The viscoelastic properties of materials are measured using DMA. For public acceptance of nanocomposites, the assessment of possible migration of nanomaterials incorporated in composites into food has to be done. For this purpose, inductively coupled plasma mass spectroscopy (ICP-MS) is very useful (Rossi et al., 2017).

3.5 CONCLUSION

As packaging materials play a crucial role in safeguarding food from farm to table, the development of polymer/clay nanocomposite-based packaging materials with improved properties is a promising strategy to keep food safe and to enable extended shelf life of the food. This chapter mainly provides details on the improved mechanical, barrier, and thermal properties of polymer/c lay nanocomposite and about the characterization methods. Polymer nanocomposites can be prepared by various routes; the major ones being solvent casting, melt intercalation, and *in situ* polymerization. After preparation, it is very essential to characterize the nanocomposites for better understanding of their structure and chemical nature. Currently, a variety of microscopic and spectroscopic analytical methods are available for characterization of polymer/clay nanocomposites. But we have to keep in mind, that each method has its own limitations. Hence, for best results, a combination of several analytical methods has to be applied for nanocomposite characterization.

KEYWORDS

- **characterization techniques**
- **mechanical properties**
- **polymer nanocomposites**
- **thermal properties**
- **solvent casting method**
- **food packaging**

REFERENCES

Anukiruthika, T.; Priyanka, S., Moses, J. A.; Anandharamakrishnan, C. Characterisation of Green Nanomaterials. In *Green Nanomaterials, Advanced Structured Materials*; Ahmed, S., Ali, W., Eds.; Springer Singapore: Singapore, 2020; pp 43–79.

Arora, A.; Padua, G. W.; Review: Nanocomposites in Food Packaging. *J. Food Sci.* **2010,** *75*, R43–R49.

Bokobza, L. Spectroscopic Techniques for the Characterization of Polymer Nanocomposites: A Review. *Polymers* **2017,** *10*, 7.

Carrado, K. A. Synthetic Organo- and Polymer–Clays: Preparation, Characterization, and Materials Applications. *Appl. Clay Sci.* **2000,** *17*, 1–23.

Dabbaghianamiri, M.; Duraia, E. M.; Beall, G. W. Self-Assembled Montmorillonite Clay-Poly Vinyl Alcohol Nanocomposite as a Safe and Efficient Gas Barrier. *Results Mater.* **2020,** *7*, 100101.

De Silva, R. T.; Mantilaka, M. M. M. G. P. G.; Ratnayake, S. P.; Amaratunga, G. A. J.; de Silva, K. M. N. Nano-MgO Reinforced Chitosan Nanocomposites for High Performance Packaging Applications with Improved Mechanical, Thermal and Barrier Properties. *Carbohydr. Polym.* **2017,** *157*, 739–747.

Farhoodi, M. Nanocomposite Materials for Food Packaging Applications: Characterization and Safety Evaluation. *Food Eng. Rev.* **2016,** *8*, 35–51.

Fu, S.; Sun, Z.; Huang, P.; Li, Y.; Hu, N. Some Basic Aspects of Polymer Nanocomposites: A Critical Review. *Nano Mater. Sci.* **2019,** *1*, 2–30.

Gong, X.; Pan, L.; Tang, C. Y.; Chen, L.; Hao, Z.; Law, W.-C.; Wang, X.; Tsui, C. P.; Wu, C. Preparation, Optical and Thermal Properties of CdSe–ZnS/Poly(Lactic Acid) (PLA) Nanocomposites. *Compos. Part B Eng.* **2014,** *66*, 494–499.

Honarvar, Z.; Hadian, Z.; Mashayekh, M. Nanocomposites in Food Packaging Applications and Their Risk Assessment for Health. *Electron. Physician* **2016,** *8*, 2531–2538.

Huang, J.-Y.; Li, X.; Zhou, W. Safety Assessment of Nanocomposite for Food Packaging Application. *Trends Food Sci. Technol.* **2015,** *45*, 187–199.

Introzzi, L.; Blomfeldt, T. O. J.; Trabattoni, S.; Tavazzi, S.; Santo, N.; Schiraldi, A.; Piergiovanni, L.; Farris, S. Ultrasound-Assisted Pullulan/Montmorillonite Bionanocomposite Coating with High Oxygen Barrier Properties. *Langmuir* **2012,** *28*, 11206–11214.

Lee, H.; Rukmanikrishnan, B.; Lee, J. Rheological, Morphological, Mechanical, and Water-Barrier Properties of Agar/Gellan Gum/Montmorillonite Clay Composite Films. *Int. J. Biol. Macromol.* **2019,** *141*, 538–544.

Li, W.; Hou, L.; Chen, Z. An NMR Investigation of Phase Structure and Chain Dynamics in the Polyethylene/Montmorillonite Nanocomposites. *J. Nanomater.* **2013,** 1–10.

Lin, P.-C.; Lin, S.; Wang, P. C.; Sridhar, R. Techniques for Physicochemical Characterization of Nanomaterials. *Biotechnol. Adv.* **2014,** *32*, 711–726.

Mallick, S.; Ahmad, Z.; Touati, F.; Bhadra, J.; Shakoor, R. A.; Al-Thani, N. J. PLA-TiO_2 Nanocomposites: Thermal, Morphological, Structural, and Humidity Sensing Properties. *Ceram. Int.* **2018,** *44*, 16507–16513.

Mathew, S.; Jayakumar, A.; Kumar, V. P.; Mathew, J.; Radhakrishnan, E. K. One-Step Synthesis of Eco-Friendly Boiled Rice Starch Blended Polyvinyl Alcohol Bionanocomposite Films Decorated with In Situ Generated Silver Nanoparticles for Food Packaging Purpose. *Int. J. Biol. Macromol.* **2019a**, *139*, 475–485.

Mathew, S.; Mathew, J.; Radhakrishnan, E. K. Polyvinyl Alcohol/Silver Nanocomposite Films Fabricated under the Influence of Solar Radiation as Effective Antimicrobial Food Packaging Material. *J. Polym. Res.* **2019b**, 26.

Mathew, S.; Sajeendra, S.; Mathew, J.; Radhakrishnan, E. K. Biodegradable and Active Nanocomposite Pouches Reinforced with Silver Nanoparticles for Improved Packaging of Chicken Sausages. *Food Packag. Shelf Life* **2019c**, *19*, 155–166.

Mathew, S.; Radhakrishnan, E. K. Polymer Nanocomposites: Alternative to Reduce Environmental Impact of Non-Biodegradable Food Packaging Materials. In *Composites for Environmental Engineering*; Ahmed, S., Chaudhry, S. A., Eds.; Wiley, 2019; pp 99–133.

Mohamed, M. A.; Jaafar, J.; Ismail, A. F.; Othman, M. H. D.; Rahman, M. A. Fourier Transform Infrared (FTIR) Spectroscopy. In *Membrane Characterization*; Elsevier, 2017; pp 3–29.

Müller, K.; Bugnicourt, E.; Latorre, M.; Jorda, M.; Echegoyen Sanz, Y.; Lagaron, J.; Miesbauer, O.; Bianchin, A.; Hankin, S.; Bölz, U.; Pérez, G.; Jesdinszki, M.; Lindner, M.; Scheuerer, Z.; Castelló, S.; Schmid, M. Review on the Processing and Properties of Polymer Nanocomposites and Nanocoatings and Their Applications in the Packaging, Automotive and Solar Energy Fields. *Nanomaterials* **2017,** *7*, 74.

Najafi, N.; Heuzey, M. C.; Carreau, P. J. Polylactide (PLA)-Clay Nanocomposites Prepared by Melt Compounding in the Presence of a Chain Extender. *Compos. Sci. Technol.* **2012,** *72*, 608–615.

Oksman, K.; Moon, R. J. Characterization of Nanocomposites Structure. In *Materials and Energy*; World Scientific, 2014; pp 89–105.

Pavlidou, S.; Papaspyrides, C. D. A Review on Polymer–Layered Silicate Nanocomposites. *Prog. Polym. Sci.* **2008,** *33*, 1119–1198.

Rossi, M.; Passeri, D.; Sinibaldi, A.; Angjellari, M.; Tamburri, E.; Sorbo, A.; Carata, E.; Dini, L. Nanotechnology for Food Packaging and Food Quality Assessment. In *Advances in Food and Nutrition Research*; Elsevier, 2017; pp 149–204.

Sadasivuni, K. K.; Cabibihan, J.-J.; Al-Maadeed, M. A. S. NMR Spectroscopy of Polymer Nanocomposites. In *Spectroscopy of Polymer Nanocomposites*; Elsevier, 2016; pp. 181–201.

Sharma, C., Dhiman, R., Rokana, N., Panwar, H. Nanotechnology: An Untapped Resource for Food Packaging. *Front. Microbiol.* **2017,** *8*.

Song, W.; Zheng, Z.; Tang, W.; Wang, X. A Facile Approach to Covalently Functionalized Carbon Nanotubes with Biocompatible Polymer. *Polymer* **2007,** *48*, 3658–3663.

Sousa, F. D. B.; de, Scuracchio, C. H. The Use of Atomic Force Microscopy as an Important Technique to Analyze the Dispersion of Nanometric Fillers and Morphology in Nanocomposites and Polymer Blends Based on Elastomers. *Polímeros* **2014,** *24*, 661–672.

Tas, C. E.; Hendessi, S.; Baysal, M., Unal, S.; Cebeci, F. C.; Menceloglu, Y. Z.; Unal, H. Halloysite Nanotubes/Polyethylene Nanocomposites for Active Food Packaging Materials with Ethylene Scavenging and Gas Barrier Properties. *Food Bioprocess Technol.* **2017,** *10*, 789–798.

Vilarinho, F.; Vaz, M. F.; Silva, A. S. The Use of Montmorillonite (MMT) in Food Nanocomposites: Methods of Incorporation, Characterization of MMT/Polymer Nanocomposites and Main Consequences in the Properties. *Recent Pat. Food Nutr. Agric.* **2020,** *11*, 13–26.

Vinyas, M.; Athul, S. J.; Harursampath, D.; Loja, M.; Nguyen Thoi, T. A Comprehensive Review on Analysis of Nanocomposites: From Manufacturing to Properties Characterization. *Mater. Res. Express* **2019,** *6*, 092002.

PART II

Types of Nanocomposite Food Packagings

CHAPTER 4

Active Nanocomposite Packaging: Functions and Applications

SOURABH SURESH KALE[1], SUNEETA PINTO[1], and MAHENDRA PAL[2*]

[1]*Dairy Technology Department, SMC College of Dairy Science, Anand Agricultural University, Anand 388110, Gujarat, India*

[2]*Narayan Consultancy on Veterinary Public Health and Microbiology, 4, Aangan, Anand 388001, Gujarat, India*

[*]*Corresponding author. E-mail: palmahendra2@gmail.com*

ABSTRACT

Active packaging is a technique of prolonging the shelf life of food products and maintaining their quality by incorporating suitable absorbers or emitters leading to improvement in the quality and shelf life of the product. One of the most studied applications of nanotechnology in the food sector is nanocomposite food packaging. The application of nanotechnology to packaging materials can improve their barrier properties and reduce the weight of food packaging materials. Applications of nanocomposites in active packaging mainly revolve around three areas: antimicrobial nanocomposites, oxygen scavenger nanocomposites, and ethylene scavenger nanocomposites. Applications of nanocomposite regarding active packaging revolves around antimicrobial and scavenging or absorbing functions. This chapter describes various nanomaterials used in nanocomposites along with their characteristics which impart them the potential to act as an active agent. Some of the major applications of nanocomposites along with their mechanism of action in the field of active packaging are also described.

4.1 INTRODUCTION

Food packaging is generally referred to as a technique of providing a barrier to safeguard the food from undesirable environmental disturbances. It plays a major role in the quality control and safety of food products throughout the supply chain (Mihindukulasuriya and Lim, 2014; Yildirim et al., 2018). The traditional role of food packaging is to protect the food from the spoilage by acting as a physical blockage between the food product and its external conditions. Developments have been made in food packaging to meet the consumer's demands for high quality, additive-free, and minimally processed foods. In this regard, various researchers have described innovations in packaging leading to improved barrier properties, improved biodegradability, active packaging (AP), and real-time quality indication through intelligent packaging (Yildirim, 2011; Pereira de Abreu et al., 2012; Dobrucka and Cierpiszewski, 2014; Realini and Marcos, 2014; Brockgreitens and Abbas, 2016; Enescu et al., 2019).

AP has redefined traditional food and beverage packaging that has a role beyond product protection and brand presentation. It impacts functions like moisture control, antioxidant activity, ethylene degradation or absorbance, and antimicrobial activity (Pal, 2017; Enescu et al., 2019). AP is a technique of prolonging the shelf life of food products and maintaining their quality by incorporating suitable absorbers or emitters leading to improvement in the quality and shelf life of the product (Yildirim et al., 2018). Active components (which includes oxygen scavengers, ethylene absorbers, carbon-di-oxide emitters, antimicrobial agents, moisture scavengers, antioxidant releasers, etc.) prevent the negative effects of deleterious factors and are good alternatives to the use of chemical treatments and disinfectants (Maneerat et al., 2003; Pal, 2017; Siripatrawan and Kaewklin, 2018). The strategy of AP excludes the addition of additives and preservatives directly into the bulk food, giving better cost-effectiveness (Panea et al., 2014; Youssef et al., 2016). The active food packaging system plays dynamic roles in the food quality, either by absorbing (Mahajan et al., 2008; Azevedo et al., 2011), degrading, and inactivating factors that affect product quality or by releasing desirable components like antimicrobial compounds (Yildirim et al., 2018).

Active systems can be classified as active-releasing systems (emitters) and active scavenging systems (absorbers). Active components, such as oxygen scavengers, moisture or ethylene absorbers, and antimicrobial or carbon dioxide releasers are added in the form of a sachet to the food packaging (Pereira de Abreu et al., 2012). Accidental breakage, which could lead to involuntary consumption of the content by the consumer is one of the major disadvantages of such sachet-based systems. Moreover, sachets cannot be added to beverages (Yildirim, 2011; Pereira de Abreu et al., 2012; Suppakul, 2015). Therefore, the recent developments in AP have focused on the direct incorporation of active components into the packaging material. The activity and capacity of active agents have to be retained when incorporated into the films along with the preservation of mechanical and physical properties of the packaging material, which is very challenging (Silvestre et al., 2013; Realini and Marcos, 2014). With advances in nanotechnology, these challenges can be overcome by creating, characterizing and using nanostructures that favor a better interaction with the polymer matrix, and the performance of the resulting material (Roco, 2003; Sorrentino et al., 2007). The application of nanotechnology to packaging materials can improve their barrier properties, and reduce the weight of food packaging materials (Sorrentino et al., 2007). As per the recommendations adopted by the EU in 2011 while defining nanotechnology, nanomaterials must have at least one dimension at the nanoscale ranging from 1 to 100 nm (Paul and Robeson, 2008; Pavlidou and Papaspyrides, 2008; Enescu et al., 2019).

Nanocomposite food packaging is one of the fastest-growing fields in nanotechnology. Nanocomposites are composite materials that are reinforced with nanoparticles (NPs). Usually, composites consist of a polymer matrix or a continuous phase and a discontinuous phase or filler (Pavlidou and Papaspyrides, 2008; Naffakh et al., 2013). The role of the matrix is keeping the NPs in place. Meanwhile, the NPs provide the composite of specific chemical or physical properties (Othman, 2014). The mechanical, thermal, and barrier properties of nanocomposites are usually widely different from those of nonreinforced synthetic or biopolymer-based materials. Due to their larger specific surface area and proportionally more surface atoms (Paul and Robeson, 2008), NPs often have different electronic properties as well as increased interfacial interactions (e.g., stronger van der Waals and electrostatic forces), which collectively contribute to their unique characteristics (Paul and Robeson, 2008) resulting in a material with improved physicochemical properties. Based on the matrix material, nanocomposites are classified as polymer matrix nanocomposites, metal matrix nanocomposites, and ceramic matrix nanocomposites (Parameswaranpillai et al., 2017).

4.2 NANOMATERIALS USED IN POLYMER NANOCOMPOSITES

Depending on the size and shape, nanomaterials are classified as 0-D (quantum dots, NPs), 1-D (carbon nanotubes, nanorods, and nanowires), 2-D (nanofilms), and 3-D nanomaterials. The performance of the nanomaterials depends more on the surface area than the material composition (Parameswaranpillai et al., 2017).

4.2.1 NANOPARTICLES

NPs have the greater surface area to the mass ratio which can alter physical and chemical properties, relative inertness at macroscopic scale, thus turns into increased reactivity and catalytic behavior at the nanoscale (Enescu et al., 2019). They are materials exhibiting all three dimensions within the nanometric scale. They can be distinguished as the natural origin (minerals), and artificially produced (Naffakh et al., 2013). Because of their resistance to harsh processing conditions (Hoseinnejad et al., 2018) NPs are predominantly used in the development of active and intelligent packages (Sinha Ray and Okamoto, 2003; Pavlidou and Papaspyrides, 2008). Chitosan-NPs, silver NPs (Ag-NPs), titanium NPs (TiO_2 NPs), magnesium oxide NPs (MgO NPs), copper NPs (Cu-NPs), zinc (ZnO-NPs), etc. (Raju, 2016) are the most used NPs for the development of active nanocomposites. Apart from its antimicrobial action, silver NPs are also able to absorb and decompose ethylene (Rhim and Ng, 2007). Nanocrystalline titania (TiO_2) can act as both UV blocker and photocatalytic oxygen scavenger under adequate working conditions (Sorrentino et al., 2007).

When irradiated with UV or visible light, the photocatalytic behavior of metal oxide NPs, such as titanium dioxide (TiO_2), magnesium oxide (MgO), copper oxide (CuO), and zinc oxide (ZnO) gives them considerable antimicrobial capabilities (Pavlidou and Papaspyrides, 2008; Kumar et al., 2011). On the other hand, NPs of some other metal oxides, such as silica (SiO_2), alumina (Al_2O_3), zirconia (ZrO_2), and antimony–tin oxide (Sb_2O_5/SnO_2), or inorganic compounds, such as calcium carbonate ($CaCO_3$) and silicon carbide (SiC) do not exhibit any significant active behavior of their own and thus can be combined with some active NPs or employed as fillers in polymer matrices aiming to improve their mechanical, barrier, thermal, and surface properties (Kumar et al., 2011; Paul and Robeson, 2008; Tjong, 2006; Rhim et al., 2013).

4.2.2 NANOFIBERS

Nanofibers are materials having two dimensions within the nanometric scale. They include nanotubes, nanowhiskers, nanorods, nanowires, nanorings, etc. Their long and thin structure is thought to puncture the microbial cells which cause irreversible damages to their cellular structure (Yuan et al., 2005; Huang et al., 2015).

Nanofibers derived from chitosan and cellulose are most studied organic fillers for the reinforcement of biopolymers, such as starch, chitosan, soy protein isolate, polylactic acid, and polyhydroxyalkanoates (Paul and Robeson, 2008; Naffakh et al., 2013). Glass and carbon nanofibers are abundantly found in nature and their relative inexpensiveness makes them ideal fillers for the reinforcement of packaging polymers (Tjong, 2006; Kiliaris and Papaspyrides, 2010; Rhim et al., 2013). Their use as a reinforcing filler is generally aimed at enhancing gas barrier or thermal resistance (Cushen et al., 2012).

4.2.3 NANOLAYERS

In the subject of nanolayers, particles of materials are having only one dimension in the nanometric scale, other two dimensions are being much larger forming sheets, flake, disk, or platelet-like structures (Pavlidou and Papaspyrides, 2008). These materials tend to be stacked by dozens and forms multi-layered agglomerates looking like decks of cards (tactoids) (Pavlidou and Papaspyrides, 2008; Paul and Robeson, 2008) due to the strong electrostatic forces, hydrogen bond, or van der Waals attractive forces around their surface (Kumar et al., 2011).

Polymers–clay nanomaterials are among the first polymer nanomaterials to emerge in the market as an improved material for food packaging applications (Sinha and Okamoto, 2003; Paul and Robeson, 2008). Clays have been used for the production of polymer nanocomposites (Pavlidou and Papaspyrides, 2008) because of their relative inexpensiveness, natural occurrence, well-known chemistry, and legal approval as food additives by various authorities (Pavlidou and Papaspyrides, 2008; Tjong, 2006). Clays, such as kaolinite, halloysite, and vermiculite have also been used because of their characteristic properties, such as the tube-forming ability of halloysite in dry conditions (Du et al., 2010), or the good optical characteristics and ethylene scavenging ability of kaolinite (Álvarez-Hernández et al., 2018). Hydrophilic characteristics and more interplatelet distance of clays allow them to yield good nanocomposite structures in highly hydrophilic polymers, such as polyethylene (PE) oxide or PVOH through solution methods (Pavlidou and Papaspyrides, 2008). However, due to the strong hydrophilic character, chemical organomodifications are required. The use of compatibilizers is necessary to improve their compatibility with low surface energy polymers, such as polyolefins (Tas et al., 2017).

Clays have a role in improving the mechanical and thermal properties of materials while silicate clays have also shown antimicrobial properties (Othman, 2014). HNTs (aluminum silicate NPs) presenting hollow tubular nanostructures are ethylene adsorbents as well as barrier properties enhancers by providing a tortuous pathway (Tas et al., 2017). Polymer/clay nanocomposites with barrier properties such as montmorillonite (MMT) nano clay for reduced gas permeability (Gorrasi et al., 2003; Morawiec et al., 2005; Picard et al., 2007; Alexandre and Dubois, 2000; Cui et al., 2015) have been reported.

4.3 NANOCOMPOSITES IN ACTIVE FOOD PACKAGING

AP based on nanocomposites has been successfully used due to the high surface-to-volume ratio and more surface reactivity of the metal/metal oxide NPs than their micro- or macroscale forms (Emamifar et al., 2010). Applications of nanocomposites in AP mainly revolve around three areas: antimicrobial nanocomposites, oxygen scavenger nanocomposites, and ethylene scavenger nanocomposites. The major applications of NPs in active food packaging are illustrated in Figure 4.1.

4.3.1 ANTIMICROBIAL NANOCOMPOSITES

Antimicrobial packaging has attracted much attention due to controlled release, reduced antimicrobial contents, and improved cost-effectiveness (Imran et al., 2010; Yildirim, 2011; Pereira de Abreu et al., 2012; Realini and Marcos, 2014). The combination of several antimicrobial agents at a single time offers a new approach to developing packaging materials with much superior antimicrobial activity or multifunctional food packaging systems (Duncan, 2011). The uniqueness of nanoantimicrobials arises from their small size resulting in higher specific surface area which makes them more reactive and efficient than their larger counterparts (Rossi et al., 2014). Mostly, microbial growth and food degradation occur at the surface. Using packaging raw materials that are inherently antimicrobial has obligated the problems associated with sachet-based applications.

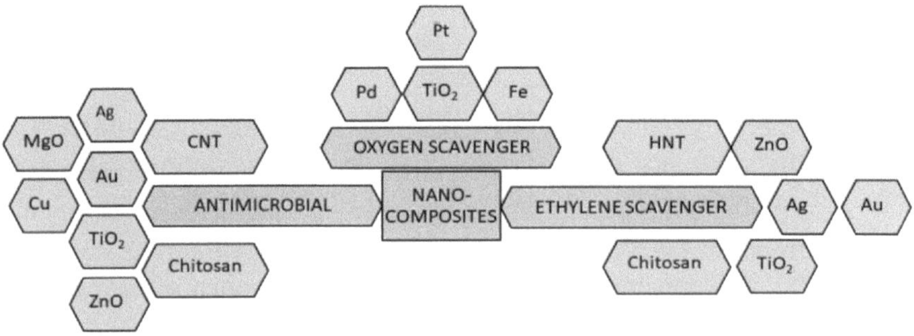

FIGURE 4.1 Major applications of active nanocomposites in food packaging.

An antimicrobial compound can be applied onto the surfaces of the polymers in contact with a foodstuff polymer by methods such as ion or covalent linkages (Khaneghah et al., 2015). Direct incorporation of antimicrobial agents can reduce their antimicrobial activity during thermal–mechanical transformation processes (i.e., extrusion) due to high temperatures used during processing (Khaneghah et al., 2018). The properties of the polymer, such as mass transport, permeability, sorption, and migration (Muriel-Galet et al., 2015), rate of addition, and the tendency of nanofillers to agglomerate also affects their activity. The storage temperature of food can also affect the release rate as well as the durability of the active antimicrobial system (Siripatrawan and Kaewklin, 2018).

Antimicrobial packaging systems are of three types: **(A)** Systems releasing antimicrobial agents that require direct food contact. **(B)** Systems that release volatile antimicrobial agents which do not require direct contact with foodstuff. **(C)** Nonmigratory antimicrobial systems in which food contact is required as the antimicrobial agent present in polymer does not intentionally migrate out. This category involves surface-modified polymeric films with antimicrobial activities or polymer films that are inherently antimicrobial (Yildirim, 2011). Active antimicrobial nanocomposites may belong to the third group. Nanomaterials like silver (Ag-NPs), gold (Au-NPs), copper (Cu-NPs), zinc oxide (ZnO-NPs), silica

(SiO_2-NPs), titanium dioxide (TiO_2-NPs), alumina (Al_2O_3-NPs), and iron oxides (Fe_3O_4, Fe_2O_3) are most commonly used as potential antimicrobials (Huang et al., 2015; Azeredo et al., 2019).

4.3.1.1 ANTIMICROBIAL POTENTIAL OF NANOMATERIALS

The potentiality of nanomaterial for its use as an antimicrobial is derived from its physicochemical characteristics which are extremely different from their macroscale counterparts. Some of them are:

1. **Size:** The size of NPs plays a great role in its antimicrobial activity. Smaller NPs having larger specific surface areas often resulted in larger permeability to the cell membranes (Gurunathan et al., 2014; Deplanche et al., 2010). The antimicrobial effectiveness of silver NPs compared to conventional NPs is a result of increased cell toxicity but more importantly a result of increased reactive surface area available for the oxidation of silver into silver ions (Hannon et al., 2015).
2. **Shape:** NPs interacting with periplasmic enzymes (enzymes in the gel-like matrix between the inner cytoplasmic membrane and outer cell membrane of Gram-negative bacteria) cause varying gradations of bacterial cell damage following the shape of NPs (Cha et al., 2015). It was found that cube-shaped Ag-NPs have stronger antibacterial activity as compared with wire-shaped and sphere-shaped Ag-NPs with similar diameters (Actis et al., 2015).
3. **Roughness:** Roughness of the surface influences antibacterial action as it promotes the adsorption of bacterial proteins, followed by a reduction in bacterial adhesion (Sukhorukova et al., 2015).
4. **Zeta Potential:** The zeta potential of NPs influences bacterial adhesion. The electrostatic attraction among NPs and the bacterial cell membrane makes NPs prone to being adsorbed on the bacterial surface. Positively charged NPs are more prone to be adsorbed, in contrast to their negatively charged counterparts (Pan et al., 2013). Positively charged NPs have been believed to enhance ROS production, which leads to interactions between the NPs and the bacterial surface (Arakha et al., 2015). Figure 4.2 depicts the various mechanisms of antimicrobial action of NPs.

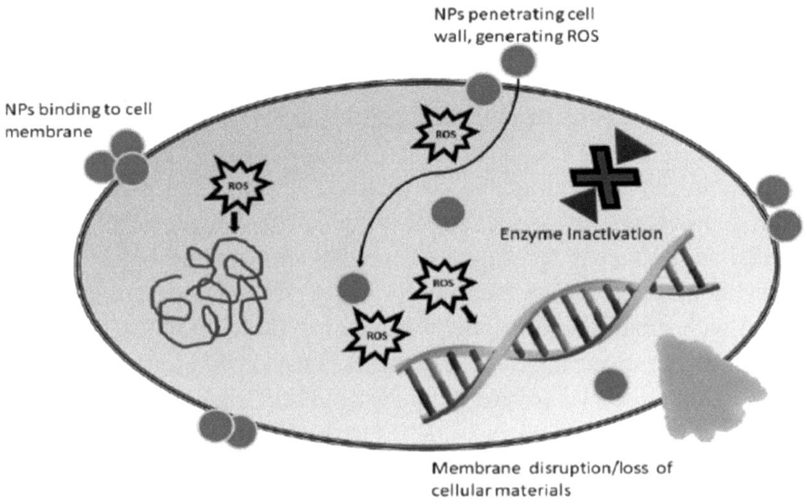

FIGURE 4.2 Antimicrobial action of nanoparticles.
Source: Adapted from Azeredo et al. (2019).

4.3.1.2 NANOCOMPOSITES CONTAINING METAL NPS WITH ANTIMICROBIAL PROPERTIES

4.3.1.2.1 Nanocomposites with Ag-NPs

Many products use some form of nanosilver (Vance et al., 2015) including textiles, cosmetics/hygiene products, electronic appliances, cleaning agents, toys, and building materials, as well as food contact materials. Nanosilver is often incorporated into coatings on food storage containers, mugs, dishes, cutlery, chopping boards, etc. (Enescu et al., 2019). Nanosilver is a potent antimicrobial against numerous species of bacteria. Ag-based nanostructures are known for strong surface plasma resonance, a large surface-to-volume ratio, efficient catalytic activity, and remarkable broad-spectrum antimicrobial activity against different strains of microorganisms (Amirsoleimani et al., 2018).

In its metal form, Ag ions and its salts as antimicrobial agents show limited potential due to interfering effects of salts in the antimicrobial mechanism, continuous release of sufficient Ag ions is not possible (Kim et al., 2007). To overcome such limitations, Ag-NPs have been applied. Ag-NPs are chemically stable, their incorporation into polymer matrices is an attractive solution as they enable controlled release over long storage periods (Duncan, 2011). A combination of the Ag-NPs and Ag+ released from Ag-NPs gives the antimicrobial effect. Various authors have proposed different mechanisms of antimicrobial behavior of Ag-NPs as shown in Table 4.1.

Among the nondegradable polymers, PE, polyvinyl chloride (PVC), polyvinylpyrrolidone, and ethylene vinyl alcohol have been intensively studied as the matrix for Ag-based nanostructures (Carbone et al., 2016; Sadeghnejad et al., 2014). Incorporation of Ag-NPs in the coating in combination with modified atmospheric packaging (MAP) (50% CO_2 and 50% N_2) resulted in strong antimicrobial effect in packaged Fiordilatte (with and without the traditional covering liquid) against *Pseudomonas* spp., Enterobacteriaceae *and Escherichia coli* (Mastromatteo et al., 2015). Mahdi et al. (2012) made PVC-PE laminate trays which were coated with Ag-NPs by inkjet printing and investigated its antimicrobial effect. The nano-Ag package was shown to significantly reduce the microbial growth (total bacteria counts, *E. coli*, and *S. aureus*) of packaged minced beef leading to a shelf life extension by 3 times (from 2 to 7 days under refrigeration).

Fernández and co-investigators (2010) developed absorbent pads consisting of cellulose–Ag-NPs for the packaging of beef-meat extrudates. It was found effective against aerobic and lactic acid bacteria, *Pseudomonas* spp. and Enterobacteriaceae (Fernández et al., 2010). It was also effective against mesophilic aerobic bacteria, psychrotrophic microorganisms, yeasts, and molds in packages containing fresh-cut melon (Fernández et al., 2010). The ripening of the fruit was also retarded along with the reduction of microorganisms, leading to an improvement in visual appearance. Azlin-Hasim et al. (2016) stated that the highest sensitivity against Ag-NPs was observed for *Pseudomonas fluorescens* among *coli, S. aureus, Bacillus cereus, and P. fluorescens.* However, when the Ag-NPs were incorporated in PVC films, their antimicrobial activity was reduced. This reduced antimicrobial activity could be because the majority of the NPs were not in direct contact with the food, as Ag-NPs were directly inserted into polymer matrices, resulting in an inhomogeneous dispersion of NPs in the film.

4.3.1.2.2 Nanocomposites with Au-NPs

The biocidal potential of gold was first recognized by Robert Koch (Glišić and Djuran, 2014). Gold has been used in the treatment in many traditional medicines. Au-NPs are reported to have antibacterial and antifungal activity (Rai et al., 2010; Zawrah et al., 2011; Lima et al., 2013; Thirumurugan et al., 2013). Very few studies are available of antimicrobial packaging using Au-NPs applied to food products. However, many in vitro studies against food-borne pathogens have shown the antimicrobial potential of gold NPs.

The interaction of Au-NPs with sulfur- or phosphorus-holding bases may lead to the inactivation of essential enzymes involved in basic metabolism and energy storage (e.g., nicotinamide adenine dinucleotide (NADH) dehydrogenases), by the generation of huge amounts of free radicals eventually results in cell death. Other proposed mechanisms of antimicrobial behavior of Au-NPs are listed in the Table 4.1.

Pagno and others (2015) prepared quinoa (*Chenopodium quinoa*) starch biofilms by incorporating gold NPs. Along with an improvement in the mechanical, optical, and structural properties, the developed biofilm exhibited strong antibacterial activity against food-borne pathogens with 99% inhibition of *E. coli* and 98% inhibition of *S. aureus* . Lima and coworkers (2013) reported that gold-NPs on the zeolite matrix are excellent biocides.

Other studies highlighted a synergistic antimicrobial effect of Au-NPs with bacteriocin-peptides produced by *Lactobacilli* and also with the commercially available nisin against different microorganisms associated with food-spoilage (Thirumurugan et al., 2013; Gomashe & Dharmik, 2014). Their findings revealed that the bacteriocins produced by *Lactobacilli* had more antimicrobial activity against *Micrococcus luteus* (Thirumurugan et al., 2013), *Klebsiella pneumonia*, *Proteus mirabilis* (Gomashe and Dharmik, 2014), *B. cereus*, *S. aureus*, and *E. coli* (Thirumurugan et al., 2013; Gomashe and Dharmik, 2014) when used as a combination with gold-NPs than alone as well as compared with nisin combined with gold-NPs. Au-NPs have good future potential for use in active antimicrobial nanocomposites.

4.3.1.2.3 Nanocomposites with Cu-NPs

Copper has been shown to possess a broad-spectrum antimicrobial activity against typical spoilage-related microorganisms (Weaver et al., 2010) as well as food-borne pathogens (Faúndez et al., 2004). The tendency of copper to alternate the oxidation state between the cuprous and the cupric form causes displacement of ions essential in the cell metabolism ultimately results in disturbed cellular functions (Quaranta et al., 2011). Even though the antimicrobial activity of copper is lower than the antimicrobial activity of silver (Shankar and Rhim, 2014), Cu-NPs are preferred over Ag-NPs due to its lower cost, ease in mixing with a polymer base, and relatively higher physicochemical stability compared with Ag (Vincent et al., 2016). The Cu^{2+} ion released from the film affects the bacterial outer membrane, causing the membrane to rupture, and killing bacteria is one of the proposed antimicrobial mechanisms of Cu-NPs. Several other mechanisms proposed by researchers are listed in Table 4.1.

Cárdenas and co-investigators (2009) developed chitosan films intended for food packaging applications with dispersed colloidal copper-NPs. Incorporation of colloidal Cu-NPs was shown to be effective against *S. aureus* and *S. Typhimurium*. Agar-based bionanocomposite film containing different types of copper salts were shown to exhibit strong antimicrobial activity against both Gram-positive (*Listeria monocytogenes*) and Gram-negative (*E. coli*) food-borne pathogen (Shankar and Rhim, 2014).

Conte and coworkers (2013) combined the bioactivity of Cu-NPs with PLA (a biodegradable polymer matrix) providing an AP for a variety of mozzarella cheese (*fior di latte*). The active, biodegradable PLA–Cu films showed delayed proliferation for *Pseudomonas* spp. (in vitro). The authors interpreted that the Cu-NPs did not affect the typical dairy flora thus preserving sensory attributes as well. Lomate and co-investigators (2018) prepared LDPE/Cu nanocomposite by incorporating Cu-NPs ranging from 0.5 to 3.0 wt.% with an average size of 50 nm. This nanocomposite showed the antimicrobial effect on both Gram-positive and Gram-negative food spoilage microorganisms.

4.3.1.3 ANTIMICROBIAL NANOCOMPOSITES CONTAINING METAL OXIDE NPS

Metal oxides primarily used as photocatalysts that derive their catalytic activity by absorbing energy from a light source. UV radiation leads to the formation of free radicals and the resulting cellular damage is considered as the dominant mechanism of their antimicrobial activity (Padmavathy and Vijayaraghavan, 2008; Xing et al., 2012; Rhim et al., 2013). Inorganic oxides as antimicrobial agents are important maybe because they contain mineral elements that are essential to humans and exhibit strong activity even applied in little amounts (Padmavathy and Vijayaraghavan, 2008).

4.3.1.3.1 Nanocomposites with TiO_2 NPs

Titanium dioxide (TiO_2) is in the top five of NPs used in consumer products (Shukla et al., 2011). TiO_2 possesses efficient photocatalytic activity making it useful as an ethylene scavenger and an antimicrobial agent (Lin et al., 2015). Titanium dioxide is an attractive photocatalyst because of its chemical stability, nontoxicity, inexpensiveness and Generally Recognized as Safe (GRAS) status (Fujishima et al., 2000; Chawengkijwanich and Hayata, 2008; Bodaghi et al., 2013; Lin et al., 2015; Yemmireddy et al., 2015). TiO_2 has been used for decades (Shukla et al., 2011), it is added to powdered foods as anticaking agents (Lomer et al., 2011), or as a pigment for food coloring and accounts for about 70% of the total production volume of pigments (Baan et al., 2006). TiO_2 NPs have been used in food packaging materials to offer UV blocking and color modification (Garcia et al., 2018).

High surface area-to-volume ratio gives the nanosized TiO_2 particles their higher photocatalytic activity than those in bulk form. The performance of TiO_2 NPs is enhanced because the nanosized TiO_2 owes greater bandgap energy under UV exposure and generates electron-hole pairs on its surface (Maneerat et al., 2003; Onda et al., 2005; Lian et al., 2016). The spontaneous agglomeration of TiO_2 NPs results in a decrease in specific surface area and consequently lessening their activity. (Lin et al., 2015; Qian et al., 2011). Paired with UV light, TiO_2 NPs had strong disinfecting activity against both Gram-negative and Gram-positive microorganisms.

The antimicrobial activity of TiO_2 NPs against various strains of bacteria is associated with the constant generation of hydroxyl radicals and superoxide ions under ultraviolet (UV) exposure (Xie and Hung, 2018). These radicals react with polyunsaturated phospholipid present in the cell membrane of microorganisms (Lin et al., 2015; de Chiara et al., 2015). In this context, Zhang and co-investigators (2014) reported that when UV rays (0.10 mW/cm^2 for 600 s) passed through the bacterial cell, the cell shrunk and the cell membrane gradually discomposed, retained the peripheral layer still its respective integrity. The introduction of 10 mg/L TiO_2 completely broken the cell structure making the outermost layer unstable which shown equal importance of UV illumination and TiO_2 NPs.Various antimicrobial mechanisms proposed by researchers are listed in Table 4.1.

The antibacterial activity of TiO_2-NPs using PE-based films as a matrix to inactivate significantly improved using UV irradiation. It resulted in an inhibition ratio of 89% for *E. coli* and 95% for *S. aureus*, respectively (Xing et al., 2012). Both TiO_2 and ZnO-NPs incorporated in poly (ethylene

TABLE 4.1 Antimicrobial Action of the Filler Materials.

Ag-NPs	• Ag+ species release reactive oxygen species (ROS) by simply dissolving or by the exchange of ions. Monovalent silver species become antibacterial agents while NPs act as a reservoir.	Rai et al. (2010), Li et al. (2010)
	• Ag+ ions have an affinity toward thiols and act as a linking agent, forming an irreversible aggregation of the molecules which contain thiols.	Li et al. (2010), Durán et al. (2016)
	• Ag+ ions have also bound sulfur-containing groups and inactivate enzymes and disturb the metabolism of bacterial cells	Pal et al. (2007), Holt and Bard (2005)
	• Adsorption and aggregation of Ag-NPs at the bacterial membrane cause perforations leading to cellular death by affecting many enzymes and pathways	Holt and Bard (2005)
Au-NPs	• Gold-NPs react with sulfur or phosphorus-holding bases leading to the inactivation of enzymes (e.g., nicotinamide adenine dinucleotide (NADH) dehydrogenases), interrupting the respiratory chains through the action of free radicals. Au-NPs also decline ATPase activities	Cui et al. (2012)
	• Interaction of nanoparticles (direct or indirect) with genetic material results in mutations and genomic instability	Lima et al. (2013)
Cu-NPs	• The tendency of alternating the oxidation state between the cuprous and the cupric form causes the displacement of essential ions in the cell, ultimately results in disturbed cellular functions.	Quaranta et al. (2011)
	• The release of Cu^{2+} ion affects the bacterial cell membrane, leading to the membrane rupture and killing of bacteria	Xue et al. (2011)
TiO_2-NPs	• On UV exposure, charge carriers (electrons (e^-) and holes (h^+)) are produced on the surface of TiO_2. h^+ reacts with adsorbed water or hydroxide ions (OH^-) to produce hydroxyl radicals (OH•) while e^- reduces oxygen and produces superoxide ions (O_2^-).	Song et al. (2017)
	• Production of ROS under UV illumination which damages the molecular components, including DNA, proteins, and lipids	Xie and Hung (2018)
	• TiO_2-NPs move into cells through damaged cell walls and cells and finally release their cellular contents	Long et al. (2014)

TABLE 4.1 *(Continued)*

ZnO-NPs	• Electrostatic interaction between ZnO-NPs and negatively charged bacterial cells, physical absorption, and Zn-containing proteins on the cell membrane collectively facilitate the binding of ZnO-NPs and their internalization into cells.	Sarwar et al. (2016), Zhang et al. (2008)
	• ROS species generated by ZnO-NPs under UV illumination including OH·, singlet oxygen, H_2O_2, and O_2– all contribute to cellular oxidative stress and damages the cellular components (e.g., DNA, proteins, and lipids), consequently leads to cell death	Chakraborti et al. (2013)
	• The dissociated Zn^{2+} from ZnO -NPs interacts with genetic material, deactivates intercellular enzymes, and causes respiratory system damage	Li et al. (2016) Tayel et al. (2011)
	• ZnO–NPs increase membrane permeability and accumulation of nanoparticles in cytoplasmic regions then leads to either inhibition or killing of the bacterial cell	Zhang et al. (2010)
Chitosan-NPs	• Under acidic conditions, the positive charge carried by the C-2 position of the glucosamine monomer enables the polymer to easily interact with microbial cell membranes (negatively charged), leading to the leakage of intracellular components	Chen et al. (1998)
Carbon Nanotubes	• The antimicrobial effect is exerted by the long and thin carbon nanotubes that puncture the microbial cell, causing irreversible damages	Huang et al. (2015)

terephthalate) (PET)/poly (butylene succinate) (PBS) blend thin films showed significant influence on the inhibitory effect of *E. coli* and *S. aureus*. TiO_2-NP-containing films were found to exhibit better inhibition performance than ZnO-NP-containing films. This could be because TiO_2-NPs were smaller than ZnO-NPs and therefore exhibiting an enhanced surface activity of antibacterial agents due to a larger surface-to-volume ratio (Threepopnatkul et al., 2014).

A combined deposit of Ag–TiO_2–SiO_2 on PE film was done to make antimicrobial plastic bags. Lettuce packaged in this film showed a considerably lower spoilage rate when stored at 20°C (Peter et al., 2015). Graphene–TiO_2 nanocomposite was reported to effectively suppress microbial growth of Gram-positive bacteria, viz., *S. aureus* and *B. anthracoides*, whereas it had a moderate effect against the Gram-negative bacteria, viz., *E. coli* and *Pasteurella multocida* (El-Shafai et al., 2019).The photocatalytic antimicrobial effect of TiO_2-NPs incorporated in LDPE films against *Pseudomonas* spp. and *Rhodotorula mucilaginosa* (in vitro) showed a significantly decreased population of mesophilic bacteria and yeast cells in packages of fresh pears (Bodaghi et al., 2013).

In another study (Chawengkijwanich and Hayata, 2008; Saraschandraa et al., 2013), TiO_2-NPs were coated on the surface of packaging films. The resulting active HDPE films (Saraschandraa et al., 2013) were observed to provide a reasonable level of antibacterial activity against *E. coli* and *S. aureus* (in vitro), whereas the active OPP films (Chawengkijwanich and Hayata, 2008) showed high antibacterial activity against *E. coli* in both in vitro and food application on fresh-cut lettuce. Hu and others (2011) investigated the combination of TiO_2-NPs and Ag-NPs. They incorporated Ag-NPs, TiO_2-NPs, and MMT particles into an LDPE film and evaluated their effect on *Botrytis cinerea* (graymold). The results revealed that the applied active film can effectively inhibit the germination of *B. cinerea.* Li et al. (2009) developed packaging films by blending PE with Ag/TiO_2/kaolin nanopowder. The resulting nanopackaging films were successfully applied to inhibit mold growth of strawberries (Yang et al., 2010), Chinese jujube (Li et al., 2009), or Chinese bayberries (Wang et al., 2010).

Nano-Ag/TiO_2-containing HDPE films were also shown to be effective against yeasts, molds, and bacteria in bread leading to a shelf life extension up to 6 days (Mihaly Cozmuta et al., 2015). Youssef et al. (2015) prepared bionanocomposites based on CS, PVA, and TiO_2-NPs with different ratios of TiO_2-NPs, viz., 2%, 4%, and 8%. The obtained CS/PVA/TiO_2 bionano composites were mechanically strong and displayed hydrophobic character. The CS/PVA/TiO_2 bionano composites exhibited good antibacterial activity against Gram-positive (*S. aureus*), Gram-negative (*P. aeruginosa, E. coli*) bacteria and fungi (*Candida albicans*). The results showed that the total bacterial counts, molds, and yeasts and coliforms reduced with an increase in storage period. Xie and Hung (2018) added TiO_2 in three biodegradable polymers (cellulose acetate (CA), polycaprolactone (PCL), and polylactic acid (PLA)) and analyzed the photocatalytic bactericidal property. Their results showed that TiO_2 embedded in PCL and PLA composite films did not show a significant bactericidal property against *E. coli* while TiO_2-embedded CA film showed remarkable antimicrobial potential.

4.3.1.3.2 Nanocomposites with ZnO-NPs

ZnO-NPs have been intensively studied because of their antimicrobial activities, good stability, and UV absorbance properties. ZnO-NPs are known for their broad-spectrum antimicrobial activity against Gram-positive and Gram-negative bacteria (Agarwal et al., 2018), fungi (Sun et al., 2018), and spores (Li et al., 2016). Many studies indicated that the incorporation of ZnO-NPs offers improved mechanical and barrier properties of the material along with broad-spectrum antimicrobial activity through cell membrane damage and generation of ROS (Li et al., 2017; Noshirvani et al., 2017; Mizielińska et al., 2018).

The release of antimicrobial ions, followed by an interaction of the ZnO-NPs with the microorganisms leading to damage to the integrity of the bacterial cell as well as the formation of ROS through light radiation has been widely proposed as the antimicrobial mechanism. A higher ZnO-NP concentration of 50 mg/L caused the complete loss of cell ammonia oxidation capacity, increased cell size, and neutralized ζ potential, all of which indicate the lethal inhibition of bacterial cells (Wu et al., 2018). Other mechanisms proposed for the antimicrobial action of ZnO-NPs are listed in Table 4.1. Synergistic effects of ZnO-NPs in combination with other antimicrobial agents, such as chitosan (Youssef et al., 2015), bacteriocins (Jin and Gurtler, 2011), or other inorganic NPs (Jin and He, 2011) have also been reported.

Emamifar and co-investigators (2010) prepared packages using nanocomposite LDPE films containing ZnO-NPs (0.25% or 1%, about 70 nm) or TiO_2 powder doped with Ag-NPs (1.5% or 5%, about 10 nm). The active packages were used for packaging fresh orange juice which had an initial load of 4.8 log CFU/mL for total aerobic bacteria and 4.9 log CFU/mL for yeasts and molds. All packages kept the microbial load below the desired limit of microbial shelf life for fresh orange juice (6 log CFU/mL) during the storage period, except the pouches having 1% ZnO-NPs.The pouches having 0.25% ZnO-NPs were found to have very low ascorbic acid degradation and browning development. Compared with the control film (pure LDPE), 0.25% ZnO-NP films showed a similar color loss, browning index, and ascorbic acid degradation. However, ranking in terms of sensory attributes (i.e., odor, taste, and overall acceptability) was remarkedly higher for the nanocomposite films, and overall, it was highest for the 0.25% ZnO-NP films. Although the antimicrobial activity of the Ag-NP films increased by increasing the nano-Ag concentration, ZnO-NPs showed the opposite results. This was explained by the tendency of agglomeration of ZnO-NPs during the processing of the films which increased with increasing ZnO-NPs content forming nanoscale aggregates up to 200 nm.

The antimicrobial effect of the combination of ZnO-NPs and Ag-NPs was investigated by Panea and coworkers (2014). Both in vitro and in vivo effect was studied by incorporating the NPs (5 or 10 wt.%) in LDPE films. In vitro tests (5 wt.% of NPs) showed a reduction in 7.34 log CFU/cm^2 for *E. coli*, 6.74 log CFU/cm^2 for *P. aeruginosa*, and 4.31 log CFU/cm^2 for *L. monocytogenes.* After 15 days of storage at 4°C, LDPE control packages filled with chicken breast fillets had a significantly higher population of total mesophilic and revealed that total mesophilic and *Enterobacteriaceae* counts, compared with the active packages containing 5 or 10 wt.% NPs. Moreover, packages containing 10 wt.% NPs showed the lowest counts of test microorganisms after 21 days compared with all other samples. The antimicrobial activity is increased with the increase in the number of NPs in the nanocomposite LDPE films.

In another study (Youssef et al., 2016), a nanocomposite packaging film containing chitosan (CH), carboxymethyl cellulose (CMC), and ZnO-NPs was prepared. The antimicrobial activity of CH/CMC/ZnO film was evaluated against total bacteria, mold, yeast, and coliforms in cheese. It was found that the prepared active films exhibited significantly stronger biocidal activity against a wide range of Gram-positive and Gram-negative bacteria as well as mold and yeast compared with the control film (CH/CMC) leading to the increased shelf life of white soft cheese.

Studies focused on the combination of ZnO-NPs with essential oils that are well known for their antimicrobial effects are also available. Arfat et al. (2015) developed a fish protein isolate/fish skin gelatin films containing 3% ZnO-NPs (ranging 20–40 nm) and basil leaf essential oil (BEO). During 12 days storage at 4°C, the sea bass samples wrapped in this active film showed the lowest count of lactic acid bacteria, psychrophilic bacteria, and spoilage microorganisms including *Enterobacteriaceae*, H_2S-producing bacteria, and *Pseudomonas,* compared with the films containing basil leaf essential oil and without ZnO-NPs, films having ZnO-NP and no BEO, and the control films (fish protein isolate/fish skin gelatin). Moreover, sensory evaluation of the fish shown the longest shelf life for the ZnO-NP/BEO films (12 days) compared with the control (6 days). The incorporation of BEO also provided antioxidant properties which reduced lipid oxidation.

ZnO-NPs embedded chitosan coatings completely inhibited the growth of *Salmonella enterica, E.coli*, and *S. aureus* after 24-h incubation (Al-Naamani et al., 2017). Another nanocomposite of PLA and ZnO-NPs was applied to paper as an antimicrobial coating, a 0.5% loading rate resulted in a 3.14 log reduction of *E. coli* as well as a considerable antimicrobial effect against *S. aureus* (Heydari-Majd et al., 2019).

Under acidic conditions, chitosan carries a positive charge. Therefore, it easily interacts with negatively charged microbial cell membranes, leading to the leakage of proteins and other intracellular constituents (Chen et al., 1998). Priyadarshi and Negi (2017) observed that the antimicrobial activity of the film against *Bacillus subtilis* and *E. coli* increased 2-fold and 1.5-fold, respectively after embedding ZnO-NPs into chitosan films.

Jafarzadeh and co-investigators (2019) reported the increased antimicrobial activity with an increase in the content of ZnO-NPs. Further, the molds were the most sensitive to the ZnO incorporated films followed by yeast and bacteria. Between the two food-borne pathogenic bacteria, *S. aureus* (Gram-positive) was more susceptible to the ZnO incorporated composite films than *E. coli* (Gram-negative). It also exhibited stronger antibacterial activity against Gram-positive bacteria than Gram-negative bacteria.

Some other nanocomposites, which have been shown to have antimicrobial activity are described in the following paragraphs.

Swaroop and Shukla (2019) produced bio-based films by reinforcing MgO NPs in PLA biopolymer using the solvent casting method and found 2 wt.% amount of MgO NPs in PLA films exhibited an improvement in the oxygen barrier and tensile strength properties (up to 29% and 25%, respectively), as well as a superior antibacterial efficacy (the 2 wt.% films caused progressive damage and death of nearly 46% of *E. coli* bacterial culture after 12 h treatment). Carbon nanotubes are cylinders with nanoscale diameters and exert powerful antimicrobial effects. They have good potential to be used in nanocomposites as an active component because the long and thin carbon nanotubes puncture the microbial cells, causing irreversible damages (Huang et al., 2015).

4.3.2 ETHYLENE SCAVENGER NANOCOMPOSITES

Ethylene is a low-molecular-weight volatile organic compound that acts as a plant hormone (Sisler and Yang, 1984). The ethylene gas deteriorates the fresh produce and reduces its shelf life, on one hand, but on the other hand, it can lead to an appropriate ripening process and prepare fresh produce for the market. However, in the case of shelf life during storage and retail, the presence of ethylene is considered as a contaminant, thereby it should be controlled or minimized.

Even at a very low concentration ranging from 10 to 100 mL/L (Aghdam et al., 2019), it can adversely affect the quality of fruits and vegetables by accelerating respiration. Accumulation of ethylene in the headspace of the package can be responsible for several undesirable reactions, such as yellowing of green vegetables, development of bitter flavors, and chlorophyll degradation. Deleterious effects of ethylene can be reduced by either its removal, degradation, or by blocking ethylene-binding sites (Hussain et al., 2011; Cao et al., 2015; Chopra et al., 2017; Sadeghi et al., 2019).

Conventionally, ethylene oxidizing materials such as potassium permanganate which oxidizes ethylene to acetate and ethanol has been used in ethylene scavenging packaging systems for climacteric fruits (Terry et al., 2007; Wills and Warton, 2004). This ethylene capturing material is usually used in the form of sachets that are placed inside the packaging, but it finds limited acceptance due to toxicity-based concerns and regulations against their food-contact applications (Tas et al., 2017). The major drawback of using scavenging agents is that they are mostly unstable and it is necessary to maintain their scavenging capacity during storage (Suppakul et al., 2003).

AP based on nanocomposites can solve the problem. Active surfaces, such as zeolite, clays, or activated carbon (Brody et al., 2008), which may be incorporated in packaging materials can remove ethylene by physical adsorption. The scavenging compound is an active agent that can be embedded in the food packaging matrix in the form of AP. The micro- and nanoscavenging materials at similar concentrations showed the highest ethylene oxidizing performance (Spricigo et al., 2017). The highly porous NPs can be applied as a delivery agent to take advantage of the scavenging capacity of materials that cannot be used directly in the packaging system (Sadeghi et al., 2019). Along with certain metal NPs, clays have also shown the ethylene scavenging ability. From among the different nanoclays vermiculite showed the strongest ethylene scavenging ability during the storage of banana fruit, while kaolinite showed the weakest performance (Álvarez-Hernández et al., 2018).

The use of potassium permanganate ($KMnO_4$) with nanomaterials like SiO_2 NPs demonstrated a stronger scavenging system (Bhattacharjee and Dhua, 2017). The main drawback of SiO_2 is related to its low saturation capacity in ethylene removal. Another matrix for the preparation of the $KMnO_4$-based ethylene scavenging system is activated aluminum oxide (Al_2O_3). Such a compound possesses a high surface area, porous matrix, high adsorption capacity, and thermal stability. Moreover, Al_2O_3 is inexpensive and nontoxic (Mallakpour and Khadem, 2015; Spricigo et al., 2017).

Metallic NPs which are commonly used as ethylene oxidizing agents to prevent deterioration effects of ethylene on fresh produce are silver (Ag), titanium oxide (TiO_2) (Wang et al., 2010), zinc oxide (ZnO) (Arfat et al., 2015), copper (Cu), and palladium (Pd) (Terry et al., 2007). Gold and cobalt oxide also have shown ethylene scavenging activity but their use is limited due to high temperature requirements.

4.3.2.1 ETHYLENE SCAVENGING MECHANISM OF NANOPARTICLES

The deleterious effect of ethylene can be reduced either by its oxidation or by inhibiting its binding to working sites. Some of the possible mechanisms of action of ethylene scavenger nanocomposites are enlisted in Table 4.2.

4.3.2.2 ETHYLENE SCAVENGING POLYMER NANOCOMPOSITES

PVC films coated with nano-ZnO powder were evaluated for their effectiveness in preserving the quality of fresh-cut "Fuji" apples. The

TABLE 4.2 Proposed Mechanism of Action of Ethylene Scavenger Nanocomposites.

Ethylene Oxidation	• In the first step, the attack of photons on the NPs surface promotes an electron from its present state (i.e., valence band) to the conductance band	Sadeghi et al. (2019)
	• Water or oxygen can be adsorbed on the surface of NPs, followed by the generation of hole (h^+)–electron (e^-) pairs.	
	• At last, the oxidizing agents (•OH and •O_2^-) can be generated which oxidize ethylene to H_2O and CO_2	
	• The large surface area of the nanosized TiO_2 helps in achieving high efficiency through higher hydroxyl groups adsorbed on the surface of TiO_2 in the photocatalytic oxidation of the ethylene	Maneerat and Hayata (2008)
Ethylene Absorption	• Nanocomposite film containing halloysite nanotubes (HNT) has proven the adsorption of 0.56 mL headspace ethylene per 1 g of film	Tas et al. (2017)
Absorption + Oxidation	• Use of potassium permanganate ($KMnO_4$) with nanomaterials like SiO_2 NPs or aluminum oxide (Al_2O_3)	Bhattacharjee and Dhua (2017)
	• These nanomaterials give a porous matrix with high adsorption capacity and act as a delivery agent to take advantage of the scavenging capacity of ($KMnO_4$)	Sadeghi et al. (2019)

use of nano-ZnO film resulted in the retardation of ethylene production compared with control (PVC film) (Li et al., 2011). In another study involving LDPE-TiO_2 film, it was reported that the ethylene production of the strawberries in the package was significantly inhibited. Changes associated with ethylene production like fruit decay, softening, weight loss, and titratable acid content were also reduced (Luo et al., 2013).

Fernández et al. (2010) developed cellulose-Ag-NP nanocomposite to control the spoilage-related microflora. These Ag-NP pads, located in trays of fresh-cut melon showed dual benefits. Apart from the antimicrobial effects, these pads indicated blocking of the ethylene and due to that the senescence of the melon pieces was retarded significantly. Lessening of ethylene-mediated effects on the ripening resulted in a less ripen and juicier product. In another study, Li and co-investigators (2009) synthesized packaging films by blending nanopowder of Ag, TiO_2, and kaolin with PE. This nanopackaging material resulted in beneficial effects like reduction in fruit softening, weight loss, and browning in Chinese Jujube compared with control PE packaging material.

TiO_2-NPs, Ag-NPs, and MMT particles were incorporated into an LDPE film and assessed for their effect on postharvest quality of ethylene-treated kiwifruit during cold storage. The application of the LDPE–Ag–TiO_2 nanocomposite packaging resulted in reduced ethylene production, extended sensory score, lowered degradation of nutritional components, reduced physiological changes, delayed ripening, and extended shelf life of the harvested kiwifruit (Hu et al., 2011). In another study involving bionanocomposite, Siripatrawan and Kaewklin (2018) synthesized the chitosan–TiO_2 nanocomposites and interpreted that ethylene photodegradation was increased with increase in TiO_2 concentration. Ethylene photocatalytic degradation properties along with the tensile strength of the chitosan film having 1% TiO_2 was found to be optimum among different films with varying concentrations of TiO_2. In addition to this, both Gram-positive (*S. aureus*) and Gram-negative (*E. coli*, *S. Typhimurium*, and *P. aeruginosa*) bacterial and fungal (*Aspergillus* and *Penicillium*) population was reduced remarkably (Siripatrawan and Kaewklin, 2018).

Yang et al. (2010) fabricated an advanced ethylene oxidizing system through the impregnation of $AgNO_3$ with SiO_2/Al_2O_3 (ZSM-5 (X) catalyst where X represents SiO_2/Al_2O_3 ratio) to produce an Ag/SiO_2/Al_2O_3 system. Among the SiO_2/Al_2O_3 ratios of 20, 25, 38, 50, and 80 the ZSM-5 catalyst with SiO_2/Al_2O_3 at a ratio of 38 exhibits the strongest oxidizing capacity.

They reported that ethylene can be completely oxidized into CO_2 by all the Ag/ZSM-5 (aluminosilicate) catalysts at 25°C. Ag/ZSM-5 with a SiO_2/Al_2O_3 ratio of 38 exhibited good catalytic stability.

The scavenging systems involving Ag, TiO_2, and kaolin NPs incorporated into PE matrix successfully maintained food quality (sensory and appearance) during storage. These NPs tend to oxidize the ethylene emitted from the fruit before it binds to target sites in the fresh produce (Wang et al., 2010). Terry and co-investigators (2007) highlighted the importance of relative humidity and demonstrated that the Pd-impregnated zeolite can oxidize ethylene gas at 100% RH. Fresh-cut apples were stored in the PVC as a packaging matrix with added zinc oxide NPs. Results showed that this system reduced the ethylene production rate by 50%, that is, from a 70 μL/kg × day (control) to 40 μL/kg × day after the 9 days storage period.

The temperature at which nanocomposites show activity is a very important factor. Jiang and others (2013) reported the oxidation of ethylene by platinum (Pt) NPs supported on MCM-41. It was discovered that the catalytic activity of ethylene oxidation has reached 100% at 0°C but its performance as an active agent for the low-temperature catalysis decreased rapidly and found not suitable for practical applications. Xue and co-investigators (2011) developed gold (Au) catalysts supported on Co_3O_4 for the removal of ethylene at low temperatures. Its decreased performance for catalytic oxidation of ethylene was supposed to be a barrier to its practical applications.

TiO_2 NPs showed higher oxidizing ability at cold storage conditions (3°C) and in the presence of humidity. This may be explained by the balance between •OH and $•O^{-2}$ groups that resulted in a higher surface activity (Hussain et al., 2011). Terry et al. (2007) developed a palladium-promoted powder-based scavenger that has ethylene adsorption capacity around 4500 μL/gadsorber at the room. Results showed that the efficiency of the palladium-promoted scavenger was far superior to $KMnO_4$ even when utilized in small quality. The ethylene scavenging action of NPs is illustrated in Figure 4.3.

4.4 OXYGEN SCAVENGER NANOCOMPOSITES

The presence of oxygen in food packages elevates the rate of oxidation of fats or vitamins present in food or promotes the growth of microorganisms, such as aerobic bacteria, yeasts, and molds which leads to rapid spoilage of food. Most of the foods encounter enzymatic browning and nutrition loss due to the presence of oxygen in the headspace (Ayranci and Tunc, 2003). MAP can reduce the residual oxygen level in the headspace up to 0.5–2.0 vol.% (Gibis and Rieblinger, 2011) at it is enough to start the deleterious changes in foodstuffs. Whereas oxygen scavengers can reduce the oxygen level to less than 0.1 vol.% (Mills et al., 2006). The use of oxygen scavengers can efficiently reduce the adverse effects that residual oxygen creates, such as oxidative rancidity, darkening, and browning of fresh meats, assisting the growth of aerobic spoilage-related microorganisms, acceleration of fresh fruit and vegetable respiration ultimately leading to the shortening of food shelf life (Brody et al., 2008). Absorption kinetics, absorption capacity, and time–temperature behavior differentiate the oxygen scavengers from one another (Gibis and Rieblinger, 2011).

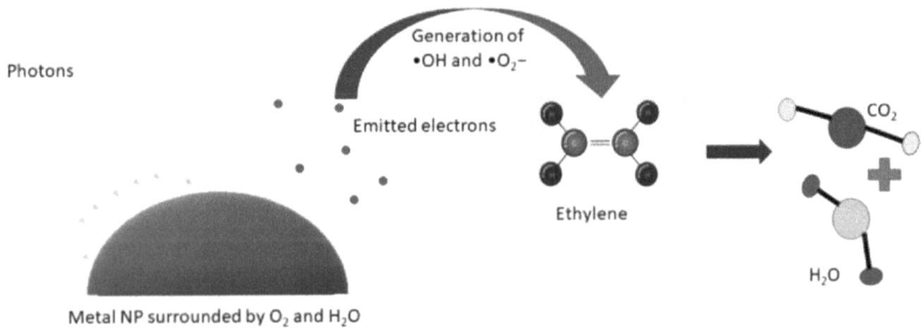

FIGURE 4.3 Ethylene scavenger nanocomposites.

Oxygen scavenger systems belong to the absorber category of AP techniques (Gill and Ahvenainen, 2003). Most of the commercially available oxygen scavengers are iron-based oxygen absorbers available in a sachet form (Brody et al., 2001; Miltz and Perry, 2005). Sachet form of the oxygen scavenger systems can be replaced by oxygen scavenging films offering several advantages like potential use with retort packaging, elimination of food product damage that may occur when a sachet contacts the food, and potential cost savings due to increased production efficiency and convenience (Brody et al., 2001).

The working mechanism of the iron-based oxygen scavenging system involves the reaction of iron with oxygen in the container to form ferric oxides. Temperature, relative humidity, and oxygen partial pressure in the package are the extrinsic factors that affect the activity of the scavenging system. They often depend on the internal environment of food packaging. Intrinsic factors are influenced by properties of iron powder, such as the particle size and specific surface area (Polyakov and Miltz, 2010; Tewari et al., 2002).

Iron powder impregnation in polymeric films also results in an increased surface area, leading to enhanced scavenging capacity. Even though iron is not a noble metal, it remains inert in dry conditions. Passivating layer in the thickness of several nanometers forms on the metallic iron surface under dry conditions (Ohtsuka, 2012; McCafferty, 2010). Nano iron showed better scavenging capacity than its micrometric counterpart (Mu et al., 2013; Foltynowicz et al., 2017). The increased activity can be explained by the increased number of reactive sites. Compared with micrometric powder form, the specific surface area of nanometric iron powder is much greater (Mu et al., 2013) resulting in excellent adsorption properties and high reduction activities.

The use of nanocomposites having MMT and iron NPs impregnated in polymeric films can affect the physical and mechanical properties of the packaging material. Addition of 2 wt.% MMT to polypropylene (PP) films reduced the oxygen permeability by 22% and water vapor permeability by 33%, respectively while the addition of only 0.2 wt.% of iron NPs resulted in a reduction in oxygen permeability by 55% and increase in the scavenging capacity by 77% (Khalaj et al., 2016). Mu and co-investigators (2013) synthesized a Fe-NPs-based oxygen scavenger. The mixture of Fe-NPs, activated carbon, NaCl, and $CaCl_2$ was filled into sachets. Storage tests of roasted sunflower seeds and walnuts showed that the oxygen scavenging capacity of the nanosized scavenger (110 nm average particle size) was almost 1.4 times higher compared with that of its microsized counterpart (about 20 μm).

Busolo and Lagaron (2012) developed and tested synthetic nano-iron-containing kaolinite for its potential as an oxygen scavenger. The authors observed that the iron kaolinite exhibited faster oxygen uptake at 100% RH. Active HDPE and LLDPE films with 10 wt.% filler content were manufactured by the extrusion method. Oxygen uptake at 24°C and 100% RH was observed. Films showed uptake of oxygen in the range between 2.4 and 4.3 mL/g of composite. At 5°C and 50% RH activity were significantly decreased (almost 50%). Byun and coworkers (2012) demonstrated that incorporation of α-tocopherol-loaded NPs and Fe-(II) chloride into a fish gelatin film was able to scavenge oxygen effectively within 50 days with the capacity of 1969 cm^3 O_2/m^2/mil film thickness.

Xiao-e and others (2005) developed packaging films by adding TiO_2-NPs into different polymers. The TiO_2 pastes were deposited on glass microscope slides and plastic (acetate) substrates. Pastes were deposited either with or without the prior addition of organic sacrificial electron donors (SEDs). Methanol, PVC, and PEG as SEDs were added to the films following film deposition. Oxygen scavenging initiated by UV illumination revealed that PVC-TiO_2 films exhibited the highest activity.

Nanocrystalline titania particles were incorporated in a flexible ethylcellulose polymer film. The nanocrystalline titanium dioxide and sacrificial electron donor were supported within the polymer matrix. The oxygen scavenging rates (0.017 cm^3 O_2/h cm^2 over 24 h) exhibited by these films were comparable to those of traditional oxygen scavengers (Mills et al., 2006), while the requirement of continuous UV illumination limits to drive the scavenging process forward (Mills et al., 2006).

The catalytic role of Pd and Pt results in a highly effective oxygen scavenging system. Yu and co-investigators (2004) used an in situ infusion method to incorporate small amounts of Pd or Pt-NPs into most popular thermoplastic polymer matrices, such as fluorinated ethylene-propylene copolymers (FEP), PP, LLDPE, PET, and nylon 6,6. They found that trace amounts (2%) of hydrogen increased the efficiency of the scavenging films.

Yildirim and others (2015) used magnetron sputtering technology to develop a Pd-based oxygen scavenging film. An additional silicon oxide (SiOx) layer was also applied to the films before Pd deposition to improve the substrate surface. The PET film with the nanoscale Pd layer was able to remove up to 2.5 vol.% residual oxygen in food packages within a few minutes. The optimal Pd layer thickness for the oxygen scavenging films ranged from 0.7 to 3.4 nm.

Oxygen scavenging films could be a blessing to the meat industry to solve the problems associated with discoloration. Hutter and coworkers (2016) found that when Pd-based oxygen scavenging film was used for packaging cooked cured ham, the red color was preserved and discoloration was prevented for 21 days even though the packages were exposed to light 24 h/day. On the other hand, ham packaged without OS film lost its redness within the first 2 h after packaging.

In another study, Pd-based OS film when used for packaging of bakery products resulted in a three- to four-fold longer shelf life of partially baked buns, toast bread slices, and gluten-free bread slices as mold growth was retarded up to 8–10 days (Rüegg et al., 2016). Whereas in samples without the OS film, mold growth was detected in all bread samples after 2 days. Gohil and Wysock (2014) mentioned that the addition of TiO_2 or Al_2O_3 as catalysts and an edible oil to the ascorbate and laccase-based formulations induced a synergistic effect upon the rate of scavenging oxygen from the packages. The addition of nano-based catalysts such as nano-TiO_2 or nano alumina (Al_2O_3) extended the OS capacity of the film. Khalaj and co-investigators (2016) developed laminated food packaging films, that is, PP/OMMT/iron nanoparticle nanocomposites with 4 wt.% OMMT and 0.2 wt.% nanoparticle. The use of OMMT reduced the oxygen and water vapor permeability while the use of 0.2 wt.% iron NPs resulted in scavenging oxygen.

4.5 CONCLUSION

Nanocomposites are the most studied and commercialized application of nanotechnology in the food sector. Nanocomposites can significantly extend the shelf life of food by improving the barrier properties of conventional as well as biodegradable packaging along with cost-effectiveness at the same time. The identification, characterization, and quantification of these NPs, their characteristic properties proved that nanomaterials can provide active properties, such as antimicrobial activity, oxygen, and ethylene scavenging capacity, offering great potential to maintain or improve the quality of food products. Synergistic effects were observed in several cases. Nanocomposites have the potential to overcome limitations involved in AP like health hazards associated with the sachet-based application, and limitations in the regeneration of active components.

At present, most of the applications of nanocomposite regarding AP revolves around antimicrobial and scavenging or absorbing functions. Although migration of the nanomaterials and its health effects are still required a lot of research, the nanocomposites as an AP material seem to have a very bright future in the upcoming era of food packaging.

KEYWORDS

- **active packaging**
- **nanotechnology**
- **nanocomposites**
- **nanomaterials**
- **antimicrobial**
- **ethylene scavenger**
- **oxygen scavenger**

REFERENCES

Actis, L.; Srinivasan, A.; Lopez-Ribot, J. L.; Ramasubramanian, A. K.; Ong, J. L. Effect of Silver Nanoparticle Geometry on Methicillin Susceptible and Resistant *Staphylococcus aureus*, and Osteoblast Viability. *J. Mater. Sci.: Mater. Med.* **2015,** *26* (7), 215.

Agarwal, H.; Menon, S.; Kumar, S. V.; Rajeshkumar, S. Mechanistic Study on Antibacterial Action of Zinc Oxide Nanoparticles Synthesized Using Green Route. *Chemico-Biol. Interact.* **2018,** *286*, 60–70.

Aghdam, M. S.; Luo, Z.; Jannatizadeh, A.; Sheikh-Assadi, M.; Sharafi, Y.; Farmani, B.; Razavi, F. Employing Exogenous Melatonin Applying Confers Chilling Tolerance in Tomato Fruits by Upregulating ZAT2/6/12 Giving Rise to Promoting Endogenous Polyamines, Proline, and Nitric Oxide Accumulation by Triggering Arginine Pathway Activity. *Food Chem.* **2019,** *275*, 549–556.

Alexandre, M.; Dubois, P. Polymer-Layered Silicate Nanocomposites: Preparation, Properties and Uses of a New Class of Materials, *Mater. Sci. Eng.: R: Rep.* **2000,** *28*, 1–63.

Al-Naamani, L.; Dobretsov, S.; Dutta, J.; Burgess, J. G. Chitosan-Zinc Oxide Nanocomposite Coatings for the Prevention of Marine Biofouling. *Chemosphere* **2017,** *168*, 408–417.

Álvarez-Hernández, M. H.; Artés-Hernández, F.; Ávalos-Belmontes, F.; Castillo-Campohermoso, M. A.; Contreras-Esquivel, J. C.; Ventura-Sobrevilla, J. M.; Martínez-Hernández, G. B. Current Scenario of Adsorbent Materials Used in Ethylene Scavenging Systems to Extend Fruit and Vegetable Postharvest Life, *Food Bioprocess Technol.* **2018,** *11*, 511–525.

Amirsoleimani, M.; Khalilzadeh, M. A.; Sadeghifar, F.; Sadeghifar, H. Surface Modification of Nanosatrch Using Nano Silver: A Potential Antibacterial for Food Package Coating. *J. Food Sci. Technol.* **2018,** *55*, 899–904.

Arakha, M.; Pal, S.; Samantarrai, D.; Panigrahi, T. K.; Mallick, B. C.; Pramanik, K.; Bibekanand M.; Jha, S. Antimicrobial Activity of Iron Oxide Nanoparticle Upon Modulation of Nanoparticle-Bacteria Interface. *Sci. Rep.* **2015,** *5*, 14813.

Arfat, Y. A.; Benjakul, S.; Vongkamjan, K.; Sumpavapol, P.; Yarnpakdee, S. Shelf-Life Extension of Refrigerated Sea Bass Slices Wrapped with Fish Protein Isolate/Fish Skin Gelatin-ZnO Nanocomposite Film Incorporated with Basil Leaf Essential Oil. *J. Food Sci. Technol.* **2015,** *52* (10), 6182–6193.

Ayranci, E.; Tunc, S. A Method for the Measurement of the Oxygen Permeability and the Development of Edible Films to Reduce the Rate of Oxidative Reactions in Fresh Foods. *Food Chem.* **2003,** *80*, 423–431.

Azeredo, H. M.; Otoni, C. G.; Corrêa, D. S.; Assis, O. B.; de Moura, M. R.; Mattoso, L. H. C. Nanostructured Antimicrobials in Food Packaging—Recent Advances. *Biotechnol. J.* **2019,** *14* (12), 1900068.

Azevedo, S.; Cunha, L. M.; Mahajan, P. V.; Fonseca, S. C. Application of Simplex Lattice Design for Development of Moisture Absorber for Oyster Mushrooms. *Procedia Food Sci.* **2011,** *1*, 184–189.

Azlin-Hasim, S.; Cruz-Romero, M. C.; Morris, M. A., Padmanabhan, S. C.; Cummins, E.; Kerry, J. P. The Potential Application of Antimicrobial Silver Polyvinyl Chloride Nanocomposite Films to Extend the Shelf-Life of Chicken Breast Fillets. *Food Bioprocess Technol.* **2016,** *9* (10), 1661–1673.

Baan, R.; Straif, K., Grosse, Y.; Secretan, B.; El Ghissassi, F.; Cogliano, V. WHO International Agency for Research on Cancer Monograph Working Group. Carcinogenicity of Carbon Black, Titanium Dioxide, and Talc, 2006.

Bhattacharjee, D.; Dhua, R. Ethylene Absorbents Improve the Shelf Life of Pointed Gourd (Trichosanthes dioica Roxb.) Fruits. *Int. J. Pure Appl. Biosci.* **2017,** *5*, 64–71.

Bodaghi, H.; Mostofi, Y.; Oromiehie, A.; Zamani, Z.; Ghanbarzadeh, B.; Costa, C.; Del Nobile, M. A. Evaluation of the Photocatalytic Antimicrobial Effects of a TiO_2 Nanocomposite Food Packaging Film by In Vitro and In Vivo Tests. *LWT-Food Sci. Technol.* **2013,** *50* (2), 702–706.

Brockgreitens, J.; Abbas, A. Responsive Food Packaging: Recent Progress and Technological Prospects. *Comprehensive Rev. Food Sci. Food Safet.* **2016,** *15*, 3–15.

Brody, A. L.; Bugusu, B.; Han, J. H.; Sand, C. K.; McHugh, T. H. Innovative Food Packaging Solutions. *J. Food Sci.* **2008,** *73*, 107–116.

Busolo, M. A.; Lagaron, J. M. Oxygen Scavenging Polyolefin Nanocomposite Films Containing an Iron Modified Kaolinite of Interest in Active Food Packaging Applications. *Innov. Food Sci. Emerg. Technol.* **2012,** *16*, 211–217.

Byun, Y.; Bae, H. J.; Whiteside, S. Active Warm-Water Fish Gelatin Film Containing Oxygen Scavenging System. *Food Hydrocoll.* **2012,** *27*, 250–255.

Cao, J.; Li, X.; Wu, K.; Jiang, W.; Qu, G. Preparation of a Novel PdCl2–CuSO4–Based Ethylene Scavenger Supported by Acidified Activated Carbon Powder and Its Effects on Quality and Ethylene Metabolism of Broccoli During Shelf-Life, *Postharvest Biol. Technol.* **2015,** *99*, 50–57.

Carbone, M.; Donia, D. T.; Sabbatella, G.; Antiochia, R. Silver Nanoparticles in Polymeric Matrices for Fresh Food Packaging. *J. King Saud Univ.-Sci.* **2016,** *28* (4), 273–279.

Cárdenas, G.; Meléndrez, M. F.; Cancino, A. G. Colloidal Cu Nanoparticles/Chitosan Composite Film Obtained by Microwave Heating for Food Package Applications. *Polym. Bull.* **2009,** *62*, 511–524.

Cha, S. H.; Hong, J.; McGuffie, M.; Yeom, B.; VanEpps, J. S.; Kotov, N. A. Shape-Dependent Biomimetic Inhibition of Enzyme by Nanoparticles and Their Antibacterial Activity. *ACS Nano* **2015,** *9*, 9097–9105.

Chakraborti, S.; Bhattacharya, S.; Chowdhury, R.; Chakrabarti, P. The Molecular Basis of Inactivation of Metronidazole-Resistant Helicobacter Pylori Using Polyethyleneimine Functionalized Zinc Oxide Nanoparticles. *PLoS One* **2013,** *8* (8).

Chawengkijwanich, C.; Hayata, Y. Development of TiO_2 Powder-Coated Food Packaging Film and Its Ability to Inactivate *Escherichia coli* in Vitro and in Actual Tests. *Int. J. Food Microbiol.* **2008,** *123* (3), 288–292.

Chen, C. S.; Liau, W. Y.; Tsai, G. J. Antibacterial Effects of N-Sulfonated and N-Sulfobenzoyl Chitosan and Application to Oyster Preservation. *J. Food Protect.* **1998,** *61*, 1124–1128.

Chopra, S.; Dhumal, S.; Abeli, P.; Beaudry, R.; Almenar, E. Metal-Organic Frameworks Have Utility in Adsorption and Release of Ethylene and 1-Methylcyclopropene in Fresh Produce Packaging. *Postharvest Biol. Technol.* **2017,** *130*, 48–55.

Conte, A.; Longano, D.; Costa, C.; Ditaranto, N.; Ancona, A.; Cioffi, N.; Del Nobile, M. A. A Novel Preservation Technique Applied to Fiordilatte Cheese. *Innov. Food Sci. Emerg. Technol.* **2013,** *19*, 158–165.

Cui, Y.; Kumar, S.; Kona, B. R.; van Houcke, D. Gas Barrier Properties of Polymer/Clay Nanocomposites. *Res. Adv.* **2015,** *5*, 63669–63690.

Cui, Y.; Zhao, Y.; Tian, Y.; Zhang, W.; Lü, X.; Jiang, X. The Molecular Mechanism of Action of Bactericidal Gold Nanoparticles on *Escherichia coli*. *Biomaterials* **2012,** *33*, 2327–2333.

Cushen, M.; Kerry, J.; Morris, M.; Cruz-Romero, M.; Cummins, E. Nanotechnologies in the Food Industry–Recent Developments, Risks and Regulation. *Trends Food Sci. Technol.* **2012,** *24*, 30–46.

Deplanche, K.; Caldelari, I.; Mikheenko, I. P.; Sargent, F.; Macaskie, L. E. Involvement of Hydrogenases in the Formation of Highly Catalytic Pd (0) Nanoparticles by Bioreduction of Pd (II) Using *Escherichia coli* Mutant Strains. *Microbiology* **2010,** *156*, 2630–2640.

Dobrucka, R; Cierpiszewski, R. Active and Intelligent Packaging Food–Research and Development–a Review. *Polish J. Food Nutr. Sci.* **2014,** *64*, 7–15.

Du, M.; Guo, B.; Jia, D. Newly Emerging Applications of Halloysite Nanotubes: A Review. *Polym. Int.* **2010,** *59*, 574–582.

Duncan, T. V. Applications of Nanotechnology in Food Packaging and Food Safety: Barrier Materials, Antimicrobials and Sensors. *J. Colloid Onterf. Sci.* **2011,** *363*, 1–24.

Durán, N.; Durán, M.; De Jesus, M. B.; Seabra, A. B.; Fávaro, W. J.; Nakazato, G. Silver Nanoparticles: A New View on Mechanistic Aspects on Antimicrobial Activity. *Nanomed.: Nanotechnol. Biol. Med.* **2016,** *12*, 789–799.

El-Shafai; N., El-Khouly, M. E.; El-Kemary, M.; Ramadan, M., Eldesoukey, I.; Masoud, M. Graphene Oxide Decorated with Zinc Oxide Nanoflower, Silver and Titanium Dioxide Nanoparticles: Fabrication, Characterization, DNA Interaction, and Antibacterial Activity. *RSC Adv.* **2019,** *9*, 3704–3714.

Emamifar, A.; Kadivar, M.; Shahedi, M.; Soleimanian-Zad, S. Evaluation of Nanocomposite Packaging Containing Ag and ZnO on Shelf Life of Fresh Orange Juice. *Innov. Food Sci. Emerg. Technol.* **2010,** *11*, 742–748.

Enescu, D.; Cerqueira, M. A.; Fucinos, P.; Pastrana, L. M. Recent Advances and Challenges on Applications of Nanotechnology in Food Packaging. A Literature Review. *Food Chem. Toxicol.* **2019,** 110814.

Fang, B.; Jiang, Y.; Nüsslein, K., Rotello, V. M.; Santore, M. M. Antimicrobial Surfaces Containing Cationic Nanoparticles: How Immobilized, Clustered, and Protruding Cationic Charge Presentation Affects Killing Activity and Kinetics. *Colloids Surf. B: Biointerf.* **2015,** *125*, 255–263.

Faúndez, G.; Troncoso, M.; Navarrete, P.; Figueroa, G. Antimicrobial Activity of Copper Surfaces against Suspensions of *Salmonella enterica* and *Campylobacter jejuni*. *BMC Microbiol.* **2004,** *4* (1), 19.

Fernández, A.; Picouet, P.; Lloret, E. Cellulose-Silver Nanoparticle Hybrid Materials to Control Spoilage-Related Microflora in Absorbent Pads Located in Trays of Fresh-Cut Melon. *Int. J. Food Microbiol.* **2010a,** *142*, 222–228.

Fernández, A.; Picouet, P.; Lloret, E. Reduction of the Spoilage-Related Microflora in Absorbent Pads by Silver Nanotechnology during Modified Atmosphere Packaging of Beef Meat. *J. Food Protect.* **2010b,** *73*, 2263–2269.

Foltynowicz, Z.; Bardenshtein, A.; Sängerlaub, S.; Antvorskov, H.; Kozak, W. Nanoscale, Zero Valent Iron Particles for Application as Oxygen Scavenger in Food Packaging. *Food Packag. Shelf Life* **2017,** *11*, 74–83.

Foster, H. A.; Ditta, I. B.; Varghese, S.; Steele, A. Photocatalytic Disinfection Using Titanium Dioxide: Spectrum and Mechanism of Antimicrobial Activity. *Appl. Microbiol. Biotechnol.* **2011,** *90* (6), 1847–1868.

Fujishima, A.; Rao, T. N.; Tryk, D. A. Titanium Dioxide Photocatalysis. *J. Photochem. Photobiol. C: Photochem. Rev.* **2000,** *1*, 1–21.

Gibis, D.; Rieblinger, K. Oxygen Scavenging Films for Food Application. *Procedia Food Sci.* **2011,** *1*, 229–234.

Gill, C. O.; Ahvenainen, R. Active Packaging in Practice: Meat. *Novel Food Packag. Techn.* **2003,** 365–383.

Glišić, B. Đ.; Djuran, M. I. Gold Complexes as Antimicrobial Agents: An Overview of Different Biological Activities in Relation to the Oxidation State of the Gold Ion and the Ligand Structure. *Dalton Trans.* **2014,** *43*, 5950–5969.

Gohil, R. M.; Wysock, W. A. Designing Efficient Oxygen Scavenging Coating Formulations for Food Packaging Applications. *Packag. Technol. Sci.* **2014,** *27*, 609–623.

Gomashe, A. V.; Dharmik, P. G. Synergistic Effect of Gold Nanoparticles and Bacteriocin against Food Blemishing Microbes: A Novel Approach for Food Packaging Material Preparation. *Global J. Res. Anal.* **2014,** *3* (5), 1–3.

Gurunathan, S.; Han, J. W., Kwon, D. N.; Kim, J. H. Enhanced Antibacterial and Anti-Biofilm Activities of Silver Nanoparticles against Gram-Negative and Gram-Positive Bacteria. *Nanoscale Res. Lett.* **2014,** *9*, 373.

Hannon, J. C.; Kerry, J.; Cruz-Romero, M.; Morris, M.; Cummins, E. Advances and Challenges for the Use of Engineered Nanoparticles in Food Contact Materials. *Trends Food Sci. Technol.* **2015,** *43*, 43–62.

Heydari-Majd, M.; Ghanbarzadeh, B.; Shahidi-Noghabi, M., Najafi, M. A.; Hosseini, M. A New Active Nanocomposite Film Based on PLA/ZnO Nanoparticle/Essential Oils for the Preservation of Refrigerated Otolithesruber Fillets. *Food Packag. Shelf Life* **2019,** *19*, 94–103.

Holt, K. B.; Bard, A. J. Interaction of Silver (I) Ions with the Respiratory Chain of *Escherichia coli*: An Electrochemical and Scanning Electrochemical Microscopy Study of the Antimicrobial Mechanism of Micromolar Ag+. *Biochemistry* **2005,** *44*, 13214–13223.

Hoseinnejad, M.; Jafari, S. M.; Katouzian, I. Inorganic and Metal Nanoparticles and Their Antimicrobial Activity in Food Packaging Applications. *Crit. Rev. Microbiol.* **2018,** *44,* 161–181.

Hu, Q.; Fang, Y.; Yang, Y.; Ma, N.; Zhao, L. Effect of Nanocomposite-Based Packaging on Postharvest Quality of Ethylene-Treated Kiwifruit (*Actinidia deliciosa*) during Cold Storage. *Food Res. Int.* **2011,** *44,* 1589–1596.

Huang, J. Y.; Li, X.; Zhou, W. Safety Assessment of Nanocomposite for Food Packaging Application. *Trends Food Sci. Technol.* **2015,** *45,* 187–199.

Hussain, M.; Bensaid, S.; Geobaldo, F.; Saracco, G.; Russo, N. Photocatalytic Degradation of Ethylene Emitted by Fruits with TiO_2 Nanoparticles. *Indust. Eng. Chem. Res.* **2011,** *50,* 2536–2543.

Hutter, S.; Rüegg, N.; Yildirim, S. Use of Palladium Based Oxygen Scavenger to Prevent Discoloration of Ham. *Food Packag. Shelf Life* **2016,** *8,* 56–62.

Jafarzadeh, S.; Rhim, J. W.; Alias, A. K.; Ariffin, F.; Mahmud, S. Application of Antimicrobial Active Packaging Film Made of Semolina Flour, Nano Zinc Oxide and Nano-Kaolin to Maintain the Quality of Low-Moisture Mozzarella Cheese during Low-Temperature Storage. *J. Sci. Food Agric.* **2019,** *99,* 2716–2725.

Jiang, C.; Hara, K.; Fukuoka, A. Low-Temperature Oxidation of Ethylene over Platinum Nanoparticles Supported on Mesoporous Silica. *Angew. Chem. Int. Ed.* **2013,** *52,* 6265–6268.

Jin, T.; Gurtler, J. B. Inactivation of Salmonella in Liquid Egg Albumen by Antimicrobial Bottle Coatings Infused with Allyl Isothiocyanate, Nisin and Zinc Oxide Nanoparticles. *J. Appl. Microbiol.* **2011,** *110,* 704–712.

Jin, T.; He, Y. Antibacterial Activities of Magnesium Oxide (MgO) Nanoparticles against Food-Borne Pathogens. *J. Nanopart. Res.* **2011,** *13,* 6877–6885.

Jung, J.; Kasi, G.; Seo, J. Development of Functional Antimicrobial Papers Using Chitosan/Starch-Silver Nanoparticles. *Int. J. Biol. Macromol.* **2018,** *112,* 530–536.

Khalaj, M. J.; Ahmadi, H.; Lesankhosh, R.; Khalaj, G. Study of Physical and Mechanical Properties of Polypropylene Nanocomposites for Food Packaging Application: Nano-Clay Modified with Iron Nanoparticles. *Trends Food Sci. Technol.* **2016,** *51,* 41–48.

Khaneghah, A. M.; Hashemi, S. M. B.; Limbo, S. Antimicrobial Agents and Packaging Systems in Antimicrobial Active Food Packaging: An Overview of Approaches and Interactions. *Food Bioprod. Process.* **2018,** *111,* 1–19.

Kiliaris, P.; Papaspyrides, C. D. Polymer/Layered Silicate (Clay) Nanocomposites: An Overview of Flame Retardancy. *Progress Polym. Sci.* **2010,** *35,* 902–958.

Kim, J. S.; Kuk, E.; Yu, K. N.; Kim, J. H.; Park, S. J.; Lee, H. J.; Kim, Y. K. Antimicrobial Effects of Silver Nanoparticles. *Nanomed.: Nanotechnol. Biol. Med.* **2007,** *3,* 95–101.

Kumar, P.; Sandeep, K. P.; Alavi, S.; Truong, V. D. A Review of Experimental and Modeling Techniques to Determine Properties of Biopolymer-Based Nanocomposites. *J. Food Sci.* **2011,** *76* (1), E2–E14.

Li, H.; Li, F.; Wang, L.; Sheng, J.; Xin, Z.; Zhao, L.; Hu, Q. Effect of Nano-Packing on Preservation Quality of Chinese Jujube (Ziziphus jujuba Mill. var. inermis (Bunge) Rehd). *Food Chem.* **2009,** *114,* 547–552.

Li, J.; Zhang, Q.; Xu, M.; Wu, C.; Li, P. Antimicrobial Efficacy and Cell Adhesion Inhibition of In Situ Synthesized ZnO Nanoparticles/Polyvinyl Alcohol Nanofibrous Membranes. *Adv. Condensed Matter Phys.* **2016.**

Li, W. R.; Xie, X. B.; Shi, Q. S.; Zeng, H. Y.; You-Sheng, O. Y.; Chen, Y. B. Antibacterial Activity and Mechanism of Silver Nanoparticles on *Escherichia coli*. *Appl. Microbiol. Biotechnol.* **2010,** *85*, 1115–1122.

Lima, E.; Guerra, R.; Lara, V.; Guzmán, A. Gold Nanoparticles as Efficient Antimicrobial Agents for *Escherichia coli* and *Salmonella typhi*. *Chem. Central J.* **2013,** *7* (1), 11.

Lin, B.; Luo, Y.; Teng, Z.; Zhang, B.; Zhou, B.; Wang, Q. Development of Silver/Titanium Dioxide/Chitosan Adipate Nanocomposite as an Antibacterial Coating for Fruit Storage. *LWT-Food Sci. Technol.* **2015,** *63*, 1206–1213.

Lomate, G. B.; Dandi, B.; Mishra, S. Development of Antimicrobial LDPE/Cu Nanocomposite Food Packaging Film for Extended Shelf Life of Peda. *Food Packag. Shelf Life* **2018,** *16*, 211–219.

Long, M.; Wang, J.; Zhuang, H.; Zhang, Y.; Wu, H.; Zhang, J. Performance and Mechanism of Standard Nano-TiO_2 (P-25) in Photocatalytic Disinfection of Food-Borne Microorganisms–*Salmonella typhimurium* and *Listeria monocytogenes*. *Food Control* **2014,** *39*, 68–74.

Mahajan, P. V.; Rodrigues, F. A.; Motel, A.; Leonhard, A. Development of a Moisture Absorber for Packaging of Fresh Mushrooms (*Agaricus bisporous*). *Postharvest Biol. Technol.* **2008,** *48* (3), 408–414.

Mahdi, S. S.; Vadood, R.; Nourdahr, R. Study on the Antimicrobial Effect of Nanosilver Tray Packaging of Minced Beef at Refrigerator Temperature. *Global Vet.* **2012,** *9*, 284–289.

Mallakpour, S.; Khadem, E. Recent Development in the Synthesis of Polymer Nanocomposites Based on Nano-Alumina. *Progress Polym. Sci.* **2015,** *51*, 74–93.

Maneerat, C.; Hayata, Y. Gas-Phase Photocatalytic Oxidation of Ethylene with TiO_2-Coated Packaging Film for Horticultural Products. *Trans. ASABE* **2008,** *51*, 163–168.

Mastromatteo, M.; Conte, A.; Lucera, A.; Saccotelli, M. A.; Buonocore, G. G.; Zambrini, A. V.; Del Nobile, M. A. Packaging Solutions to Prolong the Shelf Life of Fiordilatte Cheese: Bio-Based Nanocomposite Coating and Modified Atmosphere Packaging. *LWT-Food Sci. Technol.* **2015,** *60*, 230–237.

Mihaly Cozmuta, A.; Peter, A.; Mihaly Cozmuta, L.; Nicula, C.; Crisan, L.; Baia, L.; Turila, A. Active Packaging System Based on Ag/TiO_2 Nanocomposite Used for Extending the Shelf Life of Bread. Chemical and Microbiological Investigations. *Packag. Technol. Sci.* **2015,** *28*, 271–284.

Mihindukulasuriya, S. D. F.; Lim, L. T. Nanotechnology Development in Food Packaging: A Review, *Trends Food Sci. Technol.* **2014,** *40*, 149–167.

Mills, A.; Doyle, G.; Peiro, A. M.; Durrant, J. Demonstration of a Novel, Flexible, Photocatalytic Oxygen-Scavenging Polymer Film. *J. Photochem. Photobiol. A: Chem.* **2006,** *177*, 328–331.

Miltz, J.; Perry, M. Evaluation of the Performance of Iron-Based Oxygen Scavengers, with Comments on Their Optimal Applications. *Packag. Technol. Sci.: Int. J.* **2005,** *18*, 21–27.

Mizielińska, M.; Kowalska, U.; Jarosz, M.; Sumińska, P.; Landercy, N.; Duquesne, E. The Effect of UV Aging on Antimicrobial and Mechanical Properties of Pla Films with Incorporated Zinc Oxide Nanoparticles. *Int. J. Environ. Res. Public Health* **2018,** *15* (4), 794.

Morawiec, J.; Pawlak, A.; Slouf, M.; Galeski, A.; Piorkowska, E.; Krasnikowa, N. Preparation and Properties of Compatibilized LDPE/Organo-Modified Montmorillonite Nanocomposites. *Eur. Polym. J.* **2005,** *41*, 1115–1122.

Mu, H.; Gao, H.; Chen, H.; Tao, F.; Fang, X.; Ge, L. A Nanosized Oxygen Scavenger: Preparation and Antioxidant Application to Roasted Sunflower Seeds and Walnuts. *Food Chem.* **2013,** *136*, 245–250.

Muriel-Galet, V.; Cran, M. J.; Bigger, S. W.; Hernández-Muñoz, P.; Gavara, R. Antioxidant and Antimicrobial Properties of Ethylene Vinyl Alcohol Copolymer Films Based on the Release of Oregano Essential Oil and Green Tea Extract Components. *J. Food Eng.* **2015,** *149*, 9–16.

Naffakh, M.; Diez-Pascual, A. M.; Marco, C.; Ellis, G. J.; Gomez-Fatou, M. A. Opportunities and Challenges in the Use of Inorganic Fullerene-Like Nanoparticles to Produce Advanced Polymer Nanocomposites. *Progress Polym. Sci.* **2013,** *38*, 1163–1231.

Noshirvani, N.; Ghanbarzadeh, B.; Mokarram, R. R.; Hashemi, M. Novel Active Packaging Based on Carboxymethyl Cellulose-Chitosan-ZnO NPs Nanocomposite for Increasing the Shelf Life of Bread. *Food Packag. Shelf Life* **2017,** *11*, 106–114.

Othman, S. H. Bio-Nanocomposite Materials for Food Packaging Applications: Types of Biopolymer and Nano-Sized Filler. *Agric. Agric. Sci. Procedia* **2014,** *2*, 296–303.

Padmavathy, N.; Vijayaraghavan, R. Enhanced Bioactivity of ZnO Nanoparticles—an Antimicrobial Study. *Sci. Technol. Adv. Mater.* **2008,** *9* (3), 035004.

Pagno, C. H.; Costa, T. M.; de Menezes, E. W.; Benvenutti, E. V.; Hertz, P. F.; Matte, C. R.; Flôres, S. H. Development of Active Biofilms of Quinoa (Chenopodium quinoa W.) Starch Containing Gold Nanoparticles and Evaluation of Antimicrobial Activity. *Food Chem.* **2015,** *173*, 755–762.

Pal, M. Nanotechonlogy: A New Approach in Food Packaging. *J. Food: Microbiol. Safet. Hygiene* **2017,** *2*, 1–2.

Pal, S.; Tak, Y. K.; Song, J. M. Does the Antibacterial Activity of Silver Nanoparticles Depend on the Shape of the Nanoparticle? A Study of the Gram-Negative Bacterium *Escherichia coli. Appl. Environ. Microbiol.* **2007,** *73*, 1712–1720.

Pan, X.; Wang, Y.; Chen, Z.; Pan, D.; Cheng, Y.; Liu, Z.; Lin Z; Guan, X. Investigation of Antibacterial Activity and Related Mechanism of a Series of Nano-Mg (OH) 2. *ACS Appl. Mater. Interf.* **2013,** *5*, 1137–1142.

Panea, B.; Ripoll, G.; González, J.; Fernández-Cuello, Á.; Albertí, P. Effect of Nanocomposite Packaging Containing Different Proportions of ZnO and Ag on Chicken Breast Meat Quality. *J. Food Eng.* **2014,** *123*, 104–112.

Parameswaranpillai, J.; Hameed, N., Kurian, T.; Yu, Y. Introduction to Nanomaterials and Nanocomposites. In *Nanocomposite Materials: Synthesis, Properties and Applications*; CRC Press: Florida, 2017; pp 1–5.

Paul, D. R.; Robeson, L. M. Polymer Nanotechnology: Nanocomposites. *Polymer* **2008,** *49*, 3187–3204.

Pavlidou, S.; Papaspyrides, C. D. A Review on Polymer–Layered Silicate Nanocomposites. *Progress Polym. Sci.* **2008,** *33*, 1119–1198.

Pereira de Abreu, D. A.; Cruz, J. M.; PaseiroLosada, P. Active and Intelligent Packaging for the Food Industry. *Food Rev. Int.* **2012,** *28*, 146–187.

Peretyazhko, T. S.; Zhang, Q.; Colvin, V. L. Size-Controlled Dissolution of Silver Nanoparticles at Neutral and Acidic pH Conditions: Kinetics and Size Changes. *Environ. Sci. Technol.* **2014,** *48*, 11954–11961.

Peter, A.; Tegla, D.; Giurgiulescu, L.; Cozmuta, A. M.; Nicula, C.; Cozmuta, L. M.; Vagelas, I. Development of Ag/TiO_2-SiO_2-coated Food Packaging Film and Its Role in

Preservation of Green Lettuce during Storage. *Carpathian J. Food Sci. Technol.* **2015,** *7* (4).

Picard, E.; Vermogen, A.; Gérard, J. F.; Espuche, E. Barrier Properties of Nylon 6-Montmorillonite Nanocomposite Membranes Prepared by Melt Blending: Influence of the Clay Content and Dispersion State: Consequences on Modelling. *J. Membr. Sci.* **2007,** *292*, 133–144.

Polyakov, V. A.; Miltz, J. Modeling of the humidity effects on the oxygen absorption by iron-based scavengers, *Journal of Food Science*, **2010,** *75*(2), E91-E99.

Priyadarshi, R.; Negi, Y. S. Effect of Varying Filler Concentration on Zinc Oxide Nanoparticle Embedded Chitosan Films as Potential Food Packaging Material. *J. Polym. Environ.* **2017,** *25*, 1087–1098.

Quaranta, D.; Krans, T.; Santo, C. E.; Elowsky, C. G.; Domaille, D. W.; Chang, C. J.; Grass, G. Mechanisms of Contact-Mediated Killing of Yeast Cells on Dry Metallic Copper Surfaces. *Appl. Environ. Microbiol.* **2011,** *77*, 416–426.

Rai, A.; Prabhune, A.; Perry, C. C. Antibiotic Mediated Synthesis of Gold Nanoparticles with Potent Antimicrobial Activity and Their Application in Antimicrobial Coatings. *J. Mater. Chem.* **2010,** *20*, 6789–6798.

Realini, C. E.; Marcos, B. Active and Intelligent Packaging Systems for a Modern Society. *Meat Sci.* **2014,** *98*, 404–419.

Rhim, J. W.; Ng, P. K. Natural Biopolymer-Based Nanocomposite Films for Packaging Applications. *Crit. Rev. Food Sci. Nutr.* **2007,** *47*, 411–433.

Rhim, J. W.; Park, H. M.; Ha, C. S. Bio-Nanocomposites for Food Packaging Applications. *Progress Polym. Sci.* **2013,** *38*, 1629–1652.

Roco, M. C. Broader Societal Issues of Nanotechnology. *J. Nanopart. Res.* **2003,** *5*, 181–189.

Rüegg, N.; Blum, T.; Röcker, B.; Kleinert, M.; Yildirim, S. Application of Palladium-Based Oxygen Scavenger to Extend the Shelf Life of Bakery Products. In *Book of Abstracts of 6th International Symposium on Food Packaging*, 2016.

Sadeghi, K.; Lee, Y.; Seo, J. Ethylene Scavenging Systems in Packaging of Fresh Produce: A Review. *Food Rev. Int.* **2019,** 1–22.

Sadeghnejad, A.; Aroujalian, A.; Raisi, A.; Fazel, S. Antibacterial Nano Silver Coating on the Surface of Polyethylene Films Using Corona Discharge. *Surf. Coat. Technol.* **2014,** *245*, 1–8.

Saliani, M.; Jalal, R.; Goharshadi, E. K. Effects of pH and Temperature on Antibacterial Activity of Zinc Oxide Nanofluid against *Escherichia coli* O157: H7 and *Staphylococcus aureus*. *Jundishapur J. Microbiol.* **2015,** *8* (2).

Saraschandraa, N.; Pavithrab, M.; Sivakumar, A. Antimicrobial Applications of TiO_2 Coated Modified Polyethylene (HDPE) Films. *Arch. Appl. Sci. Res.* **2013,** *5*, 189–194.

Sarwar, S.; Chakraborti, S.; Bera, S.; Sheikh, I. A.; Hoque, K. M.; Chakrabarti, P. The Antimicrobial Activity of ZnO Nanoparticles against Vibrio Cholerae: Variation in Response Depends on Biotype. *Nanomed.: Nanotechnol. Biol. Med.* **2016,** *12*, 1499–1509.

Shankar, S.; Rhim, J. W. Effect of Copper Salts and Reducing Agents on Characteristics and Antimicrobial Activity of Copper Nanoparticles. *Mater. Lett.* **2014,** *132*, 307–311.

Shukla, R. K.; Kumar, A.; Pandey, A. K.; Singh, S. S.; Dhawan, A. Titanium Dioxide Nanoparticles Induce Oxidative Stress-Mediated Apoptosis in Human Keratinocyte Cells. *J. Biomed. Nanotechnol.* **2011,** *7* (1), 100–101.

Silvestre, C.; Pezzuto, M.; Cimmino, S.; Duraccio, D. Polymer nanomaterials for Food Packaging: Current Issues and Future Trends. In *Ecosustainable Polymer Nanomaterials for Food Packaging: Innovative Solutions, Characterization Needs, Safety and Environmental Issues*; Silvestre, C., Cimmino, S., Eds.; CRC Press: Florida, 2013; pp 1–25.

Sinha Ray, S.; Okamoto, M. New Polylactide/Layered Silicate Nanocomposites. *Macromol. Mater. Eng.* **2003,** *288*, 936–944.

Siripatrawan, U.; Kaewklin, P. Fabrication and Characterization of Chitosan-Titanium Dioxide Nanocomposite Film as Ethylene Scavenging and Antimicrobial Active Food Packaging. *Food Hydrocoll.* **2018,** *84*, 125–134.

Sisler, E. C.; Yang, S. F. Ethylene, the Gaseous Plant Hormone. *Bioscience* **1984,** *34*, 234–238.

Song, X.; Li, Y.; Wei, Z.; Ye, S.; Dionysiou, D. D. Synthesis of BiVO4/P25 Composites for the Photocatalytic Degradation of Ethylene under Visible Light. *Chem. Eng. J.*, **2017,** *314*, 443–452.

Sorrentino, A.; Gorrasi, G.; Vittoria, V. Potential Perspectives of Bio-Nanocomposites for Food Packaging Applications. *Trends Food Sci. Technol.* **2007,** *18*, 84–95.

Spricigo, P. C.; Foschini, M. M.; Ribeiro, C.; Corrêa, D. S.; Ferreira, M. D. Nanoscaled Platforms Based on SiO_2 and Al_2O_3 Impregnated with Potassium Permanganate Use Color Changes to Indicate Ethylene Removal. *Food Bioprocess Technol.* **2017,** *10*, 1622–1630.

Sukhorukova, I. V.; Sheveyko, A. N.; Kiryukhantsev-Korneev, P. V.; Zhitnyak, I. Y.; Gloushankova, N. A.; Denisenko, E. A.; Filippovich, S. Y.; Ignatov, S. G.; Shtansky, D. V. Toward Bioactive Yet Antibacterial Surfaces. *Colloids Surf. B: Biointerf.* **2015,** *135*, 158–165.

Sun, Q., Li, J.; Le, T. Zinc Oxide Nanoparticle as a Novel Class of Antifungal Agents: Current Advances and Future Perspectives. *J. Agric. Food Chem.* **2018,** *66*, 11209–11220.

Suppakul, P.; Miltz, J.; Sonneveld, K.; Bigger, S. W. Active Packaging Technologies with an Emphasis on Antimicrobial Packaging and Its Applications. *J. Food Sci.* **2003,** *68*, 408–420.

Swaroop, C.; Shukla, M. Development of Blown Polylactic Acid-MgO Nanocomposite Films for Food Packaging. *Comp. Part A: Appl. Sci. Manuf.* **2019,** *124*, 105482.

Tas, C. E.; Hendessi, S.; Baysal, M.; Unal, S.; Cebeci, F. C.; Menceloglu, Y. Z.; Unal, H. Halloysite Nanotubes/Polyethylene Nanocomposites for Active Food Packaging Materials with Ethylene Scavenging and Gas Barrier Properties. *Food Bioprocess Technol.* **2017,** *10*, 789–798.

Tayel, A. A.; EL-TRAS, W. F.; Moussa, S.; EL-BAZ, A. F.; Mahrous, H.; Salem, M. F.; Brimer, L. Antibacterial Action of Zinc Oxide Nanoparticles against Food-Borne Pathogens. *J. Food Safe.* **2011,** *31*, 211–218.

Terry, L. A.; Ilkenhans, T.; Poulston, S.; Rowsell, L.; Smith, A. W. Development of New Palladium-Promoted Ethylene Scavenger. *Postharvest Biol. Technol.* **2007,** *45*, 214–220.

Tewari, G.; Jayas, D. S.; Jeremiah, L. E.; Holley, R. A. Absorption Kinetics of Oxygen Scavengers. *Int. J. Food Sci. Technol.* **2002,** *37*, 209–217.

Thirumurugan, A.; Ramachandran, S.; Shiamala Gowri, A. Combined Effect of Bacteriocin with Gold Nanoparticles against Food Spoiling Bacteria-an Approach for Food Packaging Material Preparation. *Int. Food Res. J.* **2013,** *20* (4).

Threepopnatkul, P.; Wongnarat, C.; Intolo, W.; Suato, S.; Kulsetthanchalee, C. Effect of TiO_2 and ZnO on Thin Film Properties of PET/PBS Blend for Food Packaging Applications. *Energy Procedia* **2014,** *56*, 102–111.

Tjong, S. C. Structural and Mechanical Properties of Polymer Nanocomposites. *Mater. Sci. Eng.: R: Rep.* **2006,** *53*, 73–197.

Vance, M. E.; Kuiken, T.; Vejerano, E. P.; McGinnis, S. P.; Hochella Jr, M. F.; Rejeski, D.; Hull, M. S. Nanotechnology in the Real World: Redeveloping the Nanomaterial Consumer Products Inventory. *Beilstein J. Nanotechnol.* **2015,** *6*, 1769–1780.

Vincent, M.; Hartemann, P.; Engels-Deutsch, M. Antimicrobial Applications of Copper. *Int. J. Hygiene Environ. Health* **2016,** *219* (7), 585–591.

Wang, K.; Jin, P.; Shang, H.; Li, H.; Xu, F.; Hu, Q.; Zheng, Y. A Combination of Hot Air Treatment and Nano-Packing Reduces Fruit Decay and Maintains Quality in Postharvest Chinese Bayberries. *J. Sci. Food Agric.* **2010,** *90*, 2427–2432.

Weaver, L.; Michels, H. T.; Keevil, C. W. Potential for Preventing Spread of Fungi in Air-Conditioning Systems Constructed Using Copper Instead of Aluminium. *Lett. Appl. Microbiol.* **2010,** *50*, 18–23.

Wills, R. B. H.; Warton, M. A. Efficacy of Potassium Permanganate Impregnated into Alumina Beads to Reduce Atmospheric Ethylene. *J. Am. Soc. Horticult. Sci.* **2004,** *129*, 433–438.

Wu, Z.; Huang, X.; Li, Y. C.; Xiao, H.; Wang, X. Novel Chitosan Films with Laponite Immobilized Ag Nanoparticles for Active Food Packaging. *Carbohydr. Polym.* **2018,** *199*, 210–218.

Xiao-e, L.; Green, A. N.; Haque, S. A.; Mills, A.; Durrant, J. R. Light-Driven Oxygen Scavenging by Titania/Polymer Nanocomposite Films. *J. Photochem. Photobiol. A: Chem.* **2005,** *162*, 253–259.

Xie, J.; Hung, Y. C. UV-A Activated TiO_2 Embedded Biodegradable Polymer Film for Antimicrobial Food Packaging Application. *Lwt* **2018,** *96*, 307–314.

Xing, Y.; Li, X.; Zhang, L.; Xu, Q.; Che, Z.; Li, W.; Li, K. Effect of TiO_2 Nanoparticles on the Antibacterial and Physical Properties of Polyethylene-Based Film. *Progress Org. Coat.* **2012,** *73*, 219–224.

Xue, B.; Jiang, Y.; Liu, D. Preparation and Characterization of a Novel Anticorrosion Material: Cu/LLDPE Nanocomposites. *Mater. Trans.* **2011a,** *52*, 96–101.

Xue, W. J.; Wang, Y. F.; Li, P.; Liu, Z. T.; Hao, Z. P.; Ma, C. Y. Morphology Effects of Co_3O_4 on the Catalytic Activity of Au/Co_3O_4 Catalysts for Complete Oxidation of Trace Ethylene. *Cataly. Commun.* **2011b,** *12*, 1265–1268.

Yang, F. M.; Li, H. M.; Li, F.; Xin, Z. H.; Zhao, L. Y.; Zheng, Y. H.; Hu, Q. H. Effect of Nano-Packing on Preservation Quality of Fresh Strawberry (Fragaria ananassaDuch. cv Fengxiang) during Storage at 4 C. *J. Food Sci.* **2010,** *75*, C236–C240.

Yemmireddy, V. K.; Farrell, G. D.; Hung, Y. C. Development of Titanium Dioxide (TiO_2) Nanocoatings on Food Contact Surfaces and Method to Evaluate Their Durability and Photocatalytic Bactericidal Property. *J. Food Sci.* **2015,** *80*, N1903–N1911.

Yildirim, S. Active Packaging for Food Biopreservation. In *Protective Cultures, Antimicrobial Metabolites and Bacteriophages for Food and Beverage Biopreservation*; Lacroix, C., Ed.; Woodhead Publishing: Cambridge, 2011; pp 460–489.

Yildirim, S.; Röcker, B.; Pettersen, M. K.; Nilsen-Nygaard, J.; Ayhan, Z.; Rutkaite, R.; Coma, V. Active Packaging Applications for Food. *Comprehen. Rev. Food Sci. Food Safe.* **2018,** *17*, 165–199.

Youssef, A. M.; El-Sayed, S. M.; El-Sayed, H. S.; Salama, H. H.; Dufresne, A. Enhancement of Egyptian Soft White Cheese Shelf Life Using a Novel Chitosan/Carboxymethyl Cellulose/Zinc Oxide Bionanocomposite Film. *Carbohydr. Polym.* **2016,** *151*, 9–19.

Youssef, A. M.; El-Sayed, S. M.; Salama, H. H.; El-Sayed, H. S.; Dufresne, A. Evaluation of Bionanocomposites as Packaging Material on Properties of Soft White Cheese during Storage Period. *Carbohydr. Polym.* **2015,** *132*, 274–285.

Yu, J.; Liu, R. Y. F.; Poon, B.; Nazarenko, S.; Koloski, T.; Vargo, T.; Baer, E. Polymers with Palladium Nanoparticles as Active Membrane Materials. *J. Appl. Polym. Sci.* **2004,** *92*, 749–756.

Yuan, Q.; Jiang, W.; An, L.; Christiansen, J. D.; Li, R. K. Y. Competing Effect between Filled Glass Bead and Induced β Crystal on the Tensile Properties of Polypropylene/Glass Bead Blends. *J. Appl. Polym. Sci.* **2005,** *96*, 1729–1733.

Zawrah, M. F.; El-Moez, S. A.; Center, D. Antimicrobial Activities of Gold Nanoparticles against Major Food-Borne Pathogens. *Life Sci. J.* **2011,** *8*, 37–44.

Zhang, L.; Ding, Y.; Povey, M.; York, D. ZnO Nanofluids–a Potential Antibacterial Agent. *Progress Nat. Sci.* **2008,** *18*, 939–944.

Zhang, L.; Jiang, Y.; Ding, Y.; Daskalakis, N.; Jeuken, L.; Povey, M.; York, D. W. Mechanistic Investigation into Antibacterial Behaviour of Suspensions of ZnO Nanoparticles against *E. coli*. *J. Nanopart. Res.* **2010,** *12*, 1625–1636.

Zhu, Z.; Cai, H.; Sun, D. W. Titanium Dioxide (TiO_2) Photocatalysis Technology for Nonthermal Inactivation of Microorganisms in Foods. *Trends Food Sci. Technol.* **2018,** *75*, 23–35.

CHAPTER 5

Intelligent/Smart Nanocomposite Packaging: Functions and Applications

C. VIBHA[1], JYOTISHKUMAR PARAMESWARANPILLAI[2*], SUCHART SIENGCHIN[1], K. SENTHILKUMAR[3], G. L. PRAVEEN[4], NISA SALIM[5], and NISHAR HAMEED[6]

[1]*Department of Mechanical and Process Engineering, The Sirindhorn International Thai–German Graduate School of Engineering (TGGS), King Mongkut's University of Technology North Bangkok, Bangsue, Bangkok 10800, Thailand*

[2]*Department of Science, Faculty of Science & Technology, Alliance University, Chandapura-Anekal Main Road, Bengaluru 562106, Karnataka, India*

[3]*Center of Innovation in Design and Engineering for Manufacturing (CoI-DEM), King Mongkut's University of Technology North Bangkok, Wongsawang, Bangsue, Bangkok 10800, Thailand*

[4]*Wimpey Laboratories, Ras Al Khor, Industrial Area 2, Dubai*

[5]*Swinburne University of Technology, Faculty of Science, Engineering and Technology, Hawthorn, VIC, 3122, Australia*

[6]*Factory of the Future, Swinburne University of Technology, Hawthorn, VIC, Australia*

**Corresponding author. E-mail: jyotishkumarp@gmail.com*

ABSTRACT

The need for sustainable, healthy, and good quality food products has led to the development of intelligent/smart packaging. Innovative concepts in the food packaging sector paved the growth of smart packaging systems. The smart packaging offers numerous solutions for extending the shelf-life, improving the food safety and quality, and reducing food wastage. The smart packaging system is composed of sensors, indicators, bar code, and radiofrequency identification. These components can effectively communicate with the consumer and provides information on the goodness of the food products. The recent innovations of smart packaging technologies have resulted in the effective detection of pathogens, toxins, adulterants, allergen, freshness, and storage life of a variety of food products. The innovation such as the usage of smartphones for the real-time testing analysis of the food packaging is worth noting. The safety regulations for the use of these polymer food packaging, precise detection of the food quality, recyclability of the sensors, and indicators are still needing more attention.

5.1 INTRODUCTION

Food packaging is a conciliator linking producers and consumers that guarantee high quality, enhance safety, and extend the shelf life of the food products by preserving it from unfavorable environmental changes and other contaminations (Dobrucka and Robert, 2019; Nemes et al., 2020). The traditional packaging is less competent to meet the high demand of shell life and environmental friendliness (Erin et al., 2020). The growing desire for inventive packaging materials to satisfy consumers' requisite for good quality food evolves novel technology in the food packaging industry, and these advancements enhanced food quality, its protection, preservation, and safety devoid of affecting the environment (Han et al., 2018; Yildirim et al., 2018). An assortment of resources including polymers, nanomaterials, and biopolymers emerged as potential candidates to reverse the aforementioned challenges (Idumah et al., 2020; Mohana et al., 2020; Wicochea et al., 2019). Several advances in packaging technology occurred in the 20th century, including active and intelligent packaging based on nanoparticles, essential oil, etc. (Jayakumar et al., 2020).

"Smartness or intelligence" in food packaging is a familiar term that enfolds numerous functionalities (Jayakumar et al., 2020). In smart nanocomposite packaging, active agents are integrated within the packaging or at the headspace to scrutinize the quality and environment of the packaged food (Anukiruthika et al., 2020; Varghese et al., 2020). Unlike traditional packaging, these innovative packaging communicates information with the consumer regarding package integrity, product security, and worth of packaged food (Robertson, 2006). The communication in intelligent packaging is facilitated by employing a collection of sensors, indicators, radiofrequency identification, and barcode technologies (https://www.frontiersin.org/research-topics/13221/smart-food-packaging-active-and-intelligent-functions). This will help to track product availability, genuineness, and anti-theft (Shoue et al., 2020).

Smart packaging is a great tool for examining probable mishandling occurred at any stage of the food supply chain that will be updated to the customer. Thus, the innovations in food packaging materials steered the path for the advancements of intelligent or smart packaging (IOSP), which gives information on food products while transiting and storage.

5.2 COMPONENTS OF IOSP

The pictorial representation of smart packaging components is shown in Figure 5.1. The three intelligent components in IOSP are sensors, indicators, and radiofrequency identification. These components (1) scrutinize the system, (2) process data, and (3) provide information regarding the freshness of the packed food product. The package will switch "ON" and "OFF" accordingly to the varying environmental conditions; also, it communicates necessary information regarding the condition of the product to the customers. The indicators and sensors provide the information on the freshness of the food packaging, while the barcodes/radiofrequency identification plates store data (Ahmed et al., 2018; Saroat and Pimonpan, 2018).

5.2.1 SENSORS

The sensors comprise of two components: (1) the receptor, which generates the physiochemical signal, and (2) the transducer, which measures and converts the physiochemical signal as electrical signals (Biji et al., 2015; Fuertes et al., 2016; Gregor-Svetec, 2018). Sensors are of many types such as biosensors, gas sensors, chemical sensors, colorimetric sensors, printed electronics, and electric noses.

5.2.1.1 BIOSENSORS

Microbial growth especially bacteria deteriorates food quality, which may ensue in every phase of the food chain due to improper handling.

Immunological analysis, amplification in nucleic acid sequence, bacterial culture, and colony counting are the conventional techniques for detecting bacteria (Hamdy et al., 2018; Kucherenko et al., 2020; Mustafa and Andreescu, 2018; Rajapaksha et al., 2019). Recently, biosensors are preferred over the conventional methods for the identification of pathogens.

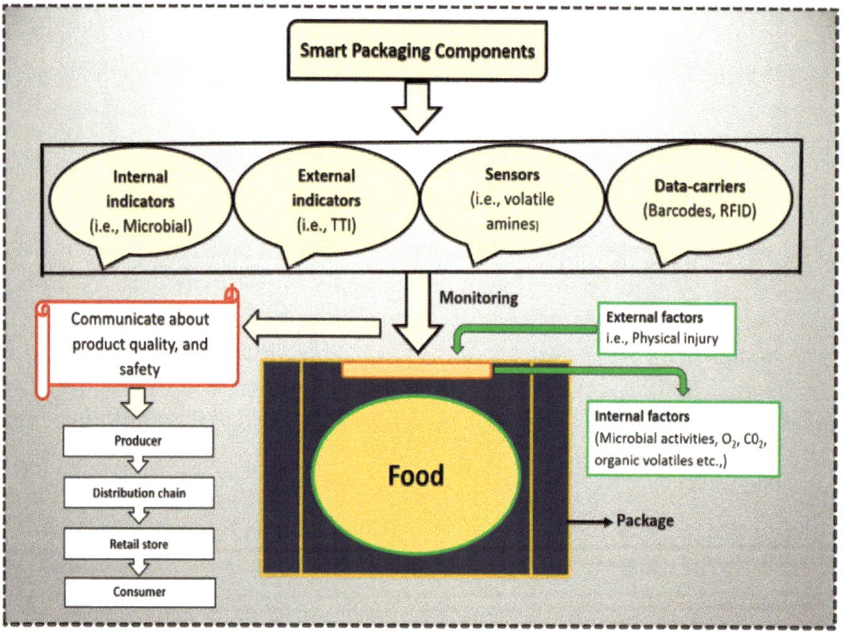

FIGURE 5.1 Pictorial representation of smart packaging components.

Source: Reprinted with permission from Ahmed et al., 2018. © 2018 John Wiley.

Biosensors are organic or biological materials used to identify, record, and communicate the data based on biological reactions. In biosensors, a bioreceptor (enzyme, hormone, nucleic acid, antigen, and microbes) is connected within a transducer (optical, electrochemical, calorimetric, or piezoelectric materials) (Kucherenko et al., 2020; Mustafa and Andreescu, 2018).

The working of the biosensor is as follows:

i) First, the biological sensor/biological material reacts with the analyte.
ii) Second, the resulting biological response is translated into an electrical signal by the transducer.

Nanomaterials, including gold nanoparticles (Hamdy et al., 2018), carbon nanotubes, and graphene (Kotsilkov et al., 2018), silver nanoparticles (Fatima and Silvana, 2020), and magnetic nanoparticles including Fe_2O_3 were generally employed for the manufacturing of efficient biosensors (Wang et al., 2016). These biosensors are incorporated into polymer films for identifying allergens, pathogens, temperature changes, residual oxygen, pesticides, toxins, and leakages (Mustafa and Andreescu, 2020; Neethirajan et al., 2018; Wang et al., 2016). The bioreceptor in the biosensor is specific in detecting the intended analyte, and the transducer (optical/acoustic/electrochemical) converts biochemical signals into measurable electronic response (Ahmed et al., 2018). The type of biosensors and their applications in food analysis are shown in Table 5.1.

The two commercially available biosensors for detecting food pathogens are Food Sentinel System® and Toxinc Guard®. In Food Sentinel System® (SIRA Technologies Inc.), from the change in the color of the barcode, one can identify whether the packed food is contaminated. On the other hand, from Toxin Guard® (Toxin Alert, Canada), one can visually identify pathogens like *Escherichia coli* and *Salmonella* (Biji et al., 2015).

5.2.1.2 GAS SENSORS

Gas sensors are employed for identifying gaseous analytes such as oxygen, carbon dioxide, hydrogen sulfide, ethylene, and water vapor in the packaging goods (Chowdhury and Morey, 2019). The device works based on the changes in the sensor's potential difference, which depends on the concentration of the gas analyte. Some of the traditional gas sensor types are metal oxide semiconductor, catalytic, electrochemical, optical, and calorimetric sensors. The gas sensors based on calorimetric have been preferred over the other traditional gas sensors in the food packing industry because they are inexpensive and have good sensitivity. It has been reported that the calorimetric gas sensors could identify the goodness of the food products by detecting carbon dioxide, hydrogen sulfide, and ammonia (Ghaani et al., 2016; Matindoust et al., 2016; Nguyen et al., 2019; Wolfbeis and List, 1995).

TABLE 5.1 Application of Biosensors in Food Analysis.

Detection	Type of biosensor	Application	References
Immune-based biosensors for allergen detection	Surface plasma resonance-based immune biosensor	Peanut allergens	Pollet et al. (2011)
	Voltammetric immunosensor based on electropolymerization	Allergens in milk	Kilic et al. (2020)
	Graphene oxide/copper nanoflower modified glassy carbon electrode	Melamine in milk	Daizy et al. (2019)
	β-Lactoglobulin sensor with surface plasmon resonance (SPR) detection	Casein in milk	Ashley et al. (2018)
	Optical thin-film biosensor having PCR amplification based on silicon	Allergens in food including meat products, fish, shrimp, nuts, and soybeans	Wang et al. (2011)
	Two-enzyme lactose biosensor; and glucose oxidase and β-galactosidase-based conductometric biosensor	Lactose in milk	Ammam et al. (2010)
	Two sensor system with layer by layer technique. First layer: Non-enzymatic sensor with anionic sulfonated copper phthalocyanine. Second layer: Cationic one with mixture of 1-butyl-3-methylimidazolium tetrafluoroborate and chitosan	Lactose in milk	Salvo-Comino et al. (2018)
Bacterial-pathogen detection	Peptides immobilized colorimetric biosensor	*Listeria monocytogenes*	Alhogail et al. (2016)
	Microfluidic based fully automated electrochemical biosensor	*Escherichia coli*	Altintas et al. (2018)
	Fluorescent DNAzyme probe with cyclo-olefin polymer package	*E. coli*	Yousefi et al. (2018)
	Lateral flow test strip immunoassays	*Klebsiella Pneumoniae, Raoultella Ornithinolytica, Klebsiella Oxytoca.*	Tominaga et al. (2018)
	CdS quantum dots conjugated with zeolotic-imidazole framework	*E. coli*	Zhong et al. (2019)

TABLE 5.1 *(Continued)*

Detection	Type of biosensor	Application	References
Toxin detection	Glassy carbon electrode modified with multiwalled carbon nanotubes	Bisphenol A	Anirudhan et al. (2018)
	Cerium dioxide nanoparticle-based sensor	Ochratoxin A	Bülbül et al. (2016)
	Localized surface plasmon resonance (LSPR) with the morphological change of gold nanocage to gold@silver nanobox	Gallic acid	Wang et al. (2018b)
	Poly (vinyl alcohol) embedded silver nanoparticles-based sensor	Gallic acid	Teerasong et al. (2017)
Adulterants detection	Combination of Pd/Au core–shell nanocrystallites and polyamidoamine dendrimers embedded with CdS quantum dots	Sudan I in food	Wang et al. (2018a)
	AuNPs immobilized LSPR sensor chip	Melamine in milk	Oh et al. (2019)

5.2.1.3 CHEMICAL SENSORS

Chemical sensors are employed for detecting concentration, composition, and activity of gas via surface adsorption (Biji et al., 2015; Gregor-Svetec, 2018). These chemical receptors adsorb the specific chemicals, and with the aid of active or passive transduce, the information collected by the chemical receptors such as a change in pH, temperature, and redox potential is converted into measurable signals (Ghaani et al., 2016). Recently, nanomaterials have been used for the manufacturing of chemical sensors because of their large surface area. It has been reported that the nano-based sensors can detect freshness, toxins, adulterants, pathogens, and chemical contaminants (Biji et al., 2015; Gregor-Svetec, 2018; Ding et al., 2020).

5.2.1.4 COLORIMETRIC SENSORS

The colorimetric sensor is a non-destructive technique used for the assessment of the shelf-life of different food products. They are fabricated by printing chemical responsive dye over the colorimetric substrate (Ahmed et al., 2018; Mukdasai et al., 2019). The interaction of the dye with the volatile organic compounds generated from the food materials results in the color change of the colorimetric sensor. The selection of chemical dye is critical, as it highly influences the sensitivity of the volatile organic compounds. Currently, natural pigments are also employed for the identification of vapors of organic compounds in calorimetric sensors. The integration of these natural pigments is attaining greater attention due to its nonhazardous nature to the food packaging. Some examples of chemical dyes/natural pigments used for monitoring the freshness of fish products are roselle anthocyanins (Zhai et al., 2017), *Arnebia euchroma* root extracts (Huang et al., 2019), and purple sweet potato anthocyanins (Jiang et al., 2020).

5.2.1.5 PRINTED ELECTRONICS

In printed electronics, inkjet printing, screen printing, gravure printing, and lithography are used to fabricate electronic circuits on the polymeric substrate. Lightweight, flexible, and portable substrates are often preferred to develop printed electronics (Biji et al., 2015; Vanderroost et al., 2014; Yu et al., 2019). They are widely used to assess the changes in the internal environment of the packaging. Figure 5.2 shows the mechanism of action of printed electronics. The change in chemical signal is converted to an electrical signal by the printed electronics. Recently, Zhou et al. (2020) developed smart, flexible, low cost, and solution printed [poly (3,4-ethylene dioxythiophene)/polystyrene sulfonate] polymer films by integrating 13.56 MHz RFID chip. Ammonia and anti-open sensors were also integrated with the printed polymer film. The printed polymer films were used for pork packaging, and the real-time freshness of the pork was monitored using a smartphone.

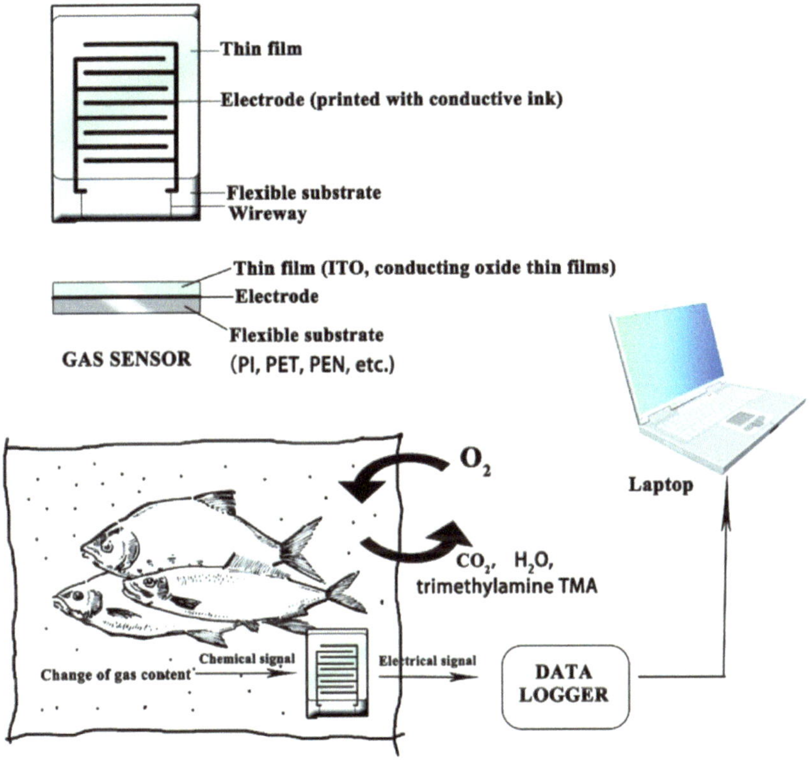

FIGURE 5.2 Mechanism of action of the printed electronics.

Source: Reprinted from Liao et al. 2009. © The Royal Society of Chemistry 2019. https://creativecommons.org/licenses/by-nc/3.0/

5.2.1.6 *ELECTRONIC NOSE*

The electronic nose is an instrument that imitates the olfactory system of mammals (Yasamin et al., 2020). They are designed to detect and classify simple or complex mixture of aromas present in food products. These electronic nose systems are capable of validating the quality of a variety of food products such as fish, meat, fruits, and vegetables (Hu et al., 2019). Wijaya et al. (2019) successfully developed a noise filtering framework for electronic nose signals, and it has superior performance compared to conventional methods. A schematic of the principle of the human nose and electronic nose is illustrated in Figure 5.3

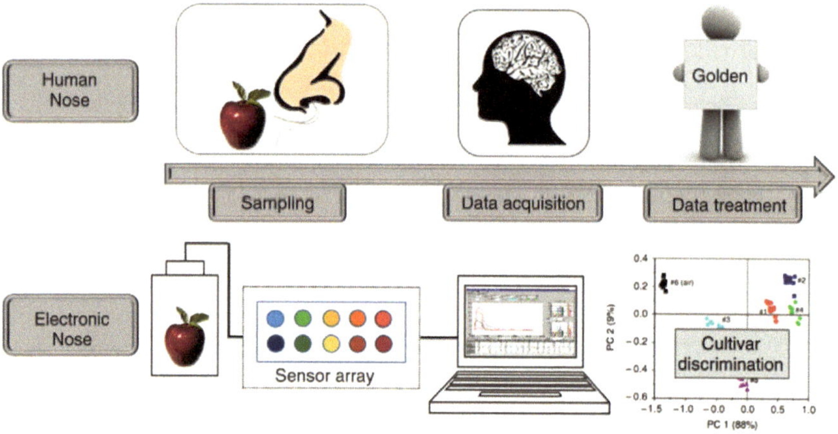

FIGURE 5.3 Schematic of the principle of the human nose and electronic nose

Source: Reprinted from Beghi et al., 2017. ©2017 Walter de Gruyter GmbH, Berlin/Boston. https://creativecommons.org/licenses/by-nc/3.0/

5.2.2 INDICATORS

Indicators indicate the presence and concentration of a material, extent of food reactions, growth of microorganisms, temperature variations, and freshness by visualizing the color changes based on the indicators placed within or out of the package (Müller and Schmid, 2019). Food indicators are classified mainly as an internal and external indicator. The internal indicators are placed within the food package, while external indicators are placed outside the food package. Different microbial indicators and atmospheric leakage indicators would come under the internal indicators, whereas the time–temperature indicator belongs to external indicators (Ahmed et al., 2018).

5.2.2.1 TIME–TEMPERATURE INDICATORS (TTIS)

For ensuring the quality and safety of the foods such as refrigerated egg, meat, fish, etc., the TTIs are significantly helpful by monitoring their temperatures (Corradini, 2018; Müller and Schmid, 2019; Tingting et al., 2020;). At present, the TTIs response based on color is the most popular. The temperature effect on chemical reactions and the biological processes can be effectively modeled with the Arrhenius equation (Peleg et al., 2012; Tingting et al., 2020). Some recent innovations in the development of TTIs and the substrates used to develop the same are summarized in Table 5.2.

TABLE 5.2 Research Works Related to the Development of Various TTIs.

Time–temperature indicators	Materials used for the development of TTIs	References
Flexible TTI	Press lamination of thermoplastic films	Jafry et al. (2017)
Plasmonic thermal history indicator	SPR of gold nanoparticles integrated into alginate	Wang et al. (2017)
Tyrosinase-based TTI	Tyrosine/tyrosinase reaction systems	Xu et al. (2017)
Lipase-based enzymatic TTI	Tricaprylin	Jaiswal et al. (2020)
Self-healing nanofiber-based self-responsive TTI	Aromatic disulfide-based TPU	Choi et al. (2020)
Enzymatic TTI	Laccase was immobilized on the electrospun zein fiber	Jhuang et al. (2020)
Non-enzymatic TTI	Fructose/glycine NEB system	Uddin et al. (2019)
Bi enzyme TTI	glucose oxidase/horseradish peroxidase reaction system	Li et al. (2019)

The three main categories of commercially available TTIs include critical temperature, partial history, and full history indicators. The critical temperature indicator responds if the food product was heated beyond or cooled below the reference temperature. The partial history indicator responds if the food product was exposed to a temperature that causes food degradation. The full history indicator provides the full time–temperature history of the packed food (Fang et al., 2017). The commercial TTIs are small labels used to track chemical, physical, or biological responses in a food product with respect to time and temperature. The quantified information was often expressed as a color change or mechanical deformation (Yousefi et al., 2019). A schematic diagram of the working of the TTIs is shown in Figure 5.4.

FIGURE 5.4 A schematic diagram of the working of the TTIs.

5.2.2.2 FRESHNESS INDICATORS

They are labels in the container that directly indicate the quality of food products while transiting and storage. These indicators mainly focus on the detection regarding microbiological growth and chemical reactions in the food products. The gas formed during the microbial growth reacts with pH-sensitive dye and results in color variation that provides the information regarding the freshness of the food products (Fang et al., 2017; Fuertes et al., 2016; Mustafa and Andreescu, 2018). RipeSense® (RipeSense, Auckland, New Zealand) introduced a ripeness sensor, in such a way that the color of the sensor changes as the fruit ripens. The customers can thereby judge the ripeness and can pick their right choice of the fruit (http://www.ripesense.co.nz).

5.2.2.3 GAS INDICATORS/INTEGRITY INDICATORS

As the food starts degradation due to enzymatic or chemical reaction, the gases will be generated within the food package. The indicators detect any gas formation within the food package and can be visualized in the form of color change (Fuertes et al., 2016). These indicators will prevent the food from decomposition and protect the consumers from health hazards. These gas indicators detect oxygen and carbon dioxide mostly, while some monitor other gases including ethanol, water vapor, and hydrogen sulfide (Lee and Ko, 2014; Meng et al., 2014; Müller and Schmid, 2019). Caroline et al. (2020) developed phosphorescent-based oxygen sensors recently via the extrusion process, and the system comprised Pt-benzoporphyrin dye integrated polystyrene-divinyl microspheres dispersed in low-density polyethylene/polylactic acid. Melnikov et al. (2018) developed fluorescent indicators for detecting molecular oxygen in the gas transmission line. In this work, Pt (II) 5,10,15,20-tetrakis (2,3,4,5,6-pentafluorophenyl)-porphyrin has been adsorbed on the nanoporous silica particle in a fluorinated polymer solution (fluoroplast 42 in acetone).

Ageless Eye® is an example of an oxygen indicator tablet manufactured by Mitsubishi Gas Chemical Company. These tablets indicate the existence of oxygen through a color change. The oxygen-free package is pink in color and the tablet becomes blue with an increase in oxygen level (AGELESS EYE, Oxygen Indicator. https://www.mgc.co.jp/eng/products/sc/ageless-eye.html).

5.2.3 DATA CARRIERS

5.2.3.1 RADIOFREQUENCY IDENTIFICATION (RFID)

RFID is an automatic recognition technique that uses wireless sensors to detect and congregate data devoid of a human interference (Fang et al., 2017). RFID comprised three essential components:

i) RFID tag, which emits radio waves
ii) Reader, which collects information from RFID tag
iii) Middleware, which is composed of hardware and software

Figure 5.5 demonstrates the schematic diagram of the working of the RFID. These superior data information systems gather, accumulate, and pass on the real-time information to consumer information system (Fuertes et al., 2016). RFID monitors numerous parameters concomitantly and piles up assorted data, including commercial information, file tracking, and real-time location. The RFID is not included either in the classifications of indicators or as a sensor; also, it is not an alternative for the barcode due to their robust electronic monitoring set up and high cost (Shoue et al., 2020). It offers abundant benefits like security, traceability, minimized food loss, labor reduction, and superior quality of food products (Bibia et al., 2017). RFID tags can be made economical by developing it through the printed ink-jet method, which utilizes conducting polymers, metallic inks, or even carbon nanotubes (Hrytsenko et al., 2018).

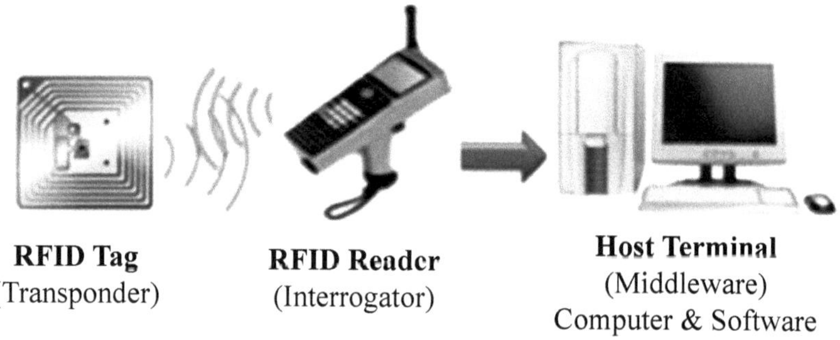

FIGURE 5.5 Schematic diagram of the working of the radiofrequency identification.

Source: Reprinted from Sardroud et al., 2012 © 2012 Sharif University of Technology. https://creativecommons.org/licenses/by-nc-nd/4.0/

Currently, smartphones are used for conducting a real-time testing analysis of the food packaging (Hussain et al., 2017; Rateni et al., 2017; Zhu et al., 2012). This platform can perform the detection of pathogens, antibodies in milk, fluoride detection, and minute traces of food allergens. By summing up, the smart food packaging provides numerous advantages to the food industry and consumers as well.

5.3 CHALLENGES AND CONCLUSION

From the brief review, it is clear that the smart packaging provides numerous advantages for the food industry. For instance, consumers are getting benefitted by ensuring safety and providing healthy and tasty foods. Besides, food waste can be reduced possible by employing sensors and indicators. The smart packaging systems also helps to identify, examine, record, track, and communicate throughout the supply chain. Since the packaging system touches the food products, the sensors and indicators should pass through strict safety regulations for precise detection of the food quality, and hence more research on the quality check of these products is required. Moreover, the sensors or indicators are very specific and sensitive for each type of food product. So, it must be efficiently sorted out and elucidate which indicator or sensor is suitable for particular food products. Besides, these intelligent packaging should be ingenious enough to identify and monitor the contaminants at the initial stage of decomposition, by assisting appropriate responses. At present, a smart packaging material with versatile responses is costly with some having less sensitivity. Also, the nondegradability and recyclability of the sensors and indicators is a concern to be addressed. More research should be done on the development of eco-friendly biodegradable substrates for manufacturing smart food packaging purposes.

KEYWORDS

- **smart packaging**
- **shelf-life**
- **intelligent component**
- **sensors**
- **indicators**
- **radiofrequency identification**

REFERENCES

Ageless Eye, Oxygen Indicator. https://www.mgc.co.jp/eng/products/sc/ageless-eye.html

Ahmed, I.; Lin, H.; Zou, L.; Li, Z.; Brody, A. L.; Qazi, I. M.; Liangtao, L. v.; Pavase, T. R.; Khan, M. U.; Khan, S.; Sun, L. An Overview of Smart Packaging Technologies for Monitoring Safety and Quality of Meat and Meat Products. *Packag. Technol. Sci.* **2018,** *31*, 449–471.

Alhogail, S.; Suaifan, G.; Zourob, M. Rapid Colorimetric Sensing Platform for the Detection of Listeria Monocytogenes Foodborne Pathogen. *Biosens. Bioelectron.* **2016,** *86*, 1061–1066.

Altintas, Z.; Akgun, M.; Kokturk, G.; Uludag, Y. A fully Automated Microfluidic-Based Electrochemical Sensor for Real-Time Bacteria Detection. *Biosens. Bioelectron.* **2018,** *100*, 541–548.

Ammam, M.; Fransaer, J. Two-Enzyme Lactose Biosensor Based on β-Galactosidase and Glucose Oxidase Deposited by AC-Electrophoresis: Characteristics and Performance for Lactose Determination in Milk. *Sens. Actuat. B Chem.* **2010,** *148*, 583–589.

Anirudhan, T. S.; Athira, V. S.; Chithra, V. S. Electrochemical Sensing and Nano Molar Level Detection of Bisphenol-A with Molecularly Imprinted Polymer Tailored on Multiwalled Carbon Nanotubes. *Polymer.* **2018,** *146*, 312–320.

Anukiruthika, T.; Priyanka,S.; Wilson, A.; Kashampur, K.; Jeyan, A. M.; Chinnaswamy, A. Multilayer Packaging: Advances in Preparation Techniques and Emerging Food Applications. *Compr. Rev*. **2020,** *19*, 1156–1186.

Ashley, J.; Roberta, D.; Monika, P.; Jeff, T.; Mike, P.; Linda, Trinh.; Thomas, L. R.; Ibtisam, E. T. Development of a β-Lactoglobulin Sensor Based on SPR for Milk Allergens Detection. *Biosensors* **2018,** *8 (2)*, 32.

Beghi, R.; Buratti, S.; Giovenzana, V.; Benedetti, S.; Guidetti, R. Electronic Nose and Visible-Near Infrared Spectroscopy in Fruit and Vegetable Monitoring. *Rev. Analyt. Chem*. **2017,** *36* (4), 20160016.

Bibia, F.; Guillaume, C.; Gontard, N.; Brice, S. A Review: RFID Technology Having Sensing Aptitudes for Food Industry and Their Contribution to Tracking and Monitoring of Food Products. *Trends Food Sci. Technol.* **2017,** *62*, 91–103.

Biji, K. B.; Ravishankar, C. N.; Mohan, C. O. et al. Smart Packaging Systems for Food Applications: A Review. *J. Food Sci. Technol.* **2015,** *52*, 6125–6135.

Bülbül, G.; Hayat, A.; Andreescu, S. ssDNA-Functionalized Nanoceria: A Redox-Active Aptaswitch for Biomolecular Recognition. *Adv. Healthc. Mater.* **2016,** *5* (7), 822–828.

Caroline, K.; Dilidaer, Y.; Irina, O.; Swagata, B.; Joe, P. K.; Andrew, M. Extruded Phosphorescence-Based Oxygen Sensors for Large-Scale Packaging Applications. *Sens. Actuat. B Chem.* **2020,** *304*, 127357.

Choi, S.; Eom, Y.; Kim, S.; Jeong, D.; Han, J.; Koo, J. M.; Hwang, S. Y.; Park, J.; Oh, D. X. A Self-Healing Nanofiber-Based Self-Responsive Time-Temperature Indicator for Securing a Cold-Supply Chain. *Adv. Mater.* **2020,** *32*, 1907064.

Chowdhury, E. U.; Morey, A. Intelligent Packaging for Poultry Industry. *J. Appl. Poult. Res.* **2019,** *28*, 791–800.

Corradini, M. G. Shelf Life of Food Products: From Open Labeling to Real-Time Measurements. *Annu. Rev. Food. Sci. Technol.* **2018,** *9*, 251–269.

Daizy, M.; Tarafder, C.; Al-Mamun, M. R.; Liu, X.; Aly Saad Aly, M.; Khan, M. Electrochemical Detection of Melamine by Using Reduced Graphene Oxide-Copper Nanoflowers Modified Glassy Carbon Electrode. *ACS Omega* **2019,** *4* (23), 20324–20329.

Ding, L.; Li, X.; Hu, L.; Zhang, Y.; Jiang, Y.; Mao, Z.; Xu, H.; Wang, B.; Feng, X.; Sui, X. A Naked-Eye Detection Polyvinyl Alcohol/Cellulose-Based pH Sensor for Intelligent Packaging. *Carbohydr. Polym.* **2020,** *233*, 115859.

Dobrucka, R.; Robert, P. New Perspectives in Active and Intelligent Food Packaging. *J. Food Process. Preserv.* **2019,** *43*, e14194.

Erin, Y.; Miranda, M.; Phil, B. A Systematic Review of Consumer Perceptions of Smart Packaging Technologies for Food. *Front. Sustain. Food Syst.* **2020,** *4*, 63.

Fang, Z.; Zhao, Y.; Robyn, D. W.; Stuart, K. J. Active and Intelligent Packaging in Meat Industry. *Trends Food Sci. Technol.* **2017,** *61*, 60–71.

Fatima, M.; Silvana, A. Nanotechnology-Based Approaches for Food Sensing and Packaging Applications. *RSC Adv.* **2020,** *10*, 19309.

Fuertes, G.; Ismael, S.; Raúl, C.; Manuel, V.; Jorge, S.; Carolina, L. Intelligent Packaging Systems: Sensors and Nanosensors to Monitor Food Quality and Safety. *J. Sens.* **2016,** *2016*, 4046061.

Ghaani, M.; Carlo, A. C.; Giulia, C.; Stefano, F. An Overview of the Intelligent Packaging Technologies in the Food Sector. *Trends Food Sci. Technol.* **2016,** *51*, 1–11.

Gregor-Svetec, D. Intelligent Packaging. In *Nanomaterials for Food Packaging*; Elsevier, **2018**; pp 203–247.

Hamdy, M. E.; Del Carlo, M.; Hussein, H. A. *et al.* Development of Gold Nanoparticles Biosensor for Ultrasensitive Diagnosis of Foot and Mouth Disease Virus. *J. Nanobiotechnol.* **2018,** *16*, 48.

Han, J. W.; Luis, R. G.; Qian, J. P.; Yang, X. T. Food Packaging: A Comprehensive Review and Future Trends. *Compr. Rev.* **2018,** *17*, 860–877.

Hrytsenko, O.; Hrytsenko, D.; Shvalagin, V.; Grodziuk, G.; Kompanets, M. The Use of Carbon Nanoparticles for Inkjet-Printed Functional Labels for Smart Packaging. *J. Nanomater.* **2018,** *2018*, 1–10.

http://www.ripesense.co.nz

https://www.frontiersin.org/research-topics/13221/smart-food-packaging-active-and-intelligent-functions

Hu, W.; Wan, L.; Jian, Y.; Ren, C.; Jin, K.; Su, X.; Bai, X.; Haick, H.; Yao, M.; Wu, W. Electronic Noses: From Advanced Materials to Sensors Aided with Data Processing. *Adv. Mater. Technol.* **2019,** *4*, 1800488.

Huang, S.; Xiong, Y.; Zou, Y.; Dong, Q.; Ding, F.; Liu, X.; Li, H. A Novel Colorimetric Indicator Based on Agar Incorporated with *Arnebia euchroma* Root Extracts for Monitoring Fish Freshness. *Food Hydrocoll.* **2019,** *90*, 198–205.

Hussain, I.; Ahamad, K. U.; Nath, P. Low-Cost, Robust, and Field Portable Smartphone Platform Photometric Sensor for Fluoride Level Detection in Drinking Water. *Anal. Chem.* **2017,** *89* (1), 767–775.

Idumah, C. I.; Zurina, M.; Ogbu, J.; Ndem, J. U.; Igba, E. C. A Review on Innovations in Polymeric Nanocomposite Packaging Materials and Electrical Sensors for Food and Agriculture. *Compos. Interf.* **2020,** *27*, 1–72.

Jafry, A. T.; Lim, H.; Sung, W.; Lee, J. Flexible Time–Temperature Indicator: A Versatile Platform for Laminated Paper-Based Analytical Devices. *Microfluid Nanofluid* **2017,** *21*, 57.

Jaiswal, R. K.; Mendiratta, S. K.; Talukder, S.; Soni, A.; Chand, S.; Saini, B. L. Application of Lipase Based Enzymatic Time Temperature Indicator (TTI) as Quality Marker for Frozen Chicken Meat. *Food Sci. Technol Res.* **2020,** *26*, 9–16.

Jayakumar, A.; Radoor, S.; Karayil, J.; Radhakrishnan, E. K.; Sanjay, M. R.; Siengchin, S.; Parameswaranpillai, J. Applications of Nanotechnology in Food Packaging. In *Food Packaging: Advanced Materials, Technologies, and Innovations*; Rangappa, M. R., Parameswaranpillai, J., Thiagamani, S. M. K., Krishnasamy, S., Siengchin, S., Eds.; CRC Press, 2020. ISBN 9780367335380.

Jhuang, J. R.; Lin, S. B.; Chen, L. C.; Lou, S. N.; Chen, S. H.; Chen, H. H. Development of Immobilized Laccase-Based Time Temperature Indicator by Electrospinning Zein Fiber. *Food Packag. Shelf Life* **2020,** *23*, 100436.

Jiang, G.; Hou, X.; Zeng, X.; Zhang, C.; Wu, H.; Shen, G.; Li, S.; Luo, Q.; Li, M.; Liu, X.; Chen, A.; Wang, Z.; Zhang, Z. Preparation and Characterization of Indicator Films from

Carboxymethyl-Cellulose/Starch and Purple Sweet Potato (*Ipomoea batatas* (L.) lam) Anthocyanins for Monitoring Fish Freshness. *Int. J. Biol. Macromol.* **2020,** *143*, 359–372.
Kilic, T.; Philipp, P. J.; Giavedoni, P. et al. Milk Allergen Detection: Sensitive Label-Free Voltammetric Immunosensor Based on Electropolymerization. *BioNanoSci.* **2020,** *10*, 512–522.
Kotsilkov, S.; Ivanov, E.; Vitanov, N. K. Release of Graphene and Carbon Nanotubes from Biodegradable Poly(Lactic Acid) Films during Degradation and Combustion: Risk Associated with the End-of-Life of Nanocomposite Food Packaging Materials. *Materials* **2018,** *11* (12), 2346.
Kucherenko, I. S.; Soldatkin, O. O.; Dzyadevych, S. V.; Soldatkin, A. P. Electrochemical Biosensors Based on Multienzyme Systems: Main Groups, Advantages and Limitations—A Review. *Analyt. Chim. Acta.* **2020,** *1111*, 114–131.
Lee, K.; Ko, S. Proof-of-Concept Study of a Whey Protein Isolate Based Carbon Dioxide Indicator to Measure the Shelf-Life of Packaged Foods. *Food Sci. Biotechnol.* **2014,** *23* (1),115–120.
Li, Q.; Duana, M.; Liu, W.; Yue, D.; Ren, L. Kinetic Modeling of a Bi-Enzyme Time-Temperature Indicator (TTI) Based on Different Paramcters for Monitoring Food Product Quality. *IOP Conf. Ser.: Mater. Sci. Eng.* **2019,** *612*, 022042.
Liao, Y.; Zhang, R.; and Qian, J. Printed Electronics Based on Inorganic Conductive Nanomaterials and Their Applications in Intelligent Food Packaging. *RSC Adv.* **2019,** *9*, 29154–29172.
Matindoust, S.; Baghaei-Nejad, M.; Abadi, M. H.; Zou, Z.; Zheng, L. R. Food Quality and Safety Monitoring Using Gas Sensor Array in Intelligent Packaging. *Sens. Rev.* **2016,** *36*, 169–183. https://doi.org/10.1108/SR-07-2015-0115
Melnikov, P. V.; Naumova, A. O.; Alexandrovskaya, A. Y. *et al.* Optimizing Production Conditions for a Composite Optical Oxygen Sensor Using Mesoporous SiO_2. *Nanotechnol.* **2018,** *13*, 602–608.
Meng, X.; Kim, S.; Puligundla, P.; Ko, S. Carbon Dioxide and Oxygen Gas Sensors-Possible Application for Monitoring Quality, Freshness, and Safety of Agricultural and Food Products with Emphasis on Importance of Analytical Signals and Their Transformation. *J. Korean Soc. Appl. Biol. Chem.* **2014,** *57*, 723–733.
Mohana, N. C.; Mithun, P. R.; Rao, H. C. Y.; Mahendra, C.; Satish, S. Nanoparticle Applications in Sustainable Agriculture, Poultry, and Food: Trends and Perspective. In *Nanotoxicity; Prevention and Antibacterial Applications of Nanomaterials Micro and Nano Technologies*; Rajendran, S., Mukherjee, A., Nguyen, T. A., Godugu, C., Shukla, R. K., Eds.; Elsevier: UK, 2020; pp 341–353.
Mukdasai, S.; Uppachai, P.; Srijaranai, S. A Novel Colorimetric Paper Sensor Based on the Layer-by-Layer Assembled Multilayers of Surfactants for the Sensitive and Selective Determination of Total Antioxidant Capacity. *RSC Adv.* **2019,** *9* (49), 28598–28608.
Müller, P.; Schmid, M. Intelligent Packaging in the Food Sector: A Brief Overview. *Foods* **2019,** *8* (1), 16.
Mustafa, F.; Andreescu, S. Nanotechnology-Based Approaches for Food Sensing and Packaging Applications. *RSC Adv.* **2020,** *10* (33), 19309–19336.
Mustafa, F.; Andreescu, S. Chemical and Biological Sensors for Food-Quality Monitoring and Smart Packaging. *Foods* **2018,** *7* (10), 168.

Neethirajan, S.; Ragavan, V.; Weng, X.; Chand, R. Biosensors for Sustainable Food Engineering: Challenges and Perspectives. *Biosensors* **2018,** *8* (1), 23.

Nemes, S. A.; Szabo, K.; Vodnar, D. C. Applicability of Agro-Industrial By-Products in Intelligent Food Packaging. *Coatings* **2020,** *10*, 550.

Nguyen, L. H.; Naficy, S.; McConchie, R.; Fariba, D.; Chandrawati, R. Polydiacetylene-Based Sensors to Detect Food Spoilage at Low Temperatures. *J. Mater. Chem. C.* **2019,** *7*, 1919–1926.

Oh, S. Y.; Lee, M. J.; Heo, N. S.; Kim, S.; Oh, J. S.; Lee, Y.; Jeon, E. J.; Moon, H.; Kim, H.S.; Park, T. J.; Moon, G.; Chun, H. S.; Huh, Y. S. Cuvette-Type LSPR Sensor for Highly Sensitive Detection of Melamine in Infant Formulas. *Sensors* **2019,** *19*, 3839.

Peleg, M.; Normand, M. D.; Corradini, M. G. The Arrhenius Equation Revisited. *Crit. Rev. Food Sci. Nutr.* **2012,** *52* (9), 830–851.

Pollet, J.; Delport, F.; Janssen, K. P. et al. Fast and Accurate Peanut Allergen Detection with Nanobead Enhanced Optical Fiber SPR Biosensor. *Talanta* **2011,** *83* (5), 1436–1441

Rajapaksha, P.; Elbourne, A.; Gangadoo, S.; Brown, R.; Cozzolino, D. Chapman, J. A Review of Methods for the Detection of Pathogenic Microorganisms. *Analyst* **2019,** *144* (2), 396–411.

Rateni, G.; Dario, P.; Cavallo, F. Smartphone- Based Food Diagnostic Technologies: A Review. *Sensors* **2017,** *17* (6), 1453.

Robertson, G. L. *Food Packaging: Principles and Practice*, 2nd ed.; CRC Press: Boca Raton, 2006.

Salvo-Comino, C.; García-Hernández, C.; García-Cabezón, C.; Rodríguez-Méndez, M. L. Discrimination of Milks with a Multisensor System Based on Layer-by-Layer Films. *Sensors* **2018,** *18* (8), 2716.

Sardroud, J. M. Influence of RFID Technology on Automated Management of Construction Materials and Components. *Scientia Iranica* **2012,** *19* (3), 381–392.

Saroat, R.; Pimonpan, K. Valorization of Food Processing By-Products as Smart Food Packaging Materials and Its Application. In *Food Preservation and Waste Exploitation*; Thierry, A., Sonia, S., Anca, F., Jean-Claude, L. Eds.; IntechOpen Books, 2018; pp 1–28.

Shoue, C.; Sandrayee, B.; Jonathon, M.; Changyong, C.; Bahar, A. The Role of Smart Packaging System in Food Supply Chain. *Concise Rev. Hypotheses Food Sci.* **2020,** *85*, 517–525.

Teerasong, S.; Jinnaraka, A.; Sumonmarn, C.; Prapin, W.; Duangjai, N. Poly (Vinyl Alcohol) Capped Silver Nanoparticles for Antioxidant Assay Based on Seed-Mediated Nanoparticle Growth. *Talanta* **2017,** *170*, 193–198.

Tingting, G.; Tian,Y.; Zhu, Z.; Da-Wen, S. Modelling, Responses and Applications of Time-Temperature Indicators (TTIs) in Monitoring Fresh Food Quality. *Trends Food Sci. Technol.* **2020,** *99*, 311–322.

Tominaga, T. Rapid Detection of *Klebsiella pneumoniae*, *Klebsiella oxytoca*, *Raoultella ornithinolytica* and Other Related Bacteria in Food by Lateral-Flow Test Strip Immunoassays. *J. Microbiol. Methods* **2018,** *147*, 43–49.

Uddin, Z.; Boonsupthip, W. Development and Characterization of a New Nonenzymatic Colored Time–Temperature Indicator. *J. Food Process Eng.* **2019,** *42*, e13027.

Vanderroost, M.; Ragaert, P.; Devlieghere, F.; Meulenaer, B. D. Intelligent Food Packaging: The Next Generation. *Trends Food Sci. Technol.* **2014,** *39*, 47–62.

Varghese, S. A.; Siengchin, S.; Parameswaranpillai, J.; Essential Oil as a Potent Antimicrobial Agent in Biopolymer-Based Food Packaging—A Comprehensive Review. *Food Biosci.* **2020,** Accepted.

Wang, C.; Liuyi, H.;Kang, Z.; Anping, D.; Jianguo, L. Multiple Signal Amplification Electrochemiluminescent Immunoassay for Sudan I Using Gold Nanorods Functionalized Graphene Oxide and Palladium/Aurum Core-Shell Nanocrystallines as Labels. *Electrochim. Acta* **2018a,** *278*, 352–362.

Wang, Y.; Zhang, P.; Fu, W.; Zhao, Y. Morphological Control of Nanoprobe for Colorimetric Antioxidant Detection. *Biosens. Bioelectron.* **2018b,** *122*, 183–188.

Wang, J.; Wu, X.; Wang, C.; Rong, Z .; Ding, H.; Li, Hui.; Li, S.; Shao, N.; Dong, P.; Xiao, R.; Wang, S. Facile Synthesis of Au-Coated Magnetic Nanoparticles and Their Application in Bacteria Detection via a SERS Method. *ACS Appl. Mater. Interf.* **2016,** *8* (31), 19958–19967.

Wang, W.; Han, J.; Wu, Y.; Yuan, F.; Chen, Y.; Ge, Y. Simultaneous Detection of Eight Food Allergens Using Optical Thin-Film Biosensor Chips. *J. Agric. Food Chem.* **2011,** *59* (13), 6889–6894.

Wang, Y. C.; Lu, L.; Gunasekaran, S. Biopolymer/Gold Nanoparticles Composite Plasmonic Thermal History Indicator to Monitor Quality and Safety of Perishable Bioproducts. *Biosens. Bioelectron.* **2017,** *92*, 109–116.

Wicochea, R. J. D.; Chalier, P.; Ruiz, T.; Gastaldi, E. Active Food Packaging Based on Biopolymers and Aroma Compounds: How to Design and Control the Release. *Front. Chem.* **2019,** *7*, 398.

Wijaya, D. R.; Sarno, R.; Zulaika, E. Noise Filtering Framework for Electronic Nose Signals: An Application for Beef Quality Monitoring. *Comput. Electron. Agric.* **2019,** *157*, 305–321.

Wolfbeis, O. S.; List, H. Method for Quality Control of Packaged Organic Substances and Packaging Material for Use with This Method. *U.S. Patent* **1995,** *5*, 407, 829.

Xu, F.; Ge, L.; Li, Z.; Lin, H.; Mao, X. Development and Application of a Tyrosinase-Based Time-Temperature Indicator (TTI) for Determining the Quality of Turbot Sashimi. *J. Ocean U. China* **2017,** *16* (5), 847–854.

Yasamin, A.; Jodat, K. K.; Daniel, V. J.; Rosakaren, L.; De la, G. H.; Wang, T.; Joshi, S.; Rezaei, Z.; Bruna, A. G. M.; David, G.; Manu, S. M.; Su, R. S. A 3D-Printed Hybrid Nasal Cartilage with Functional Electronic Olfaction. *Adv. Sci.* **2020,** *7*, 1901878.

Yildirim, S.; Röcker, B.; Pettersen, M. K.; Julie, N. N.; Zehram, A.; Ramune, R.; Tanja, R.; Patrycja, S.; Begonya, M.; Véronique, C. Active Packaging Applications for Food. *Compr. Rev.* **2018,** *17*, 165–199.

Yousefi, H.; Ali, M. M.; Su, H. M.; Filipe, C. D. M.; Didar, T. F. Sentinel Wraps: Real-Time Monitoring of Food Contamination by Printing DNA Zyme Probes on Food Packaging. *ACS Nano* **2018,** *12* (4), 3287–3294.

Yousefi, H.; Su, H. M.; Sara, M. I.; Kais, A.; Carlos, D. M. F.; Tohid, F. D. Intelligent Food Packaging: A Review of Smart Sensing Technologies for Monitoring Food Quality. *ACS Sens.* **2019,** *4* (4), 808–821.

Zhai, X.; Shi, J.; Zou, X.; Wang, S.; Jiang, C.; Zhang, J.; Huang, X.; Zhang, W.; Holmes, M. Novel Colorimetric Films Based on Starch/Polyvinyl Alcohol Incorporated with Roselle Anthocyanins for Fish Freshness Monitoring. *Food Hydrocoll.* **2017,** *69*, 308–317.

Zhong, M.; Lu,Y.; Hui,Y.; Chang, C.; Wenfang, D.; Yueming,T.; Qingji, X.; Shouzhuo, Y. An Electrochemical Immunobiosensor for Ultrasensitive Detection of *Escherichia coli* O157:H7 Using CdS Quantum Dots-Encapsulated Metal-Organic Frameworks as Signal-Amplifying Tags. *Biosens. Bioelectron.* **2019,** *126*, 493–500.

Zhou, H.; Li, S.; Chen, S.; Zhang, Q.; Liu, W.; Guo, X. Enabling Low Cost Flexible Smart Packaging System with Internet-of-Things Connectivity via Flexible Hybrid Integration of Silicon RFID Chip and Printed Polymer Sensors. *IEEE Sens. J.* **2020,** *20*, 5004–5011.

Zhu, H.; Sikora, U.; Ozcan, A. Quantum Dot Enabled Detection of *Escherichia coli* Using a Cell Phone. *Analyst* **2012,** *137* (11), 2541–2544.

CHAPTER 6

Biopolymers-Based Nanocomposites: Functions and Applications

ALKA YADAV[1], GAURAVI AGARKAR[1],
LUIZA HELENA DA SILVA MARTINS[2], and MAHENDRA RAI[1*]

[1]*Department of Biotechnology, SGB Amravati University, Amravati, Maharashtra, India*

[2]*ISPA (Institute of Animal Health and Production) Federal Rural University of Amazonia Avenue Presidente Tancredo Neves, Terra Firme, Belém, Pará, Brazil*

[*]*Corresponding author. E-mail: mahendrarai7@gmail.com*

ABSTRACT

Scientists all over the world show strong interest in developing novel materials that are eco-friendly, cost effective, and biodegradable due to the increase in environment concern. Thus, bio-based products have raised great attention in recent times. Bio-based polymers are defined as polymers that are obtained from biological origin like plants, animals, and microbes including bacteria, fungi, and algae. The simplest form of biopolymers includes cellulose and starch that have been widely used since centuries. Biopolymers show application in different fields like packaging, textiles, and water treatment, but they depict some shortcomings when compared to synthetic polymers. However, biopolymers formulated using nanomaterials depict a wide range of improved properties including permeability, thermal stability, elasticity, and crystallinity. These biopolymer-based nanocomposites provide an opportunity to replace the traditional nonbiodegradable polymers and also reduce the over dependence on petroleum-based products. Biopolymer-based nanocomposites demonstrate applications in different fields including packaging, water treatment, antimicrobials, wound dressing, wound healing, disease diagnosis, and sensors. The present chapter gives an overview on biopolymers and biopolymer-based nanocomposites, their functions, and applications in different fields of science and technology.

6.1 INTRODUCTION

Macromolecules are present in nature in the form of carbohydrates and proteins in all biological entities (Dufresne et al., 2009; Xiong et al., 2018; Kassab et al., 2019). These macromolecules occur in the form of large monomer units. Scientists after detailed study have chemically designed similar type of molecules in controlled conditions in the laboratory. These synthetically formed chemical macromolecules are known as "polymers" (Bledzki et al., 2016). Polymers are mostly derived from petroleum oil; some of the examples of polymers are polyethylene (PE), polyamides (Nylon), poly vinyl chloride (PVC), polystyrene (PS), synthetic rubber, and teflon (Bibi et al., 2019; Luzi et al., 2020). Synthetic polymers possess versatile mechanical, thermal, and degradation properties as they are synthesized under controlled temperature and pressure conditions. Synthetic polymers are used for a wide range of applications like packaging, PVC pipes, plastics, implants, and catheters (Othman, 2014). The introduction of synthetic polymers brought a boom in the global industry due to their adaptability, durability, and low cost in the past two decades (Xiong et al., 2018; Luzi et al., 2019). However, most of the synthetic polymers were derived from petroleum and coal, making them incompatible with the environment as they could not be recycled (Dufresne et al., 2009; Patra et al., 2018). To overcome the problem, polymers with biological sources were crafted and described as biopolymers. Biopolymers were extracted from natural sources by living organisms or chemically synthesized using biological entities. Figure 6.1 illustrates that the various biopolymers are derived from carbohydrates (cellulose and starch), polysaccharides (alginate and pectin), animal protein based (silk, gelatin, collagen, chitin, and gum) Bledzki et al., 2016; Kassab et al., 2019; Mellinas et al., 2020). Compared to synthetic polymers, biopolymers have simpler structure and sustainability (Bledzki et al., 2016; Fiori et al., 2019). Also, biopolymers get degraded on exposure to bacteria in soil or compost making them environment friendly (Abdul Khalil et al., 2017; Bibi et al., 2019). However, poor mechanical and barrier properties limit the use of biopolymers commercially (Dufresne et al., 2009; Luzi et al., 2019).

FIGURE 6.1 Types of biopolymers.

Source: Reproduced from Mellinas et al. (2020) under a Creative Commons Attribution (CC BY) license (http://creativecommons.org/licenses/by/4.0/).

Nanotechnology has been gaining tremendous impetus since the last few decades (Othman, 2014). Due to their relatively small structure and enhanced properties, nanoparticles have received much attention in different fields of science and technology, including medicines, therapeutics, electronics, agriculture, and textiles (Nitta and Numata, 2013; Bertolino et al., 2018). Nanoparticles combined with biopolymers form nanocomposites. Biopolymer-based nanocomposites comprise a broad class of components on a length scale of less than 100 nm (Shchipunov, 2012; Othman, 2014). These components arranged in nanosized clusters form the nanocomposites. Nanomaterials comprise high surface area to volume ratio, increased mechanical properties and high optical properties (Okpala, 2013; Bertolino et al., 2018; Bibi et al., 2019), and biopolymers forming composites with nanomaterials depict improved mechanical and thermal properties (Jamroz et al., 2019). Bionanocomposites comprise inorganic three-dimensional materials like zeolites, two-dimensional materials including clays, metal oxides, metal phosphates, and also one- and zero-dimensional materials (Bertolino et al., 2018). Experimental evaluation has depicted that all types of nanocomposites that show refined and improved properties (Bledzki et al., 2016; Ali and Ahmed, 2018; Mellinas et al., 2020). Bio-nanocomposites also show interesting functional properties that make them available for wide range of applications including sensors, food packaging, drug delivery, and tissue engineering (Nitta and Numata, 2013; Ali et al., 2016; Bertolino et al., 2018; Fiori et al., 2019).

The present chapter also showcases a detailed study on biopolymers and biopolymer-based nanocomposites. The review introduces about polymers, synthetic polymers and how biopolymers are different from synthetic polymers. It largely focuses on biopolymer-based nanocomposites and their functions and applications.

6.2 POLYMERS

Polymers form a very important part of our daily life ranging from rubber, plastics, to resins, and adhesives (Mohan et al., 2016). The word polymer is taken from a Greek word "Poly" that means "many" and "mers" are "parts or units" (Goudoulas, 2012). Thus, polymers can be explained as any class of synthetic or natural material composed of multiple assemblies of a simple structural unit termed as monomers (Beisl et al., 2017; Bertolino et al., 2018). The reaction due to which the monomers join to form a polymer is known as polymerization reaction. Polymers can be fabricated using a number of materials including ethylene, propylene, starch, cellulose, and gelatin (Bibi et al., 2019). Polymers possess different chemical structures, mechanical, physical, and thermal properties (Carvalho et al., 2019). Polymers can be divided into synthetic and natural or biologically originated polymers (Maitz et al., 2015).

6.2.1 SYNTHETIC POLYMERS

Synthetic polymers are manmade macromolecules made up of numerous repeating monomer units (Maitz, 2015). Examples of synthetic polymers include polyamides (nylon), PVC, PE, PS, synthetic rubber, and teflon (Goudoulas, 2012; Mohan et al., 2016). Synthetic polymers are usually synthesized using petroleum and are made up of carbon–carbon bonds as backbone (Mohan et al., 2016). Synthetic polymers possess high mechanical, thermal, and elastic properties that make them efficient for a number of applications (Maitz et al., 2015; Mohan et al., 2016). A wide range of industries like automobile, textile, packaging, and paper depend on synthetic polymers (Bibi et al., 2019). Synthetic polymers are generally classified based on their response to heat. Thermoplastic polymers get softened by heating and could be transformed into desired shapes whereas thermosetting polymers remain intact at high temperature and pressure conditions (Maitz et al., 2015).

6.2.2 BIOPOLYMERS

Polymers of natural origin are defined as "Biopolymers" (Sharma, 2017). Naturally originated polymers exist since the beginning of life. Polymers like DNA, RNA, and polysaccharides play a vital role in the development of plants and animals (Zakaria et al., 2017; Ali and Ahmed, 2018; Kausar, 2020). Thus, biopolymers are polymeric structures produced by living organisms and are grabbing attention due to their unique properties (Goudoulas, 2012; Nitta and Numata, 2013; Bano et al., 2017; Rouf and Kokini, 2018). Biopolymers can be classified into biodegradable and nonbiodegradable biopolymers (Ali and Ahmed, 2018). Biodegradable biopolymers are biobased and include plants, animals, and microorganisms, while nonbiodegradable biopolymers are fossil based (Ali et al., 2016; Patra et al., 2017; Luzi et al., 2019). Biopolymers obtained from biodegradable materials like plant origin (starch, cellulose, pectin, and chitosan), animal origin (gelatin and casein), and microbial origin (polyhydroxy butyrate and polyhydroxy valerate) (Rendon-Villalobor et al., 2016; Rouf and Kokini, 2018; Kassab et al., 2018; Aftab et al., 2020) (Fig. 6.2).

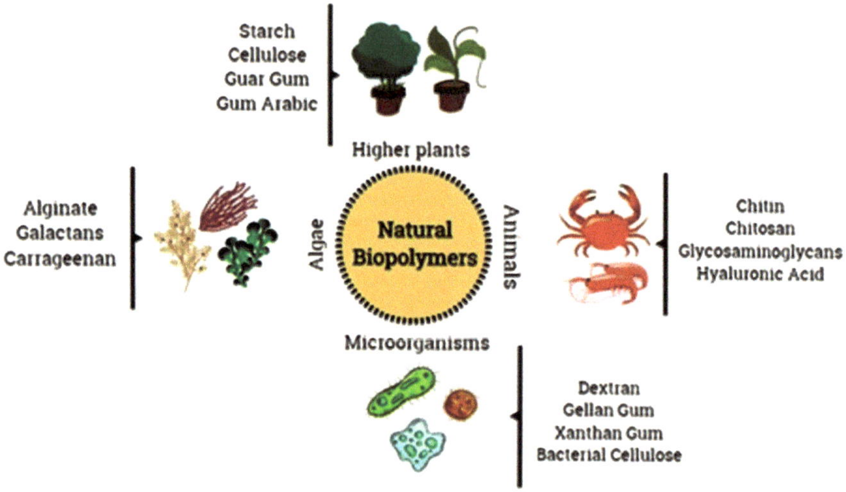

FIGURE 6.2 Natural biopolymers used in biomedical applications.
Source: Reproduced from Patra et al. (2018) under a Creative Commons Attribution 4.0 International License (http://creativecommons.org/licenses/by/4.0/).

Biopolymers have polysaccharides as prominent derivatives that are broken down into sugar monomers (Nitta and Numata, 2013; Abdul Khalil et al., 2017). Thus, due to their biodegradable nature and environment friendly synthesis process, biopolymers are used in a wide range of applications like packaging, tissue engineering, drug delivery, and dentistry (Ali et al., 2016; Bano et al., 2017; Fiori et al., 2019).

6.3 BIOPOLYMER-BASED NANOCOMPOSITES

In recent years, researchers have focused on the replacement of petroleum-based polymers with biopolymers (Carvalho et al., 2019). However, biopolymers exhibit poor thermal, mechanical, and barrier properties compared to nonbiodegradable polymers (Rendon-Villalobos et al., 2016; Bibi et al., 2019). Thus, bionanocomposites were developed as a promising application to improve the mechanical and barrier properties of biopolymers (Hossain et al., 2018; Carvalho et al., 2019). They are a novel class of advanced materials (Okpala, 2013; Jamroz et al., 2019). The addition of nanomaterials like titanium, silica, and silver to biopolymers not only improves the mechanical and barrier properties but also offers application of biopolymers in different areas like food packaging, anti-microbial agent, biosensors, and antibiofilm applications (Kashiri et al., 2017; Xiong et al., 2018; Kausar, 2020).

Most commonly used biopolymers to synthesize bionanocomposite materials include starch, cellulose, and chitosan (Xiong et al., 2018). At the nanoscale level, the size of the nanomaterials is significantly reduced leading to an enhancement in its surface area. Thus, the synthesized bionanocomposite relies on the high surface area of the nanomaterials, and this results in a large interfacial area between the biopolymer and the nanomaterials (Kashiri et al., 2017; Luzi et al., 2019). This large interface enables the modification of the mechanical, thermal, and barrier properties of the bio-nanocomposite (Kausar, 2020).

Many types of nanomaterials are used to enhance the properties of the bio-nanocomposite that include chitosan, titanium, silver, polymeric nanoparticles, and magnetite (Othman, 2014; Aftab et al., 2020). In a bio-nanocomposite, the polymer matrix forms the biological part while the nanoparticles are considered as the value-added material (Xiong et al., 2018; Kausar, 2020). These bio-nanocomposites are toward a greener approach, show good biodegradability and biocompatible properties (Abdul Khalil et al., 2017; Luzi et al., 2019). The properties and applications of the bio-nanocomposite depend on the characteristics of the nanomaterials. Some of the examples of biopolymers used as nanocomposites include cellulose, chitosan, alginate, zeolite, and clay (Okpala, 2013; Ramesh and Radhakrishnan, 2019; Kausar, 2020).

6.3.1 CELLULOSE

Cellulose is a polysaccharide obtained from both plants and animals (Ioelovich, 2008; Ramesh and Radhakrishnan, 2019). It is the building material of long fibrous cells and highly strong natural polymers (Pande and Saklecha, 2017; Rajnipriya et al., 2018). Cellulose has gained much attention for its application as nanocomposite material due to its high strength, low weight, biodegradability, and biocompatibility (Aftab et al., 2020). The potential high stiffness of cellulose can easily be employed in a nanocomposite material (Kassab et al., 2019). Cellulosic nanocomposites can be differentiated based on their synthesis process; they can be obtained as bacterial cellulose (BC) and microfibrillated cellulose (MFC) or nanofibrillated cellulose (NFC) (Carvalho et al., 2019). Nanocellulose can further be exploited in applications like medicines and cosmetics, pharmaceutical, dentistry, and automobiles (Hossain et al., 2018; Rajnipriya et al., 2018) (Fig. 6.3).

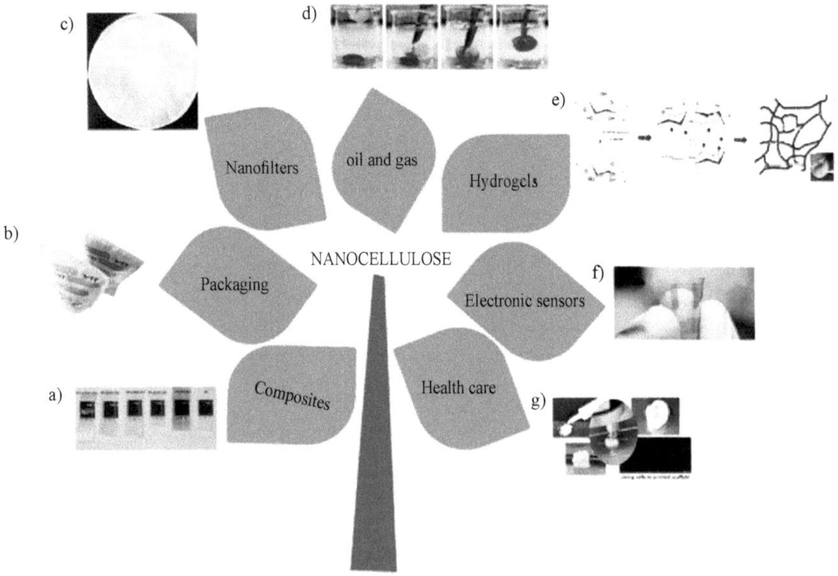

FIGURE 6.3 Importance of industrial and agricultural wastes in the field of nanocellulose and recent industrial developments of wood-based nanocellulose:

Source: Reprinted with permission from Rajinipriya et al., 2018. © 2018 American Chemical Society.

6.3.2 LIGNIN

Lignin forms the main constituent of the woody plants where it works as a lining agent. Lignin forms one of the major constituents of the lignocellulosic material found in the cell wall (Beisl et al., 2017). After cellulose, lignin is the second most abundantly harnessed biopolymer for the preparation of biocomposites (Feldman, 2016). Lignin exhibits ideal properties like low weight, abundance, high stiffness, antimicrobial, and antioxidant with highly biodegradable character (Beisl et al., 2017). Nanocomposites synthesized using lignin depict improved mechanical, thermal, and barrier properties (Feldman, 2016). These lignin-based nanocomposites are used as ultraviolet (UV) blocker, biocide, antioxidants/radical scavengers, and surfactants Beisl et al., 2017).

6.3.3 CHITOSAN

Chitosan naturally occurs as amino polysaccharide obtained as a deacetylated form of chitin (Ali and Ahmed, 2018). Chitosan is nontoxic, biodegradable, antibacterial and biocompatible (Bano et al., 2017). These properties of chitosan make it accessible to use in different applications like wound healing, drug delivery, tissue engineering, and food and paint industry (Hamed et al., 2016). Chitosan as a bio-nanocomposite material exhibits significant mechanical, thermal, and tensile properties (Bano et al., 2017; Ali and Ahmed, 2018).

6.3.4 STARCH

Starch as a biopolymer shows high biodegradable and biocompatible property, but it also depicts certain disadvantages like hydrophilic nature, brittleness, and crystallization (Dufresne et al., 2009). However, development of starch-based nanocomposites with addition of fillers shows improved physical, mechanical, and barrier properties (Zakaria et al., 2017). Starch acts as a matrix component, and the nanostructured materials acts as fillers thus enhancing the properties of starch for particular application (Goudarzi and Shahabi-Ghahfarrokhi, 2018). Starch nanocomposites are used in applications like tissue engineering, drug delivery, and thermoplastic materials (Zakaria et al., 2017; Goudarzi and Shahabi-Ghahfarrokhi, 2018).

6.3.5 CLAY

A lot of research has been conducted on clay nanocomposites as they are easily accessible, cost efficient, possess high aspect ratio, and offer enhanced properties keeping its original properties intact (Ali et al., 2016). Clay biopolymers like montmorillonite, vermiculite, saprolite, laponite, bentonite, and attapulgite are majorly used for synthesis of polymer nanocomposites (Fiori et al., 2019). Clay nanocomposites illustrate enhanced thermal, barrier, flame retardancy, and anticorrosive properties (Kashiri et al., 2017). Clay nanocomposites are exploited for applications like packaging, insulation, coatings, electrochemicals, and construction materials (Ali et al., 2016; Fiori et al., 2019).

6.3.6 ALGINATE

Alginate is a biopolymer belonging to carbohydrate family and is obtained from brown algae (Helmiyati and Dini, 2018). Alginate is nontoxic, immunogenic, biodegradable, and biocompatible polymer and hence is used in many biomedical applications (Bibi et al., 2019). To enhance the desired properties of alginate, nanocomposite materials are designed using alginate matrix (Bibi et al., 2019). Alginate-based nanocarriers are used as drug delivery carriers, enzyme immobilization, immunosensors, and scaffolds (Helmiyati and Dini, 2018; Bibi et al., 2019).

6.4 APPLICATIONS OF BIONANOCOMPOSITES

6.4.1 COSMETICS

Bionanocomposites using cellulose are used for a wide range of health-care applications involving cosmetics and biomedicine (Ioelovich, 2008). Nanocellulose can act as carrier–drug complex and penetrate through the skin pores of a person and treat topical infections. Also, bacterial cellulose incorporated with nanomaterials exhibits higher hydration property and is used as facial masks that soften the skin or as moisturizing cream (Ioelovich, 2008).

Like chitosan, cellulose is also used in cosmetics due to its antimicrobial property. Sakulwech et al. (2018) reported synthesis of nanoparticles using quaternized cyclodextrin grafted chitosan (QCD-g-CS) associated with hyaluronic acid (HA). The QCD-g-CS acts as a carrier for negatively charged hydrophilic molecules and hydrophobic drugs that help in skin penetration. Also, its antibacterial nature aids in fighting skin infections making it compatible to use in cosmetics (Sakulwech et al., 2018). Photo-aging is caused by UV radiations. Aranaz et al. (2018) has reviewed the use of chitin nanocomposites for protection against UV radiations. In the study, chitin nanofibrils and chitin nanocrystals were used to develop skin-protective formulations. It was observed that the chitin nanocrystals and nanofibrils improved the epithelial granular layer of the skin and also increased the granular density. Also, the nanocrystals and nanofibrils reduced the synthesis of transforming growth factor beta (TGF-β), thus slowing the photoaging process.

6.4.2 PACKAGING

Packaging materials are generally made from plastics specifically those using petroleum. Packaging materials made from biopolymers depict low mechanical and barrier properties. However, biopolymers incorporated with nanofillers show improved mechanical, thermal, and barrier properties. Nanofillers also enhance the antimicrobial and oxygen scavenging activity of the biopolymers. These bionanocomposites enable environment friendly and sustainable packaging material compared to the synthetic packaging material (Othman, 2014). Bionanocomposites like starch, cellulose, chitosan, carboxymethyl cellulose (CMC), and cellophane are generally used as packaging material. The nanofillers involved in the packaging material include nanoclays, zinc oxide, and metal oxides as they provide improved barrier properties against oxygen, carbon dioxide, and diffusion of flavor complexes (Oladimeji and Singh, 2018). Thus, the compatibility between nanofillers and polymer matrices lead to preparation of novel bionanocomposites that offer application in packaging of materials (Youssef and El-Sayed, 2018) (Table 6.1).

6.4.3 DENTISTRY

Bionanocomposites show enhanced mechanical properties that further supplement in polymer matrices favoring cellular behavior (Lee et al., 2016). Bio-nanocomposites can substitute native dental tissue in dentistry as bioactive nanoparticles conjugated with polymer matrix depict tremendous bioactivity with bone-derived cells and tooth-derived cells (Lee et al., 2016) (Fig. 6.4). Bio-nanocomposites due to their low toxicity, antimicrobial activity and surface–protein interactions offer diverse dental applications. They are used for regeneration of periodontal

TABLE 6.1 List of Bionanocomposites for Food Packaging Applications.

Sr. No.	Biopolymer	Nanoparticle	Applications	References
1	Guar Gum Polymer	Silver nanoparticles	Antimicrobial films	Abdullah et al. (2015)
2	Cellulose	Zinc oxide	Antimicrobial packaging	Shanshan et al. (2017)
3	Semolina	Zinc Nanorod Nano Kaolin	Biodegradable films	Jafarzadeh et al. (2017)
4	Tapioca Starch Bovine Gelatin	Zinc Nanorod	Edible films	Marvizadeh et al. (2017)
5	Starch	Titanium Dioxide	Biodegradable packaging material	Goudarzi and Shahabi-Ghahfarrokhi (2018)
6	Whey Protein	Nano Silica	Packaging solid materials	Hassannia-Kolaee et al. (2018)
7	Poly Lactic Acid	Magnesium oxide	Antimicrobial films	Swaroop and Shukla (2018)
8	Zein	Sodium Bentonite Nanoclay	Antimicrobial films	Kashiri et al. (2019)
9	Carboxy Methyl Cellulose	Sodium Bentonite Nanoclay	Edible food packaging	Fiori et al. (2019)
10	Starch Kefiran	Zinc oxide	Sanitizing and food packaging	Shahabi-Ghahfarrokhi and Babae-Ghazvini, (2019)
11	Poly Vinyl Alcohol	Cellulose nanoparticles	Biodegradable films	Ramesh and Radhakrishnan (2019)
12	Agar Carboxy Methyl Cellulose	Silver-Montmorillonite	Antimicrobial films	Makwana et al. (2020)
13	Seed Mucilage – *Ocimum basilicum*	Montmorillonite	Food packaging	Rohini et al. (2020)

apparatus that includes dentine, cementum, periodontal ligaments, and bone (Foong et al., 2020).

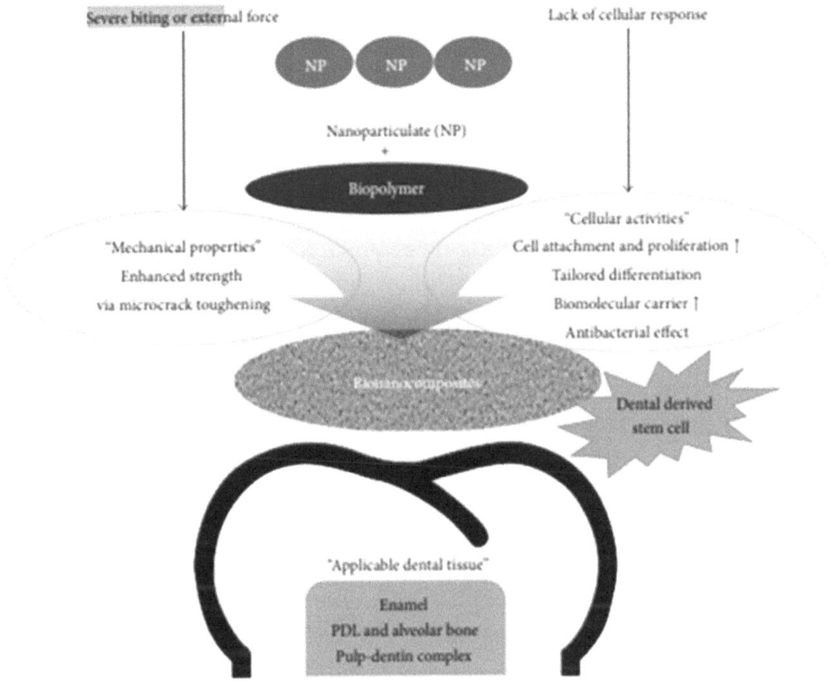

FIGURE 6.4 Use of bio-nanocomposites for dental tissue regeneration.
Source: Reprinted from Lee et al. (2016). https://creativecommons.org/licenses/by/4.0/.

Jahanizadeh et al. (2017) studied the use of curcumin loaded chitosan/carboxymethyl starch/montmorillonite biocomposites for its activity against dental biofilm formation. The anti-biofilm activity was checked against *Streptococcus mutans* using response surface technique to obtain maximum and minimum medicine loading particle sizes, entrapment efficacy, surfactant concentrations, and polysaccharide concentrations. The results of the study showed powerful reduction of biofilm by the curcumin nanocomposite. Aranaz et al. (2018) also suggested the use of fluoride, nanophosphate, and chitosan composite for the preparation of toothpaste to overcome dental abrasion. The nanophosphate combination with fluoride and chitosan was observed to prevent dental erosion. The nanocomposite denoted the ability to bind tooth enamel and function as a protective shield for the teeth.

6.4.4 TREATMENT OF CONTAMINATED WATER

Organoclays are utilized for wastewater treatment by many industries as they show synergistic effect with water treatment processing units like activated charcoal unit and reverse osmosis (Patel et al., 2006). Contaminated water contains many types of impurities including smaller particle impurities, humic acid, oil, and grease. The activated charcoal unit easily removes the smaller particle impurities but is unable to remove humic acid impurities. Organoclay can easily remove humic acid as well as oil and grease impurities from water and prevent it from turning carcinogenic. Organoclays use the partitioning mechanism for the absorption of organics as it contains alternate organic and inorganic layers. Organoclays prove to be superior that any other water treatment strategy as it cleans sufficient amount of oil and grease from the water (Patel et al., 2006). Pandey et al. (2017) reviewed the use of polymer nanocomposites for the adsorption of contaminants from drinking water. Polymer nanocomposites possess highly tunable adsorption properties as the nanoparticles increase the surface area of polymers composites. These polymer nanocomposites can be harnessed for the adsorption of contaminants like toxic metal ions, dyes, and microorganisms from contaminated water. Researchers have developed binary, tertiary, and quaternary polymer nanocomposites to increase the adsorption potential. Opoku et al. (2017) explained the use of metal oxide nanocomposites for water treatment. Metal oxide nanocomposites depict superior photodegradation activity. This property helps in the degradation of highly toxic and nonbiodegradable pollutants in the water. Nanofiber membranes like cellulose, poly vinyl acetate, and poly vinyl pyrrolidone work as a catalyst with metal oxide nanoparticles and show improved optical, mechanical, and adsorption properties. Metal oxide nanocomposites offer promising solution for wastewater treatment.

6.4.5 TISSUE ENGINEERING

Tissue engineering refers to the combination of scaffolds, biologically active molecules, cells, engineering, and assembling functional material that can construct, maintain, or improve the damaged tissues or whole organs (Jahangirian et al., 2018). Different classes of nanomaterials including ceramic, metallic, and polymer-based nanocomposites, and their combinations are being extensively studied as a scaffold for bone tissue engineering. Such nanocomposites are having great potential to be used in the treatment of critical size bone defects caused due to trauma or disease (Bramhill et al., 2017).

Wang et al. (2017) have obtained a novel tissue based on recombinant human bone morphogenetic protein-2 (rhBMP-2)-loaded calcium phosphate (Ca-P) nanoparticle/poly (l-lactic acid) (PLLA) nanocomposite by cryogenic 3D printing method. It reported good surface morphology while mechanical properties similar to human cancellous bone. This tissue showed sustained release of Ca^{2+}ions and rhBMP-2, favorable porosity, osteogenic differentiation, and sufficient cell viability suggesting it a promising material for bone regeneration. Tahriri et al. (2017) studied PLGA/nanofluoro hydroxyapatite microspheres as a candidate for bone tissue engineering. It revealed enhanced mechanical properties of nanocomposite as compared to PLGA scaffold alone when incubated in simulated body fluid. Also, biocompatibility testing by MTT cytotoxicity assays and alkaline phosphatase (ALP) activities reported increase in cell viability versus time against control. Mirza et al. (2020) attempted to fabricate ideal bone scaffold using ternary systems consisting of nano-hydroxyapatite (n-HA)/gum Arabic (GA)/κ-carrageenan (κ-CG) with different concentration ratios. One of the combination ratios of this scaffold revealed superior biocompatibility, protein adsorption, osteogenic protein expression, and maximum cell proliferation indicating its suitability for bone tissue engineering.

Liu et al. (2018) produced multilayered braided scaffold using poly(L-lactide-co-ε-caprolactone) (PLCL). A layer-by-layer coating was obtained by immersing the scaffolds into poly-L-lysine solution (polycation) and then into HA solution (polyanion) in order to promote MSCs growth, differentiation, and migration. This multilayered scaffold showed good biocompatibility and mechanical properties appropriate for ligament tissue engineering.

The treatment of peripheral nerve injury has limitations, as the neurons being a vital unit of neural systems cannot divide at all and the supporting glial cells has limited capacity of division. Recently, there are increasing reports mentioning that the graphene nanocomposites can be employed to stimulate neural stem cell adhesion, proliferation, differentiation, and neural regeneration (Bei et al., 2019). Qian et al. (2018) synthesized graphene-loaded polycaprolactone (PCL) nanoscaffolds along with dopamine (DPA) and arginyl glycylaspartic acid (RGD) by 3D-printing using a layer-by-layer technique. They reported that PCL conduits could connect damaged nerve stumps providing long-term support, whereas DPA and RGD within the inner tubing enhanced adhesion. These nanoscaffolds were implanted in rats, showed faster recovery in the Schwann cell-loaded nanoscaffolds, over 12 weeks span. The results pointed the synergistic effects of graphene, PCL, adhesion molecules, and cells as good alternative for peripheral nerve recovery.

For peripheral nerve regeneration and recovery, Pillai et al. (2020) endeavored to make scaffold with carbon nanofiber (CNF) with varying concentrations (5, 7.5, 10% w/w) dispersed in poly-ε-caprolactone (PCL) (10% w/v) nanocomposite and a unique combination of biomolecules coated on silk fibroin. Such silk threads were braided to obtain a scaffold structure. In the coating solution, as the concentration of CNF was increased, the electrical impedance was decreased up to 400 Ω, thus indicated better conductivity. It demonstrated significant enhancement in cell attachment, growth and proliferation, better mechanical properties, and nontoxicity (cytocompatibility) of the developed braided conduits.

6.4.6 DRUG DELIVERY SYSTEM

Various types of nanoparticles are being considered as vehicles for drug delivery, for example hydrophobic anticancer drugs. Most of the anticancer drugs are having limitations like poor solubility, physicochemical and pharmaceutical properties, and serious side effects due to use of adjuvant when administered intravenously. Different nano-formulations have been studied to overcome these limitations and making the drugs suitable for oral administration. Saha et al. (2016) have designed a delivery model that can carry more than one drug. It was achieved by synthesis of poly lactic-co-glycolic acid (PLGA) nanoparticles (NPs) that encapsulated hydrophobic anticancer drug Quercetin in core while the surface of NPs modified to carry hydrophilic drugs like Adriamycin (ADR) and mitoxantrone (MTX). The biopolymers used for surface modification such as bovine serum albumin (BSA) or histones (His) can play dual role, that is, carrying the drug as well as shielding the NPs against body's immune system, cellular degradation, and reticulo-endothelial system (RES). This model with multiple drugs can be useful to overcome multidrug resistance (MDR), which is a critical task in cancer treatment. Sharma et al. (2018)

effectively used sodium dodecyl sulphate-supported iron silicophosphate (SDS/FeSP) nanocomposite as a drug carrier for ondansetron. The drug release study was carried out at various pH values (pH 2.2, 5.5, 7.4, 9.4) for different time intervals (20 min–48 h), and the maximum release of nanocomposite encapsulated drug was reported about 45.38% at pH 2.2. This nanocomposite was found excellent for controlled drug release and having noncytotoxic nature.

Diabetes is one of the serious causes of death worldwide, in which the pancreas is unable to synthesize insulin and therefore needs to be injected externally to the patient. Studies have revealed that only 20% insulin reaches its target when subcutaneously administered while in case of oral route it is very less, that is, 1–2% due to enzymatic degradation and acidity of digestive system (Chen et al., 2011; Fonte et al., 2015). The oral route is most desired route owing to patient compliance, easier administration, site specificity, and immediate release of drug (Sharma et al., 2018). Thus, developing a drug delivery system for oral administration to enhance bioavailability of insulin and increasing its uptake by intestine is much anticipated. In this regard, Collado-González et al. (2020) have reported the fabrication of nanocomposites of alginate, dextran sulfate, poly-(ethylene glycol) 4000, poloxamer 188, chitosan, and bovine serum albumin for oral delivery of insulin. HPLC analysis showed that these nanocomposites could protect the insulin from physiological gastric environment where the peptide was fully released to interact with the absorptive mucosa. Moreover, all the nanocomposites showed enhanced stability over time, different ionic strength, and pH.

For ocular drug delivery, topical administration has been the most preferred route as it is painless, noninvasive and reaches anterior tissues as well as back-of-the-eye tissues (Cholkar et al., 2012; Agarwal and Rupenthal, 2016). But effective delivering a topical drug has major challenge of low ocular bioavailability that can be enhanced by expanding the retention time in precorneal area and elevating the permeability of drugs across the ocular tissues (Barar et al., 2016). Xu et al. (2018) have developed functional intercalated nanocomposites based on chitosan-glutathione-glycylsarcosine (CG-GS) and layered double hydroxides (LDH) as ocular drug carriers to treat mid-posterior diseases. Glycylsarcosine (GS), an active target ligand of peptide transporter-1 (PepT-1), distinctively connects with PepT-1 present on the cornea and leads the nanoparticles to the targeted site. In the experiments on rabbits, these CG-GS-LDH nanocomposites showed sustained release, enhanced bio-adhesion, longer precorneal retention, higher distribution of fluorescence probe/model drug, and also exhibited increased cellular uptake in comparison to pure drug solution. Further, an ocular irritation study and cytotoxicity testing had revealed good biocompatibility and no significant irritant effects.

6.4.7 CANCER THERAPY

Nanotechnology is also finding its application in cancer therapy by various strategies such as combination of photothermal therapy (PTT) and photoacoustic imaging (PAI) that shows fast recovery, preventing damage to nontargeted regions and minimal invasiveness. Manivasagan et al. (2017), in their study, synthesized chitosan-polypyrrole nanocomposites (CS-PPy NCs) and intelligently used them for photoacoustic imaging-guided photothermal ablation of cancer. These nanocomposites being biocompatible, stable and having strong near-infrared (NIR) absorbance found successful for enhancing PAI and accurately locating cancerous tissue. The tumor in mice almost disappeared within 20 days, after injecting CS-PPy NCs and NIR 808 nm laser irradiation, subsequently mice fully recovered.

6.4.8 ANTIFOULING APPLICATION

Polyhydroxyalkanoates (PHAs) have been considered as a good replacement material for traditional plastic. But this commercial PHA releases unpleasant odor due to its volatile compounds that may badly affect the final quality of the product. García-Quiles et al. (2019) have tried to resolve this issue by developing customized PHA bionanocomposites. They reinforced PHA materials with organomodified nanoclays like sepiolite and montmorillonite and obtained nanocomposites that were having high adsorbance toward volatile compounds those responsible for unpleasant odor.

6.5 CONCLUSION

Bionanocomposites are a unique group of biobased nanomaterials. They comprise a constituent of biological origin and particles with at least one dimension in the nanoscale level. Bionanocomposites are an exciting and flourishing field in the area of science and technology. Biopolymers along with nanomaterials showcase diverse group of nanocomposites depicting different structures, function, and applications. Biopolymers form the matrix whereas nanoparticles modify the matrix enhancing its mechanical, thermal, and barrier properties. These bionanocomposites are further harnessed for desired applications in biotechnology and biomedical science. This chapter also gives can overview of biopolymers and bionanocomposites mostly focusing on the different applications bionanocomposites offer in the field of science and technology.

KEYWORDS

- **polymers**
- **biopolymers**
- **nanocomposites**
- **application**
- **nanomaterials**

REFERENCES

Abdullah, M. F.; Ghosh, S. K.; Basu, S.; Mukherjee, A. Cationic Guar Gum Orchestrated Environmental Synthesis for Silver Nano-Bio-Composite Films. *Carbohydr. Poly.* **2015,** *134*, 30–37.

Aftab, M. N.; Karadag, A.; Muha-ud-din, R.; Iqbal, I. Cellulose Based Nanocomposites. *Bionanocomp. Green Synth. App.* **2020,** 207–231.

Agarwal, P.; Rupenthal, I. D. Modern Approaches to the Ocular Delivery of Cyclosporine A. *Drug Discov. Today* **2016,** *21* (6), 977–988.

Ali, A.; Ahmed, S. A Review on Chitosan and Its Nanocomposites in Drug Delivery. *Int. J. Biol. Macromol.* **2018,** *109*, 273–286.

Ali, F.; Ullah, H.; Ali, Z.; Rahim, F., Khan, F.; Rehman, Z. U. Polymer-Clay Nanocomposites, Preparations and Current Applications: A Review. *Curr. Nanomater.* **2016,** *1*, 83–95.

Aranaz, I.; Acosta, N.; Civera, C.; Elorza B.; Mingo J.; Castro, C.; Gandia, M. L.; Caballero A. H. Cosmetics and Cosmeceutical Applications of Chitin, Chitosan and Their Derivatives. *Polymers* **2018,** *10*, 213.

Barar, J.; Aghanejad, A.; Fathi, M.; Omidi, Y. Advanced Drug Delivery and Targeting Technologies for the Ocular Diseases. *BioImpacts* **2016,** *6* (1), 49–67.

Bei, H. P.; Yang, Y.; Zhang, Q.; Tian, Y.; Luo, X.; Yang, M.; Zhao, X. Graphene-Based Nanocomposites for Neural Tissue Engineering. *Molecules* **2019,** *24* (4), 658.

Beisl, S.; Friedl, A.; Miltner, A. Lignin from Micro to Nanosize: Applications. *Int. J. Mol. Sci.* **2017,** *18*, 2367.

Bertolino, V.; Cavallaro, G.; Milioto, S.; Parisi, F.; Lazzara, G. Thermal Properties of Multilayer Nanocomposites Based on Halloysite Nanotubes and Biopolymers. *J. Comp. Sci.* **2018,** *2*, 41.

Bibi, A.; Rehman, S.; Yaseen, A. Alginate-Nanoparticles Composites: Kinds, Reactions and Applications. *Mater. Res. Exp.* **2019,** *6*, 092001.

Bledzki, A. K.; Jaszkiewicz, A.; Urbaniak, M.; Stankowska-Walczak, D. Biocomposites in the Past and in the Future. *Fibers Text. Eastern Eur.* **2016,** *20* (6B), 15–22.

Bramhill, J.; Ross, S.; Ross, G. Bioactive Nanocomposites for Tissue Repair and Regeneration: A Review. *Int. J. Environ. Res. Public Health* **2017,** *14* (1), 66.

Chen, M. C.; Sonaje, K.; Chen, K. J.; Sung, H. W. A Review of the Prospects for Polymeric Nanoparticle Platformsin Oral Insulin Delivery. *Biomaterials* **2011,** *32*, 9826–9838.

Cholkar, K.; Patel, A.; Vadlapudi, A. D.; Mitra, A. K. Novel Nanomicellar Formulation Approaches for Anterior and Posterior Segment Ocular Drug Delivery. *Recent Patents Nanomed.* **2012,** *2* (2), 82–95.

Collado-González, M.; Ferreri, M. C.; Freitas, A. R.; Santos, A. C.; Ferreira, N. R.; Carissimi, G.; Sequeira, J. A. D.; DíazBaños, F. G.; Villora, G.; Veiga, F.; Ribeiro, A. Complex Polysaccharide-Based Nanocomposites for Oral Insulin Delivery. *Mar. Drugs* **2020,** *18*, 55.

Dufresne, A.; Medeiros, E. S.; Orts, W. J. Starch-Based Nanocomposites. *Starches: Charac., Prop. Appl.* **2009,** 205–246.

Feldman, D. Lignin Nanocomposites. *J. Macromol. Sci. Part A* **2016,** *53*(6), 382–387.

Fiori, A. P. S.; Camani, P. H.; Rosa, D. S.; Carastan, D. J. Combined Effects of Clay Minerals and Polyethylene Glycol in the Mechanical and Water Barrier Properties of Carboxymethyl Cellulose Films. *Indust. Crop. Prod.* **2019,** *140*, 111644.

Fonte, P.; Araújo, F.; Silva, C.; Pereira, C.; Reis, S.; Santos, H. A.; Sarmento, B. Polymer-Based Nanoparticles for Oral Insulin Delivery: Revisited Approaches. *Biotechnol. Adv.* **2015,** *33*, 1342–1354.

Foong, L. K.; Foroughi, M. M.; Mirhosseini, A. F.; Safaei, M.; Jahani, S.; Mostafavi, M.; Ebrahimpoor, N.; Sharifi, M.; Varma, R. S.; Khatami, M. Applications of Nano-Materials in Diverse Dentistry Regimes. *RSC Adv.* **2020,** *10*, 15430.

García-Quiles, L.; Valdés, A.; Cuello, A. F.; Jiménez, A.; Garrigós, M. C.; Castell, P. Reducing off-Flavour in Commercially Available Polyhydroxyalkanoate Materials by Autooxidation through Compounding with Organoclays. *Polymers* **2019,** *11*, 945.

Goudarzi, V.; Shahabi-Ghahfarrokhi, I. Photo-Producible and Photo-Degradable Starch/TiO_2 Bionanocomposite as a Food Packaging Material: Development and Characterization. *Int. J. Biol. Macromol.* **2018,** *106*, 661–669.

Goudoulas, T. B. Polymers and Biopolymers as Drug Delivery Systems in Nanomedicne. *Recent Patents Nanomed.* **2012,** *2*, 52–61.

Hassannia-Kolaee, M.; Shahabi-Ghahfarrokhi, I.; Hassannia-Kolaee, M. Development and Characterization of a Novel Ecofriendly Biodegradable Whey Protein Concentrate Film with nano-SiO_2. *Int. J. Food Eng.* **2018**. doi: 10.1515/ijfe-2017-0098

Helmiyati, H.; Dini, F. W. Synthesis and Application of Nanocomposite Based on Nano Sodium Alginate from Brown Seaweed Impregnation TiO_2 as a Catalyst for the Synthesis of 5-Hydroxymethylfurfural from Fructose. *AIP Conf. Proceed.* **2018,** *2023*, 020101.

Ioelovich, M. Cellulose as a Nano Structured Polymer: A Short Review. *BioResources* **2008,** *3* (4), 1403–1418.

Jafarzadeh, S.; Ariffin, F.; Mahmud, S.; Alias, A. K.; Hosseini, S. F.; Ahmad, M. Improving the Physical and Protective Functions of Semolina Films by Embedding a Blend Nanofillers (ZnO-nr and Nano-Kaolin). *Food Packag. Shelf Life* **2017,** *12*, 66–75.

Jahangirian, H.; Lemraski, E. G.; Rafiee-Moghaddam, R.; Webster, T. J. A Review of Using Green Chemistry Methods for Biomaterials in Tissue Engineering. *Int. J. Nanomed.* **2018,** *13*, 5953–5969.

Jahanizadeh, S.; Yazdian, F.; Marjani, A.; Omidi, M.; Rashedi, H. Curcumin-Loaded Chitosan/Carboxymethyl Starch/Montmorillonite Bio-Nanocomposite for Reduction of Dental Bacterial Biofilm Formation. *Int. J. Biol. Macromol.* **2017,** *105* (1), 757–763.

Jamroz, E.; Kulawik, P.; Kopel, P. The Effect of Nanofillers on the Functional Properties of Biopolymer-Based Films: A Review. *Polymers* **2019,** *11*, 675.

Kashiri, M.; Maghsoudlo, Y.; Khomeiri, M. Incorporating Zataria Multiflora Boiss. Essential Oil and Sodium Bentonite Nano-Clay Open a New Perspective to Use Zein Films as Bioactive Packaging Materials. *Food Sci. Technol. Int.* **2017,** *23* (7), 582–596.

Kausar, A. Progress in Green Nanocomposites for High-Performance Applications. *Mater. Res. Innov.* **2020,** 1–13.

Lee, J. H.; Kim, H. W.; Seo, S. J. Polymer-Ceramic Bionanocomposites for Dental Application. *J. Nanomater*. **2016,** 3795976.

Liu, X.; Laurent, C.; Du, Q.; Targa, L.; Cauchois, G.; Chen, Y.; Wang, X.; Isla, N. Mesenchymal Stem Cell Interacted with PLCL Braided Scaffold Coated with Poly-l-lysine/Hyaluronic Acid for Ligament Tissue Engineering. *J. Biomed. Mater. Res. Part A.* **2018,** *9999*, 1–11.

Luzi, F.; Torre, L.; Kenny, J. M.; Puglia, D. Bio and Fossil-Based Polymeric Blends and Nanocomposites for Packaging: Structure–Property Relationship. *Materials* **2019,** *12*, 471.

Maitz, M. F. Applications of Synthetic Polymers in Clinical Medicine. *Biosurf. Biotribol.* **2015,** *1*, 161–176.

Makwana, D.; Castano, J.; Somani, R. S.; Bajaj, H. C. Characterization of Agar-CMC/Ag-MMT Nanocomposite and Evaluation of Antibacterial and Mechanical Properties for Packaging Applications. *Arab. J. Chem.* **2020,** 3092–3099.

Manivasagan, P.; Bui, N. Q.; Bharathiraja, S.; Moorthy, M. S.; Oh, Y. O.; Song, K.; Seo, H.; Yoon, M.; Oh, J. Multifunctional Biocompatible Chitosan-Polypyrrolenano Composites as Novel Agents for Photoacoustic Imaging-Guided Photothermal Ablation of Cancer. *Sci. Rep.* **2017,** *7*, 43593.

Marvizadeh, M. M.; Oladzadabbasabadi, N.; Nafchi, A. F.; Jokar, M. Preparation and Characterization of Bionanocomposite Film Based on Tapioca Starch/Bovine Gelatin/Nanorod Zinc Oxide. *Int. J. Biol. Macromol.* **2017,** *99*, 1–7.

Mellinas, C.; Ramos, M.; Jimenez, A.; Garrigoz, M. C. Recent Trends in the Use of Pectin from Agro-Waste Residues as a Natural-Based Biopolymer for Food Packaging Applications. *Materials* **2020,** *13*, 673.

Mirza, S.; Jolly, R.; Zia, I.; Saad Umar, M.; Owais, M.; Shakir, M. Bioactive Gum Arabic/κ-Carrageenan-Incorporated Nano-Hydroxyapatite Nanocomposites and Their

Relative Biological Functionalities in Bone Tissue Engineering. *ACS Omega* **2020,** *5* (20), 11279–11290.

Mohan, S,; Oluwafemi, O. S.; Kalarikkal, N.; Thomas, S.; Songca, S. P. Biopolymers- Application in Nanoscience and Nanotechnology. *Rec. Adv. Biopoly.* **2016**. dx.doi.org/10.5772/62225

Nitta, S. K.; Numata, K. Biopolymer-Based Nanoparticles for Drug/Gene Delivery and Tissue Engineering. *Int. J. Mol. Sci.* **2013,** *14*, 1629–1654.

Okpala, C. C. Nanocomposites- An Overview. *Int. J. Eng. Res. Dev.* **2013,** *8* (11), 17–23.

Oladimeji, H.; Singh, S. Bio-nanocomposites in Packaging: Business Model for Products' Commercialisation. In *Bio-Based Materials for Food Packaging*; Ahmed, S., Eds.; Springer: Singapore, 2018.

Opoku, F.; Kiarii, E. M.; Govender, P. P.; Mamo, M. A. Metal Oxide Polymer Nanocomposites in Water Treatments. In *Descriptive Inorganic Chemistry Researches of Metal Compounds*; 2017; Chap 8, pp 173–199.

Othman, S. H. Bio-Nanocomposite Material for Food Packaging Applications: Types of Biopolymer and Nano-Sized Fillers. *Agri. Agric. Sci. Proceedia* **2014,** *2*, 296–303.

Pandey, N.; Shukla, S. K.; Singh, N. B. Water Purification by Polymer Nanocomposites: An Overview. *Nanocomposites* **2017,** *3* (2), 47–66.

Patel, H. A.; Somani, R. J.; Bajaj, H. C.; Jasra, R. V. Nanoclays for Polymer Nanocomposites, Paints, Inks, Greases and Cosmetic Formulations, Drug Delivery Vehicle and Waste Water Treatment. *Bull. Mater. Sci.* **2006,** *29* (2), 133–145.

Patra, J. K.; Das, G.; Fraceto, L. F.; Campos, E. V. R.; Rodriguez-Torres, M. P.; Acosta-Torres, L. S.; Diaz-Torres, L. A.; Grillo, R.; Swamy, M. K.; Sharma, S.; Habtemariam, S.; Shin, H. S. Nano Based Drug Delivery Systems: Recent Developments and Future Prospects. *J. Nanobiotechnol.* **2018,** *16*, 71.

Pillai, M. M.; Kumar, G. S.; Houshyar, S.; Padhye, R.; Bhattacharyya, A. Effect of Nanocomposite Coating and Biomolecule Functionalization on Silk Fibroin Based Conducting 3D Braided Scaffolds for Peripheral Nerve Tissue Engineering. *Nanomedicine* **2020,** *24*, 102131.

Qian, Y.; Zhao, X.; Han, Q.; Chen,W.; Li, H.; Yuan,W. An Integrated Multi-Layer 3D-Fabrication of PDA/RGD Coated Graphene Loaded PCL Nanoscaffold for Peripheral Nerve Restoration. *Nat. Commun.* **2018,** *9*, 323.

Rajnipriya, M.; Nagalakshmaiah, M.; Robert, M.; Elkoun, S. Importance of Industrial and Agricultural Waste in the Field of Nanocellulose and Recent Industrial Developments of Wood Based Nanocellulose: A Review. *ACS Sustain. Chem. Eng.* **2018**. doi: 10.1021/acssuschemeng.7b03437

Ramesh, S.; Radhakrishnan, P. Cellulose Nanoparticles from Agro-Industrial Waste for the Development of Active Packaging. *Appl. Surf. Sci.* **2019,** *484*, 1274–1281.

Rendon-Villalobos, R.; Ortiz-Sanchez, A.; Tovar-Sanchez, E.; Flores-Huicochea, E. The Role of Biopolymers in Obtaining Environmentally Friendly Materials. *Comp. Renew. Sus. Mater.* **2016**. dx.doi.org/10.5772/65265

Rohini, B.; Ishwarya, S. P.; Rajasekaran, R.; Vijayakumar, A. K. *Ocimum basilicum* Seed Mucilage Reinforced with Montmorillonite for Preparation of Bionanocomposite Film for Food Packaging Applications. *Poly. Test.* **2020,** *87*, 106465.

Rouf, T. B.; Kokini, J. L. Natural Biopolymer-Based Nanocomposite Films for Packaging Applications. In *Bionanocomposites for Packaging Applications*; Jawaid, M., Swain, S., Eds.; Springer: Cham, 2018. https://doi.org/10.1007/978-3-319-67319-6_8

Saha, C.; Kaushik, A.; Das, A.; Pal, S.; Majumder, D. Anthracycline Drugs on Modified Surface of Quercetin-Loaded Polymer Nanoparticles: A Dual Drug Delivery Model for Cancer Treatment. *PLoS ONE* **2016,** *11* (5), e0155710.

Sakulwech, S.; Lourith, N.; Ruktanonchai, U.; Kanlayavattanakul, M. Preparation and Characterization of Nanoparticles from Quaternized Cyclodextrin-Grafted Chitosan Associated with Hyaluronic Acid for Cosmetics. *Asian J. Pharma. Sci.* **2018,** *13*, 498–504.

Shahabi-Ghahfarrokhi, I.; Babae-Ghazvini, A. Using Photo-Modification to Compatibilize Nano-ZnO in Development of Starch-Kefiran-ZnO Green Nanocomposite as Food Packaging Material. *Int. J. Biol. Macromol.* **2019,** *124*, 922–930.

Shanshan, G.; Xiaoming, S.; Jianhua, W.; Shitao, Y.; Fushan, C.; Xinyu, S. Structure, Mechanical Properties, and Antimicrobial Activity of Nano-ZnO/Cellulose Composite Films. *Cellulose Chem. Technol.* **2017,** *51* (3–4), 355–361.

Sharma, A. K. Biopolymers in Drug Delivery. *Biopoly. Res.* **2017,** *1*, 1.

Sharma, G.; Naushad, M.; Thakur, B.; Kumar. A.; Negi, P.; Saini, R.; Chahal, A.; Kumar, A.; Stadler, F. J.; Aqil, U. M. H. Sodium Dodecyl Sulphate-Supported Nanocomposite as Drug Carrier System for Controlled Delivery of Ondansetron. *Int. J. Environ. Res. Public Health* **2018,** *15*, 414.

Shchipunov, Y. Bionanocomposites: Green Sustainable Materials for the Near Future. *Pure Appl. Chem.* **2012,** *84* (12), 2579–2607.

Swaroop, C.; Shukla, M. Nano-Magnesium Oxide Reinforced Polylactic Acid Biofilms for Food Packaging Applications. *Int. J. Biol. Macromol.* **2018,** *113*, 729–736.

Tahriri, M.; Moztarzadeh, F,; Tahriri, A.; Eslami, H.; Khoshroo K.; Jazayeri H.; Tayebi, L. Evaluation of the in Vitro Biodegradation and Biological Behavior of Poly (Lactic-co-Glycolic Acid)/Nano-Fluorhydroxyapatite Composite Microsphere-Sintered Scaffold for Bone Tissue Engineering. *J. Bioact. Compat. Polym.* **2017,** *33* (2), 146–159.

Wang, C.; Zhao, Q.; Wang, M. Cryogenic 3D Printing for Producing Hierarchical Porous and rhBMP-2-Loaded Ca-P/PLLA Nanocomposite Scaffolds for Bone Tissue Engineering. *Biofabrication* **2017,** *9* (2), 025031.

Youssef, A. M.; El-Sayed, S. M. Bionanocomposites Materials for Food Packaging Applications: Concepts and Future Outlook. *Carbohydr. Poly.* **2018,** *193*, 19–27.

Zakaria, N. H.; Muhammad, N.; Abdullah, M. M. A. B. Potential of Starch Nanocomposites for Biomedical Applications. *Int. Conf. Innov. Res.* **2017**. doi:10.1088/1757-899X/209/1/012087

PART III

Role of Nanotechnology in Food Preservation

CHAPTER 7

Nano-Innovations in Food Packaging to Preserve Food Flavor and Odor

AISWARYA SATHIAN[1], K. S. JOSHY[1], and SABU THOMAS[2*]

[1]*School of Energy Materials, Mahatma Gandhi University, Kottayam*

[2]*School of Chemical Sciences, Mahatma Gandhi University, Kottayam 686560, Kerala, India*

[*]*Corresponding author. E-mail: sabuthomas@mgu.ac.in*

ABSTRACT

Nanoscience and nanotechnology hold a huge potent in several fields and are envisaged as technology that can lead the way toward sustainable environment-friendly development in the coming years. These are very rapidly growing disciplines in the field of the food industry. By using this technology, the food industry is expanding the process of encapsulation of aromas and flavors and nanodelivery systems. This chapter is a compendium, which addresses important modern technologies in the field of food nanotechnology. In the given current situation that we are all in, it is a very challenging task for using and synthesis of unique, innovative materials in a system that is nano-sized for the packaging of food. The preservation of the organoleptic profile of the product is done by encapsulating the aromatic compounds until it can be used and this ensures the quality remains high as well as the commercial value. In the following chapter, those materials that are being used commercially, how the taste and odor are maintained and the various encapsulation techniques for the process of encapsulation are discussed.

7.1 INTRODUCTION

Several disciplines are integrated in a wide variety of arenas of nanotechnology, including chemistry, biotechnology, physics, and engineering, and refers to the utilization of nanomaterials whose size ranges from 1 to 100 nm. Technological progression has seen a lot of changes and nanotechnology has been looked into as an attractive source the past few decades that has transformed the food sector. The fast-paced growth of nanotechnology has revolutionized many areas of food science, mainly involving the packaging, transportation, processing, storage, and other safety aspects of food (Bajpai et al., 2018). Many industrial and scientific fields including the food industry have been transformed due to the recent innovations in nanotechnology. Nanoparticles has been used in diverse fields of food science and food microbiology, and it is inclusively used in food processing, packaged foods, development of functional food, safety of food, detection of food borne pathogens, and prolongation of the shelf-life food has led to various functions of nanotechnology as showed in Figure 7.1. The demand of nanoparticle-based materials usage increased in the food industries. As many of them contain crucial elements and they were also found to be non-toxic (Brabin et al., 2001).

From food manufacturing, processing to packaging, nanotechnology offers all variety of complete food solutions for the same. In the realm of food quality and safety, nanomaterials bring a great deal of differences including health benefits that food delivers. Novel techniques, methods, and products are coming up which include industries, organizations, and researchers which have direct applications of nanotechnology in the arena of food science (Singh et al., 2017). In the area of food packaging, applications of nanotechnology can be classified into mainly three divisions:

Improved packaging: The nanomaterials incorporated into polymers matrix to better the properties like; temperature, humidity, and gas barrier resistance of being packaging materials.

Active packaging: The usage of nanomaterials in packaging allows for the interaction of food with the environment while also playing a dynamic role in the preservation of food. Antimicrobial properties exhibited by several materials like carbon nanotubes, nano copper oxide, nano silver, nano titanium dioxide, and nano magnesium oxide. The usage of silver nanoparticles in food packaging as antibacterial agents is increasing presently.

Smart packaging: It incorporates nanoparticles that can sense the biochemical or microbial changes in food and acts as a tracking system for safety of food and to prevent food adulteration and counterfeit.

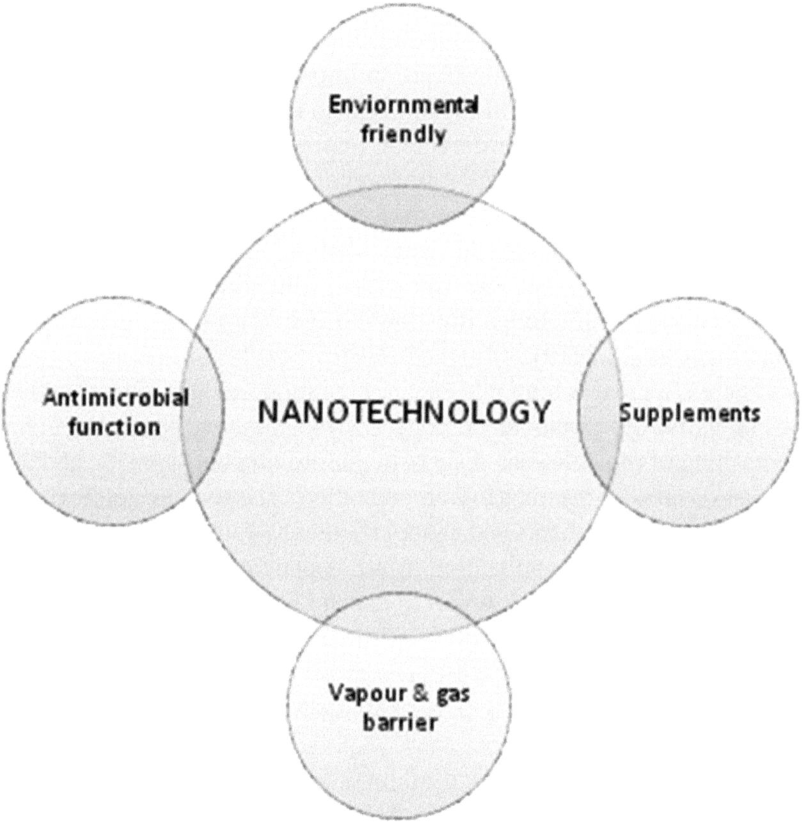

FIGURE 7.1 Functions of nanotechnology in food.

7.2 FOOD NANOTECHNOLOGY

Conventional food science and food industry has been thoroughly revolutionized by nanotechnology (He and Hwang, 2016). Nanotechnology is a leading technology which is empowering the sectors of food, agriculture, and medicine (Nile et al., 2020). As the consumer concerns about food quality and health benefits are on the rise, researchers are driven to find a path to enhance the food quality while causing minimum disturbance to the nutritional value of the product. Nanoparticle-based materials are in great demand in the food industry owing to the presence of essential elements in many of them, their non-toxic nature and stability at high temperature and pressure.

In the area of food processing, these nanoparticles can be utilized as food additives, agents of anti-caking, for the smart delivery of nutrients as carriers, agents of antimicrobial, to improve mechanical strength, and durability as fillers of the material packaged. Whereas food nano sensing is applicable in the achievement of improved food safety and quality evaluation (Singh et al., 2017). In an economy that is circular in nature, the aim is in reduction of waste by recycling and reusing materials. One great source of nature-derived sustainable polymers would be from food industry wastes, like bagasse and okara, which have higher contents of polysaccharides. Effective utilization of food waste for bioplastic production (Tsang et al., 2019).

There is a great potential to enhance the food and agricultural industries by innovative use of nanotechnology. Use of nanotools that are novel for the controlling of rapid disease diagnostic, improving the capacity, and ability of plants to absorb the nutrients among others. The specific application is in the field of nanofertilizers and nanopesticides to trail nutrients and products that have remarkable application in nanotechnology in agriculture as well as to utilize these to check levels to expand its productivity without water and soil contamination, at the same time providing protection to plants against several microbial diseases and insects. Nanosensors for monitoring soil quality, which would help to maintain the health of agricultural plants (Prasad et al., 2017).

The production and design of food through a variety of techniques like molecule shaping and that of atoms is the upcoming future of global food industries. On one hand, developments in crop DNA decoding and analyzing permit the industries to foresee, control and better the agro and farm produce. On the other hand, with technological advancements in manipulating the molecules and atoms of food, the future food industry has an effective method to design food with much more precision and with reduced expenditure. Meanwhile, the research on the combination of DNA and nanotechnology gives rise to the new nutrition delivery system, which directs active agents more accurately and effectively to the targeted parts of the tissues or cells. Foods that are functional will reap primarily from the new technologies after which standard food, nutraceuticals, and others will follow (Ravichandran, 2010).

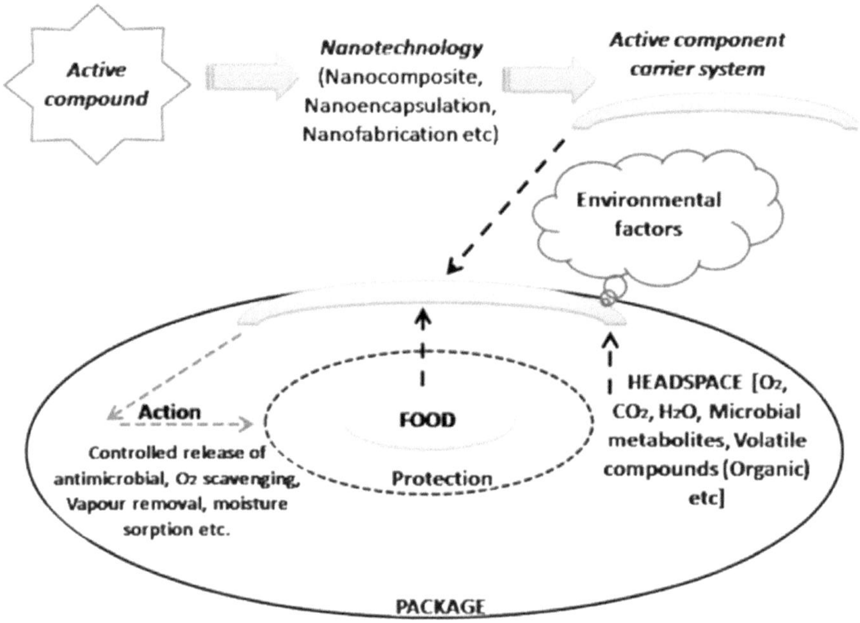

FIGURE 7.2 Active packaging and its association with nanotechnology.

Source: Reprinted from Sharma et al. (2017). © 2017 Sharma, Dhiman, Rokana and Panwar. https://creativecommons.org/licenses/by/4.0/

The issue lies in determining whether the change in the size of the materials could lead to radical, albeit useful properties, or affect other properties in particular, the potential toxicity of the material. In spite of it being intended for food consumption, nanotechnology products are expected to be classified as novel products and need testing and clearance. Concern regarding accidental release and ingestion of nanoparticles of undermined toxicity is raised especially in the area of food contact materials. Such concerns need to be addressed as the ultimate success of these products would finally depend on the consumer acceptance. The current explosion in the availability of nanoproducts showcases both the direct and indirect impacts it will have on the food industry in the near future. The sector that has generated more debate is the possible utilization of products of evolutionary nanotechnology in the food sector (Ravichandran, 2010).

The amelioration of nanotechnology signifies wide usage of nanomaterials especially in the food sector with a lot of potential benefits in the arenas of food quality and safety, distribution of ingredients which are micronutrients and bioactive, processed and in packaged food. Currently, many such studies are carried out to design innovative packaging (smart, intelligent, and active food packaging) in-order to boost the effectiveness and efficiency of packaging, as well as balance environmental issues. It is possible that these fabricated technologies will be designed for nano delivery frameworks in the future. Figure 7.2 depicts the active packaging and its association with nanotechnology. Nano delivery systems will concentrate principally on the bioavailability and its enhancement or the bioactive materials absorption and will help in precise delivery to specific parts of the gastrointestinal tract. In the market, there are a number of nanoproducts that are food-related. For example, nanoparticles of carotenoids are available, that can be dispersed in water, giving bioavailability improvement and also nano-based mineral supplements such as nano-size iron and calcium; nano-micellar systems those of which are needed to supply systems for delivery for minerals, vitamins, and phytochemicals (Chaturvedi and Dave, 2019).

7.3 FOOD TECHNOLOGY PROCESS

There are a set of chemical and physical techniques in the food nanotechnology and is used in food ingredients transformation and agricultural products into food. It encircles numerous forms of food processing, like grinding grain to make raw flour to industrial complex methods and home cooking used to make convenience foods. Merits of food processing include toxin removal, preservation, easy marketing and tasks of distribution, food consistency increase and improving the quality of food. With the aid of processing food technology, a huge profit potent persists for producers and supply agents of food products that are processed.

One of the areas under which, the general drive for improved quality has led to more important changes is called minimally processed foods. Using new technological solutions, such as modified atmosphere packaging, new packaging materials and technologies, or combinations of different treatments, a variety of products much closer to fresh and wholesome products is increasingly reaching the market. High-pressure treatment is another very promising technology, which is more severe in nature, but it has been proven to be successful in preserving quality attributes and for the same reason is sometimes included in this category, with mild referring to process impact.

Recent consumer trends turn toward a stronger emphasis on balancing food products that are healthy. Super-foods myths and hypes are only voicing of consumer's doubts for healthier options of products. Combining a pre-existing food profiling system with food processing-related descriptors permits the judgment of the trade-offs of processing of food on the design of products that are food. A novel division system was developed by Monteiro and collaborators to distinguish between processed and unprocessed food. The guidelines as justified by Monteiro states that for healthier food options by observing brought out that most works of public health and nutrition overvalued the importance of food and nutrients and has led to undervaluation and even in some cases ignored the importance of processing (Monteiro, 2009).

The NOVA categorization contains four classes of foods that are of rising processing intensities. Foods that are extremely highly processed, UPFD, are ultra-processed foods so-called industrial formulations, which, besides salt, sugar, oils, and fats, and include substances not used in culinary preparations, specifically additives, utilized to emulate sensorial attributes of minimum processed foods and their culinary preparations (Fernanda et al., 2019). The identification of ultra-processed foods can be known by the presence of salt; furthermore, it includes added sugar, high-fat content, additives, taste compounds, colors, substances from contact with packaging materials, and compounds generated during processing and storage (Monteiro et al., 2010). The definition does not refer to processing or unit operations which are made use in the process of production of the described products of food at all, but remain in the nutritional realms of compositions and put the significance on formulation and additives. Most significantly, the ability to extrapolate must be present in the consumer about the decreased nutritional features from the existence of an additive or in-process (or storage) generated compound in a product.

Consequentially, this would denote that rather than the use right processing of food, the utilization of additives (such as added sugar and salt), of agricultural raw material, would determine the qualities of ultra-processed foods. This definition would strangely and instantly transform a minimized yogurt that is processed into a product that is ultra-processed when the consumer sweetens it with extra sugar. To advance in food research and development, an adaptation that is worldwide to research that is critical in nature and the development of agendas are needed. A forceful paradigm shift is required in the field of food process technology and science, to handle the big challenges stemming from the interface processing food with nutrition. Most significantly, a paradigm change is essential for research on foods (Knorr and Watzke, 2019).

7.4 APPLICATION OF NANOTECHNOLOGY IN FOOD PACKAGING

In recent years, there has been a growing concern on research and innovation in packaging, encapsulating of food, and its materials. A wide range of nanomaterials (AgNPs, TiONPs, ZnONPs, TiO_2NPs, and nano clay) are subjected as functional additives and have been introduced to the packaging of these foods (Duncan, 2011; Lamba and Garg, 2018).

Biopolymeric carriers are composed of natural polymers like starch, pectin, etc., can form excellent complexes with other molecules. The main advantage of nanodelivery and encapsulation is to reduce the wastage of materials. These biopolymeric nanomaterials have significant applications in the field of food and agricultural sector and potential benefits. The biopolymeric nanodelivery is biocompatible, less toxic, and also delivers the precise amount of encapsulated material. Figure 7.3 shows the application of biopolymeric nanodelivery and summarizes its application in the agri-food sector. Polymeric materials are currently used in various sectors due to their tremendous application. The various types of nanomaterial applications in modern food technology process is shown in Table 7.1.

A broad range of nature-derived polymers utilizing the above-mentioned methods has been fabricated into systems of nanodelivery for the encapsulation of bioactive molecules such as nutraceuticals, antimicrobials, antioxidants, and flavors (Jafarizadeh-Malmiri et al., 2019; Jideani et al., 2017). The prime merit of utilizing nanoparticles for encapsulation is that nano-size and they do not generally affect the organoleptic properties in juxtaposition with their micron-sized counterparts. Another salient merit is that owing to their increased surface-volume ratio, nanoparticles provide higher bioavailability of the encapsulated nutrient (Sampathkumar et al., 2020).

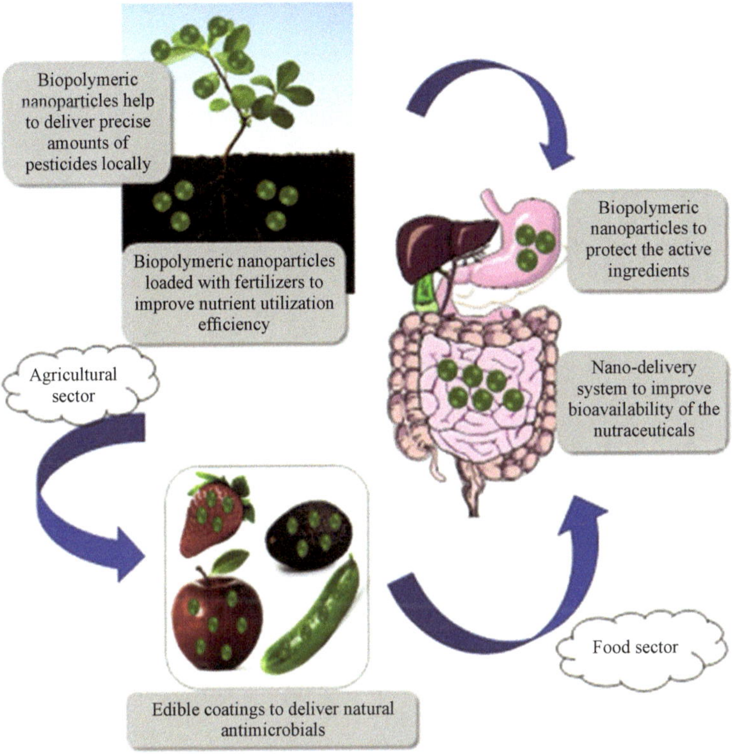

FIGURE 7.3 Applications of biopolymeric nano-delivery systems in the agri-food sector advantage.

Source: Reprinted from Sampathkumar et al. (2020). © 2020 The Authors. https://creativecommons.org/licenses/by-nc-nd/4.0/

7.5 NANOENCAPSULATION TECHNOLOGY

The encapsulation process has been developed 30 years ago. It is the method, materials one or combinations of materials is enclosed within another system or material. The material coated is called core or active material, and the material coating is called shell, wall material, carrier, or encapsulant. The microencapsulation products and their development began in 1950s in studies into pressure-sensitive coatings for the production

TABLE 7.1 Types of Nano Innovation in Food Packaging: Modern Food Technology Process.

Nanotechnique	Characteristic feature	Example	References
Edible coatings	To preserve the quality of fresh foods during extended storage	Application of novel zinc oxide reinforced xanthan gum hybrid system for edible coatings.	K. S. et al. (2020)
		• Effect of gelatin-based edible coatings incorporated with aloe-vera and green tea extracts on the shelf-life of fresh-cut apple.	Amiri et al. (2018)
		A cassava starch–chitosan edible coating enriched with lippia sidoides Cham. Essential oil and pomegranate peel extract for the preservation of Italian tomatoes (Lycopersicon esculentum Mill.) stored at room temperature	Araújo et al. (2018)
Hydrogel	Can be easily placed into capsules that aids in the protection of drugs from extreme environments, and to deliver them in response to environmental stimuli such as pH and temperature	A green strategy for preparing durable underwater superoleophobic calcium alginate hydrogel coated-meshes for oil/water separation	Wang et al. (2019)
Polymeric micelles	Solubilize water-insoluble compounds in the hydrophobic interior, high solubility, and low toxicity	Extraction and determination of flavonoids in fruit juices and vegetables using $Fe3O_4/SiO_2$ magnetic nanoparticles modified with mixed hemi/ad-micelle cetyltrimethylammonium bromide and high performance liquid chromatography	Sani et al. (2020)
Nanoemulsions	Their greater stability, higher optical clarity, increases the bioavailability and their efficiency to deliver nutraceuticals	Nanoemulsions as advanced edible coatings to preserve the quality of fresh-cut fruits and vegetables: a review	Hasan et al. (2020)
Liposomes	Liposome surrounds an aqueous solution inside a hydrophobic- membrane, it can be used as a carrier vehicle	Formation and characterization of spray dried coated and uncoated liposomes with encapsulated black carrot extract	Guldiken et al. (2019)
Inorganic NPs	It has impressive encapsulation capability and the rigid surface allows for controlled functionalization.	Palladium nanoparticles immobilized over Strawberry fruit extract coated Fe_3O_4 NPs: A magnetic reusable nanocatalyst for Suzuki-Miyaura coupling reactions.	Hemmati et al. (2020)

of carbonless paper copying (Madene et al., 2006). Nanoencapsulation is the process of encapsulating the required individual materials/particles inside nanocarriers. There can be two types of nanocarrier walls: microencapsulation and nanoencapsulation. When particle size ranges from 3 to 800 m in diameter, it is microencapsulation and the nanoparticles between 10 and 1000 nm in diameter used in nanoencapsulation.

In the microencapsulation, it is very easy to covert the state of matter in order to enhance its property. For the water release and sorption isotherm at 36°C and 80°C, microcapsules were studied in their structural level. It was seen that increase in the rate of release was not due to a temperature increase, which is indicative that microcapsules were resistant to relatively high temperatures (80°C). There has been a several presented advantages of the nanoencapsulation systems by providing of higher surface area some of which include higher bioavailability of the target compounds as well as target-release (Delshadi et al., 2020). There are certain properties of the materials, which are mandatory to be used for the encapsulation of active components and they include that it needs to be biodegradable, cost-effective, and nontoxic, an efficiency for binding of active ingredients, resistant to environmental changes which include temperature that is relatively high, humidity, and mechanical pressure, as well as have a strong barrier property.

There has been a wide application of nanoencapsulation techniques for active packaging in various food industries globally which is mainly because of their efficiency in extending shelf life, preserving of the natural properties (aroma, color, and flavor), properties of controlled release and addition of value in several food systems. In addition, it is highly efficient in the protection of food from a wide variety of contamination of microbes which are pathogenic in nature (Das et al., 2020). There are nanocarriers like nanoliposomes, nanoemulsions, solid-lipid nanoparticles, lipid nanocarriers, biopolymeric nanoparticles, and nanofibers for the delivery of encapsulated materials. Figure 7.4 shows the schematic representation of a nanocarrier with its encapsulated core material and wall material (Bahrami et al., 2019).

Nanoemulsions, formed by the combination of two immiscible liquids and these nanoemulsions are thermodynamically stable in various conditions. It is because of monodispersivity and small size (comparing 50–500 nm to 1200 nm). The stability of functional ingredients and their applicability in various food matrices can guarantee by this technology.

Encapsulation is to pack materials inside the core by using techniques like nanocomposites, nanoemulsification, etc. Production of foods that are functional in nature along with enhanced stability and functionality may be achieved using this technique, which makes use of compounds that are bioactive such as antioxidants, proteins, vitamins, lipids as well as carbohydrates. Nanoemulsions can encapsulate the required nutrients or drugs in their water/oil bond or all through the continuous phase of the system. There are different types of techniques available for the production of nanocarriers.

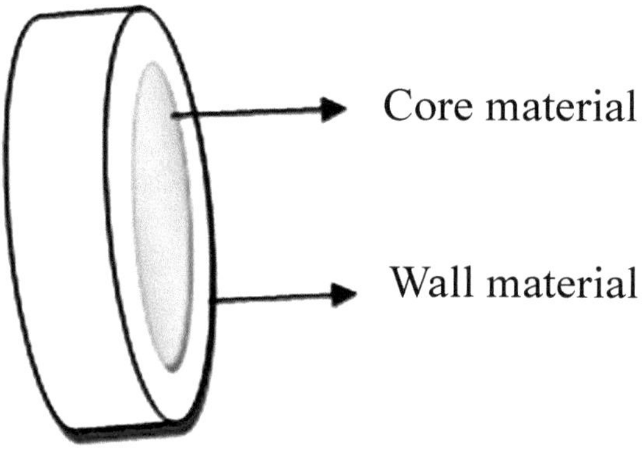

FIGURE 7.4 Schematic representation of nanocarriers/nanocapsule.

Recently, controlled-released technology has made a novel food delivery system; it has the property of releasing minimum or a reduced amount of encapsulated active compounds or formulations (Sekhon, 2010). The objective of encapsulation is to allow molecules to move along the membrane while at the same time protecting the contents from the environment. Some of the natural examples are eggshells of birds, skin, and seashells. The techniques like top-down and bottom-up are used for the synthesis of nanoparticles and nanocapsules. Figure 7.5 shows the different methods for the production of nanocarriers (Abd et al., 2020).

There are varieties of foods encapsulated with flavoring agents, acids, colorants, base, preservatives, antioxidants, and artificial sweeteners. In addition, there exists a variety of methods to releasing of the ingredients from the nanocarriers/capsules. The release of encapsulated particles can be site-specific, stage-specific, or signaled by changes in pH, temperature, and osmotic shock. The most common technique used in the food industry is by solvent-activated release. There is the development of new markets as well as current researches that are being undertaken in order to reduce the production costs, which are usually high as well as the lack of food-grade materials. The carrier wall or the capsule can be made up of sugars, proteins, synthetic polymers, and lipids. The pros of encapsulation can include an improvement of flow properties and handling which is made easier since now it is in the form of a solid rather than liquid. There can be the improvement of stability of the material that is encapsulated due to the protection from moisture or heat (Gibbs et al., 1999).

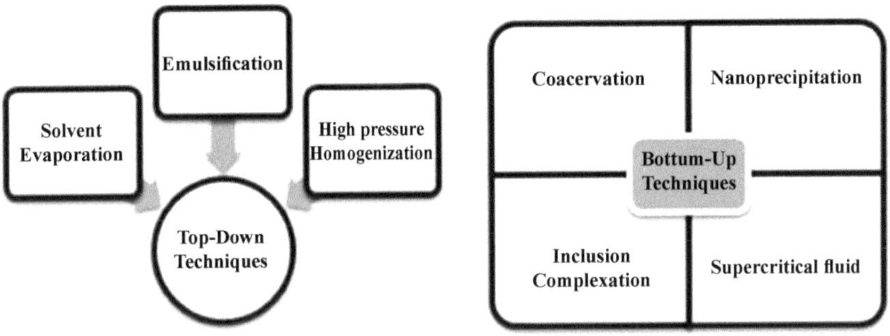

FIGURE 7.5 Top-down and bottom-up techniques for the synthesis of nanocapsules/carriers.

The main reasons for the usage of encapsulation are as follows:

1. Protection of the product from the surrounding conditions (temperature, moisture, etc.).
2. Protection against the deterioration of the ingredient that is active and limitation of the evaporation (losses) of material which is volatile.
3. Sustainable protection of the environment from hazardous and toxic materials to be safe during the handling procedures.
4. By converting liquids and sticky solids to powders (dry handling).
5. Undesired characteristics of the active components are being masked which include taste, odor, and the like.

Food products that have been cultivated, produced, packaged by using nanoscience and nanotechnological tools are so called nanofoods. Nanostructures are naturally occurring in food and these nanofoods are having variety of properties like longer shelf life, health-promoting additives, addition of flavor and aroma. There are different types of functional nanostructures like nanoliposomes, nanoparticles, nanoemulsions, and nanofibers (Press, 2010). Aroma compounds are active agents, which are widely used, as antimicrobial, anti-oxidant, insect repellent agents, etc. These are aiming on the application of extending shelf life of foods and biological products. The volatile chemical compounds present in aromas can loss at high temperature (Wicochea-Rodríguez et al., 2019). Flavor scalping can be defined as the interaction of compounds like aroma and flavor with plastic packaging may leads to the loss of flavor and aroma. An analysis done on the flavor compounds and fatty acids showed a confirmation that a good retention can be taken from these materials during the optimized process conditions.

The materials used for flavors and aromas in nanoencapsulation are polymers, lipids, polysaccharides, and proteins. Encapsulation of hydrophobic and hydrophilic active agents, like polysaccharides and proteins, are needed whereas polymers with hydrophobic characters or lipids should be made use for the encapsulation of compounds that are hydrophilic. It is quite necessary to bring to notice that the materials encapsulated should be made inclusive in the positive list marked by the current legislation for substances, which may be utilized in the plastic material manufacturing those of which are intended to come into contact with food. There has been a huge array of materials to encapsulate flavors that have been proposed, it includes proteins, lipids, polysaccharides, and other polymers. Materials that are of different types be made to use as building blocks in order to create the nanostructured carriers. There are both inorganic as well as organic substances, which are of nanomaterials based utilized in food applications. The engineered nanomaterials (ENMs) are of three main categories: inorganic, surface-functionalized materials, and organic-engineered nanomaterials (Kumar et al., 2019). The different types of carrier materials and their applications are given in Table 7.2.

There might be a requirement for using of different techniques of encapsulation since there exists functional ingredients of different types. These include drying techniques and wall materials to meet the specific physicochemical and molecular requirements, as well as its desirability. This can solve the micronutrients deficiencies that take place worldwide. The transformation of techniques that are of lab scale in nature into industrial scale poses a big challenge and need to further explore and overcome in future studies (Chew et al., 2019).

TABLE 7.2 Types of Nanocarriers and Their Applications.

	Carrier material	Applications	References
Lipids-based nanosystems	Nanoliposomes	• Used in areas like encapsulation and controlled release of food materials • It has higher chemical stability • It carries nutrients, enzymes, antimicrobial agents and food additives	Mozafari et al. (2008)
	Nanochelates	• They are nano coiled particles which is wrapped around micronutrients • It acts as a stabilizer • It helps to increase the nutritional value of food	Thangavel and Thiruvengadam (2014) Paredes et al. (2016)
	Nanoemulsions	• They are droplets with a diameter of less than 100–500 nm • It is an good carrier for poorly water soluble ingredients	Sen Gupta and Ghosh (2015)
Polymeric type nanoparticles	Nanocomposites	• They are nanoparticles incorporated with plastics • They are thermoplastic polymers • Natural biopolymer-based nanocomposites are widely used in food industry for food packaging	Alonso et al. (2010)
	Nanofibers	• It has diameter of less than 100 nm • These are produced by electrospinning • These are synthetic polymers	Weiss et al. (2006)
Carbohydrate-based delivering systems	Starch	• It is biodegradable, biocompatible and digestible polymer. • Hydrophilic nature of starch limits in its application of encapsulation. • They are sensitive to acids	Fathi et al. (2014)
	Guar gum	• It is a hydrophilic polysaccharide • It is used as thickening, retro gradation retardant agent in food industry	Paredes et al. (2016) Fathi et al. (2014)
	Dextran	• It is a bacterial polysaccharide of glucan • It can be act as water soluble and insoluble by changing its substitution.	Sen Gupta and Ghosh (2015) Fathi et al. (2014).

7.6 HOW THE TASTE AND ODOUR IS MAINTAINED

Food flavor is a sensory attribute of food material. It is the combination which is chemical in nature and of taste and aroma. Aromas and flavors are the most salient attributes of food material. They may inherently be presented in the food or during processing may develop or can be added during processing according to the consumers' demands. Aromas and flavors give the initial impression of the material of food, which can change the consumers' intention for consumption for further use. This is due to the fact that it has a powerful relationship with the acceptability and quality of food materials, but it is hard to control (Jun-xia et al., 2011; Fuciños et al., 2017).

Many food products available are sensitive to oxygen and upon prolonged exposure can lead to deterioration in the quality of food like change in taste, odor, and food color or facilitate the microbial growth in them. Nanoscale ingredients are utilized by the food industry to better the color, texture, quality, and flavor of food. The nanoscale food ingredients utilized include TiO_2 and SiO_2. The nanoparticles TiO_2 and SiO_2 and amorphous silica are used as food additives. TiO_2 is used as a coloring agent in powdered sugar coating on doughnuts. Nanocapsulation technology has been used to mask the taste and odor of tuna fish oil in order to be utilized to enrich bread with healthy Omega-3 fatty acids. Particles of fish oil are packaged into a film coating that averts the fish oil from reaction with oxygen and the release of its smell. Upon reaching the stomach the nanocapsules break so as to receive the benefits of eating them without experiencing the odor.

Micro- and nano-sized encapsulated particles help in the facilitation as well as in the flavor compound distribution in food products and aid in the release of flavoring which is also prolonged. In processing of food operations, there are several factors (e.g., temperature, pH, and enzyme), which may bring about degradation in the flavor. Therefore, flavoring ingredients at different stages are added of processing depending upon the product types made. For instance, in some processes, it is in the last step of processing that the flavor is added to lower thermal loss during processing. Further, the encapsulated flavor and aroma compounds in suspension and paste forms are also used in foods differently, depending on the processing operation and the food products. Nanoencapsulation approaches provide a bigger loading efficiency and capacity, enhanced stability, as well as finer control on flavor release profile as compared to microencapsulation (Saifullah et al., 2019).

The creation of nanoparticles that fasten onto the taste receptors on the tongue, can help in controlling how individual flavors can be experienced. By making changes to the chemical bonds that held the molecules and atoms in unison, small building blocks can be arranged in methods that make salt saltier or sweets sweeter. Thousands of nanoparticles that contain a variety of flavors or color enhancers can be food incorporated, unless specifically triggered, they would stay dormant. A new concept of food source, called nutraceuticals, is developed to minister in the treatment of illnesses, like cancer, diabetes, and heart disease. Flavors and aromas are organic molecules having low molecular weight. They are comparatively volatile and very sensitive to heat, air, light, and moisture (Bakry et al., 2016).

In the processed food products, the degradation of aroma and flavor occurs during processing and storing of the product. To lower the level of degradation and for the preservation of the originality of aroma and flavor, volatile pre-encapsulated ingredients can be utilized in foods and beverage (Madene et al., 2006). Flavor stability in distinct types of foods has been of growing interest because of the quality and acceptability of foods and its relationship, but the controlling of it usually requires a lot of effort. Processes of manufacturing and storage, packaging materials and food ingredients often cause alterations in overall flavor by lowering aroma compound intensity or off-flavor components production. In carbohydrates, some are more stable those which are soluble in water and some are more stable in the coating which are lipid-based. The overall quality of the food is affected by many factors linked to the aroma; examples are physico-chemical properties, concentration, volatile aroma molecules, and their interactions with food components. To limit aroma loss or degradation during storage and processing, it is beneficial and effective for the volatile ingredients to be encapsulated prior, to be utilized in foods or beverages.

7.7 FOOD SAFETY

The innovative nanotechnology formulation in health care, such materials like nano-calcium and nano-iron as supplements, which are more easily absorbed than the standard forms of helpful additives such as probiotics or vitamins, which needs to be encapsulated to withstand the processing, or be made very small to stop them from coming out of the solution in drinks. Innovation in food safety can be brought out by nanotech by aiding to develop new ingredients that are antimicrobial or nanocoatings that could be utilized in food preparation to coat equipment to avert microbial growth. The food industry would be benefitted largely from nanotechnology and it can be assumed that it will grow faster with time. Nanotechnology developing fast and has had strong impacts on food systems like on production of food, cultivation, packaging, their processing, transportation, nutrients that are biodiverse, and the like. There is a great impact on the utilization of nanomaterials commercially on the food industry (Chaturvedi and Dave, 2019).

Besides the above mentioned, there has been a lot of merits of making use of nanotechnology in the food industry, as well as regarding the safety issues connected with the nanomaterials that cannot be ignored. The safety concerns connected with nanomaterials was discussed by researchers gave greater emphasis on the possibility of using nanoparticles from the material packaging into the food and their impact on consumer's health (Bradley et al., 2011; Jain et al., 2018).

Although a material is being regarded as GRAS (generally regarded as safe) substance, further studies must be conducted and carried out to study the dangerousness of it, because the physiochemical properties in micro states are completely different from that that are nano state. Moreover, the nanomaterials due to their small size may add the risk for bioaccumulation within the body organs and tissues. For example, silica nanoparticles which are utilized as anti-caking agents could be cytotoxic in lung cells of humans when subjected to its higher exposure (Athinarayanan et al., 2014). There are many elements which affect dissolution inclusive of surface energy, surface morphology, adsorption, concentration, and aggregation. A prototype has been developed in order to examine the migration of particles from food packaging of food; a model was developed by Cushen et al. (2014).

The migration of copper and silver was studied from nanocomposites and it was seen that the percentage of nanofillers in the nanocomposites was one of the main essential parameters driving migration, more so than particle contact time, temperature, or size. Since every nanomaterial has its individual attribute, hence, toxicity would mainly be initiated on a basis that is case-by-case. In order to be sure of quality product, safety and health, and ecological directives; authorities must develop some standards for commercial products. Environmentalists are frightful that nanotechnology may produce contaminants, because of their nano-size, may prove to be ultra-hazardous. The reduced size of nanomaterials can extend the chance for bioaccumulation inside the body organs and tissues (Wallace Hayes and Sahu, 2017). Standard procedural tests are needed on living cells to understand the nanoparticles side effects and also the evaluation of potential hazards to health. In addition, a nanoparticles testing is mandatory before their release in the market for consumption by humans. It is popularly believed that in the near future, the food products derived from nanotechnology will be increasingly available for consumption worldwide.

The nanoparticles sometimes cause degradation of the protein because it also gets coated with proteins, resulting in the normal cellular mechanism to be disrupted. Silver nanoparticles have been identified on the human system to have adverse effects. The human lung fibroblast is affected by reducing ATP content, increase in the ROS production, and damaging mitochondria and DNA. It also leads in the aberration of chromosomes. Some of the nanoparticles are cytotoxic, carcinogenic, and genotoxic. The diminished size of the nanomaterials helps to crossover the cellular barriers and its exposure leads to free radicals formation in tissues which further leads to the oxidative destruction in the cells and tissues (Pradhan et al., 2015; Maness et al., 1999).

The technological and science advances made in developed countries for retail packaging that is nano-enabled should be adopted in developing countries as well (Bradley et al., 2011). Due to the combination of nanocomposites, biodegradable nanocomposites, nanosensors, for leakage proof, gases free, and pathogen less food packaging, packaging is revolutionized. They act as a barrier for exchange of gases and for the maintenance in the food quality using nanoclays, the creation of bacterial and fungal organisms or any kind of toxins and pathogens, which are eliminated using antimicrobial packaging that is antimicrobial, which uses silver, titanium oxide, zinc oxide, and other bionanoparticles. Biodegradable nanocomposites packaging to the environment is of considerable potential (Chellaram et al., 2014). The largest share formed in the present and short-term predicted market is of food packaging applications for nano-enabled products in the food sector. Market uptake and regulation are impeded by uncertainties in the safety of consumers and environment (Kalita and Baruah, 2019).

7.8 CONCLUSION

Nanotechnology and its applications provide a huge array of merits in the fields related to food safety and quality. From the production of food to its consumption, nanotechnology has been able to positively influence every stage of progression such as production enhancement, longer shelf life, nutrients retention, controlling quality, and smart packaging. Smart packaging helps the consumers to select the fair products, which have a better shelf life as well as indicating other features of the food and its nature. This technology also paves the way to food security and safety.

Nanoencapsulation technology bought a new era of food packaging techniques. This technology in the field of food industry and food processing bought effective alternate methods to reduce the wastage of activated core materials and to maintain odor and taste. Encapsulation technology promises advantage for the effective protection of functional materials and their proper delivery system. The strategy of this technology results in the protection of odor and flavors and also extended shelf life and quality. By experimenting with different materials to develop nanocarrier will result in new innovative nanocarriers with unique features.

There are ongoing extensive research and exploring in identifying cost-effective nanomaterials with similar properties. Human system can easily taken up nanoparticles and at the same time create toxic effects. The intensified toxicity form due to their higher bioavailability and it can further affect the immune system. Silver nanoparticles, for example, can make the cells resistant to any other antibacterial, however, nanoparticles that function within the living system is unknown. Many other nanoparticles, such as TiO_2 and ZnO showing higher toxic level, which show the impact on the environment. Eco-friendly nanoparticles require to be planned which can serve both as an antibacterial and also not to cause harmful effects to the environment.

Given the promise for encapsulation in agri-food applications of polymers, which are nature-derived, it is foreseeable that its utilization will with no doubt increase. The safe nature and biodegradable of such materials in addition, puts them in a better position for food ingestion and ecological (agriculture) utilization. The usage of these materials would, therefore, be an approach that is sustainable to agri-food purposes. With intelligent innovations, it will positively affect the food quality, safety, and security of food to meet the consumer demands. However, more research studies are required to be conducted to examine the migration behavior of nanoparticles from the materials that are packaged to food and their potential health implications.

However, nanotechnology and its usage may present potential risks not only to the health of humans, but can also affect the environment and animals. Toxicological effects on the biological systems can be affected, recent studies have shown due to which there is inhibition on the use of nanoparticles. FCMs are already in the market in some countries; therefore, data about the ENMs and safety of nanoproducts affecting human health are essential to ensure proper control and their practical administration for FCMs. There is a lack of knowledge, in comparison to its benefits and effects and potential use of ENMs, which are very well described, and about the potential ecotoxicological effects of nanoparticles. Nanotechnology has already bought significant changes in our society and this may lead to explore nanomaterials in all sectors. At the end, the materials that we use in the agricultural sector and food industry should not leave negative impact on the human body. For this, we have to do more research and experimentation. This chapter ends with an outlook on nano innovation in food packaging, nanoencapsulation technology, and concluding state on future innovations.

KEYWORDS

- **nanoencapsulation**
- **flavors**
- **aromas**
- **shelf life**
- **edible coating**
- **carrier materials**
- **food technology**

REFERENCES

Abd, A. E.; Kader, E.; Hashish, H. M. A. Encapsulation Techniques of Food Bioproduct. *Egypt. J. Chem.* **2020,** *63* (5), 1881–1909.

Alonso, D.; Gimeno, M.; Sepúlveda-Sánchez, J. D.; Shirai, K. Chitosan-Based Microcapsules Containing Grapefruit Seed Extract Grafted onto Cellulose Fibers by a Non-Toxic Procedure. *Carbohydr. Res.* **2010,** *345* (6), 854–859.

Amiri, S.; Akhavan, H. R.; Zare, N.; Radi, M. Effect of Gelatin-Based Edible Coatings Incorporated with Aloe Vera and Green Tea Extracts on the Shelf-Life of Fresh-Cut Apple. *Italian J. Food Sci.* **2018,** *30* (1), 61–74.

Araújo, J. M. S.; de Siqueira, A. C. P.; Blank, A. F.; Narain, N.; de Aquino Santana, L. C. L. A. Cassava Starch–Chitosan Edible Coating Enriched with Lippia sidoides Cham. Essential Oil and Pomegranate Peel Extract for Preservation of Italian Tomatoes (Lycopersicon esculentum Mill.) Stored at Room Temperature. *Food Bioprocess Technol.* **2018,** *11* (9), 1750–1760.

Athinarayanan, J.; Periasamy, V. S.; Alsaif, M. A.; Al-Warthan, A. A.; Alshatwi, A. A. Presence of Nanosilica (E551) in Commercial Food Products: TNF-Mediated Oxidative Stress and Altered Cell Cycle Progression in Human Lung Fibroblast Cells. *Cell Biol. Toxicol.* **2014,** *30* (2), 89–100.

Bahrami, A.; Delshadi, R.; Jafari, S. M.; Williams, L. Nanoencapsulated Nisin: An Engineered Natural Antimicrobial System for the Food Industry. *Trends Food Sci. Technol.* **2019,** *94*, 20–31.

Bajpai, V. K.; Kamle, M.; Shukla, S.; Mahato, D. K.; Chandra, P.; Hwang, S. K.; Kumar, P.; Huh, Y. S.; Han, Y. K. Prospects of Using Nanotechnology for Food Preservation, Safety, and Security. *J. Food Drug Analys.* **2018,** *26* (4), 1201–1214.

Bakry, A. M.; Abbas, S.; Ali, B.; Majeed, H.; Abouelwafa, M. Y.; Mousa, A.; Liang, L. Microencapsulation of Oils: A Comprehensive Review of Benefits, Techniques, and Applications. *Compreh. Rev. Food Sci. Food Safet.* **2016,** *15* (1), 143–182.

Brabin, B. J.; Hakimi, M.; Pelletier, D. Imported from https://academic.oup.com/jn/article/148/suppl_1/1001S/5033576. *J. Nutr.* **2001,** *131* (2), 604S–615S.

Bradley, E. L.; Castle, L.; Chaudhry, Q. Applications of Nanomaterials in Food Packaging with a Consideration of Opportunities for Developing Countries, *Trends Food Sci. Technol.* **2011,** *22* (11), 604–610.

Chaturvedi, S.; Dave, P. N. Application of Nanotechnology in Foods and Beverages. *Nanoeng. Bev. Indust. Sci. Bev.* **2019,** *20*, 137–162.

Chellaram, C.; Murugaboopathi, G.; John, A. A.; Sivakumar, R.; Ganesan, S.; Krithika, S.; Priya, G. Significance of Nanotechnology in Food Industry. *APCBEE Procedia*, **2014,** *8*, 109–113.

Chew, S-K.; Tan, C-H.; Pui, L-P.; Chong, P-N.; Gunasekaran, B.; LiN, N. K. Encapsulation Technologies: A Tool for Functional Foods Development. **2019,** *8*, 5S.

Cushen, M.; Kerry, J.; Morris, M.; Cruz-Romero, M.; Cummins, E. Evaluation and Simulation of Silver and Copper Nanoparticle Migration from Polyethylene Nanocomposites to Food and an Associated Exposure Assessment. *J. Agric. Food Chem.* **2014,** *62* (6), 1403–1411.

Das, S.; Chaudhari, A. K.; Dwivedy, A. K.; Upadhyay, N.; Singh, V. K.; Singh, A.; Dubey, N. K. Nanoencapsulation Technology: Boon to Food Packaging Industries, 2020; pp 17–40.

Delshadi, R.; Bahrami, A.; Tafti, A. G.; Barba, F. J., Williams, L. L. Micro and Nano-Encapsulation of Vegetable and Essential Oils to Develop Functional Food Products with Improved Nutritional Profiles. *Trends Food Sci. Technol.* **2020,** *104*, 72–83.

Duncan, T. V. Applications of Nanotechnology in Food Packaging and Food Safety: Barrier Materials, Antimicrobials and Sensors. *J. Colloid Interf. Sci.* **2011,** *363* (1), 1–24.

Fathi, M.; Martín, Á.; McClements, D. J. Nanoencapsulation of Food Ingredients Using Carbohydrate Based Delivery Systems. *Trends Food Sci. Technol.* **2014,** *39* (1), 18–39.

Fernanda, A. R.; Oliveira, J. C. Processing Foods: Quality Optimization and Process Assessment, 2019.

Fuciños, C.; Míguez, M.; Fuciños, P.; Pastrana, L. M.; Rúa, M. L.; Vicente, A. A. Creating Functional Nanostructures: Encapsulation of Caffeine into α-Lactalbumin Nanotubes. *Innov. Food Sci. Emerg. Technol.* **2017,** *40*, 10–17.

Gibbs, B. F.; Kermasha, S.; Alli, I.; Mulligan, C. N. Encapsulation in the Food Industry: A Review. *Int. J. Food Sci. Nutr.* **1999,** *50* (3), 213–224.

Guldiken, B.; Linke, A.; Capanoglu, E.; Boyacioglu, D.; Kohlus, R.; Weiss, J.; Gibis, M. Formation and Characterization of Spray Dried Coated and Uncoated Liposomes with Encapsulated Black Carrot Extract. *J. Food Eng.* **2019,** *246*, 42–50.

Hasan, S. M. K.; Ferrentino, G.; Scampicchio, M. Nanoemulsion as Advanced Edible Coatings to Preserve the Quality of Fresh-Cut Fruits and Vegetables: A Review. *Int. J. Food Sci. Technol.* **2020,** *55* (1), 1–10.

He, X.; Hwang, H. M. Nanotechnology in Food Science: Functionality, Applicability, and Safety Assessment. *J. Food Drug Analys.* **2016,** *24* (4), 671–681.

Hemmati, S.; Yousefi, M.; Salehi, M. H.; Amiri, M.; Hekmati, M. Palladium Nanoparticles Immobilized over Strawberry Fruit Extract Coated Fe_3O_4 NPs: A Magnetic Reusable Nanocatalyst for Suzuki-Miyaura Coupling Reactions. *Appl. Organometal. Chem.* **2020**.

Jafarizadeh-Malmiri, H.; Sayyar, Z.; Anarjan, N.; Berenjian, A. Nanobiotechnology in Food: Concepts, Applications and Perspectives. In *Nanobiotechnology in Food: Concepts, Applications and Perspectives*, 2019.

Jain, A.; Ranjan, S.; Dasgupta, N.; Ramalingam, C. Nanomaterials in Food and Agriculture: An Overview on Their Safety Concerns and Regulatory Issues. *Crit. Rev. Food Sci. Nutr.* **2018,** *58* (2), 297–317.

Jideani, A. I. O.; Anyasi, T. A.; Mchau, G. R. A.; Udoro, E. O.; Onipe, O. O. Processing and Preservation of Fresh-Cut Fruit and Vegetable Products. *Postharvest Handling*, 2017.

Jun-xia, X.; Hai-yan, Y.; Jian, Y. Microencapsulation of Sweet Orange Oil by Complex Coacervation with Soybean Protein Isolate/Gum Arabic. *Food Chem.* **2011,** *125* (4), 1267–1272.

K. S., J.; Jose, J.; Li, T.; Thomas, M.; Shankregowda, A. M.; Sreekumaran, S.; Kalarikkal, N.; Thomas, S. Application of Novel Zinc Oxide Reinforced Xanthan Gum Hybrid System for Edible Coatings. *Int. J. Biol. Macromol.* **2020,** *151*, 806–813.

Kalita, D.; Baruah, S. The Impact of Nanotechnology on Food. In Nanomaterials Applications for Environmental Matrices: Water, Soil and Air, 2019.

Knorr, D.; Watzke, H. Food Processing at a Crossroad. *Front. Nutr.* **2019,** *6*, 85.

Kumar, A.; Gupta, K.; Dixit, S.; Mishra, K.; Srivastava, S. A Review on Positive and Negative Impacts of Nanotechnology in Agriculture. *Int. J. Environ. Sci. Technol.* **2019,** *16* (4), 2175–2184.

Lamba, A.; Garg, V. Nanotechnology Approach in Food Science: A review. *Int. J. Food Sci. Nutr.* **2018,** 183–186.

Madene, A.; Jacquot, M.; Scher, J.; Desobry, S. Flavour Encapsulation and Controlled Release—a Review. *Int. J. Food Sci. Technol.* **2006,** *41* (1), 1–21.

Maness, P. C.; Smolinski, S.; Blake, D. M.; Huang, Z.; Wolfrum, E. J.; Jacoby, W. A. Bactericidal Activity of Photocatalytic TiO_2 Reaction: Toward an Understanding of Its Killing Mechanism. *Appl. Environ. Microbiol.* **1999,** *65* (9), 4094–4098.

Monteiro, C. A. Nutrition and Health. The Issue Is Not Food, Nor Nutrients, So Much as Processing. *Public Health Nutr.* **2009,** *12* (5), 729–731.

Monteiro, C. A.; Levy, R. B.; Claro, R. M.; de Castro, I. R. R.; Cannon, G. Uma nova classifi cação de alimentos baseada na extensão e propósito do seu processamento. *Cadernos de Saude Publica* **2010,** *26* (11), 2039–2049.

Mozafari, M. R.; Johnson, C.; Hatziantoniou, S.; Demetzos, C. Nanoliposomes and Their Applications in Food Nanotechnology. *J. Liposome Res.* **2008,** *18* (4), 309–327.

Nile, S. H.; Baskar, V.; Selvaraj, D.; Nile, A.; Xiao, J.; Kai, G. Nanotechnologies in Food Science: Applications, Recent Trends, and Future Perspectives. *Nano-Micro Lett.* **2020,** *12* (1), 1–34.

Paredes, A. J.; Asencio, C. M.; Manuel, L. J.; Allemandi, D. A.; Daniel Palma, S. Nanoencapsulation in the Food Industry: Manufacture, Applications and Characterization. **2016,** *1* (1).

Pradhan, N.; Singh, S.; Ojha, N.; Shrivastava, A.; Barla, A,; Rai, V.; Bose, S. Facets of Nanotechnology, Food Processing, Packaging, and Preservation Industry, 2015.

Prasad, R.; Bhattacharyya, A.; Nguyen, Q. D. Nanotechnology in Sustainable Agriculture: Recent Developments, Challenges, and Perspectives. *Front. Microbiol.* **2017,** *8* (6).

Press, D. NSA_8677_food-nanotechnology--an-overview; 2010.

Ravichandran, R. Nanotechnology Applications in Food and Food Processing: Innovative Green Approaches, Opportunities and Uncertainties for Global Market. *Int. J. Green Nanotechnol. Phys. Chem.* **2010,** *1* (2), 72–96.

Saifullah, M.; Shishir, M. R. I.; Ferdowsi, R.; Tanver Rahman, M. R.; Van Vuong, Q. Micro and Nano Encapsulation, Retention and Controlled Release of Flavor and Aroma Compounds: A Critical Review. *Trends Food Sci. Technol.* **2019,** *86*, 230–251

Sampathkumar, K.; Tan, K. X.; Loo, S. C. J. Developing Nano-Delivery Systems for Agriculture and Food Applications with Nature-Derived Polymers. *iScience* **2020,** *23* (5), 101055.

Sani, T. H.; Hadjmohammadi, M.; Fatemi, M. H. Extraction and Determination of Flavonoids in Fruit Juices and Vegetables Using Fe_3O_4/SiO_2 Magnetic Nanoparticles Modified with Mixed Hemi/Ad-Micelle Cetyltrimethylammonium Bromide and High Performance Liquid Chromatography. *J. Separation Sci.* **2020,** *43* (7), 1224–1231.

Sekhon, B. S. Food Nanotechnology–an Overview. *Nanotechnol. Sci. App.* **2010,** *3* (1), 1–15.

Sen Gupta, S.; Ghosh, M. Formulation Development and Process Parameter Optimization of Lipid Nanoemulsions Using an Alginate-Protein Stabilizer. *J. Food Sci. Technol.* **2015,** *52* (5), 2544–2557.

Sharma, C.; Dhiman, R.; Rokana, N.; Panwar, H. Nanotechnology: An Untapped Resource for Food Packaging. *Front. Microbiol.* **2017,** *8* (9), 1735.
Singh, T.; Shukla, S.; Kumar, P.; Wahla, V.; Bajpai, V. K. Application of Nanotechnology in Food Science: Perception and Overview. *Front. Microbiol.* **2017,** *8* (8).
Thangavel, G.; Thiruvengadam, S. Nanotechnology in Food Industry—a Review. *Int. J. ChemTech Res.* **2014,** *6* (9), 4096–4101.
Tsang, Y. F.; Kumar, V.; Samadar, P.; Yang, Y.; Lee, J.; Ok, Y. S.; Song, H.; Kim, K. H.; Kwon, E. E.; Jeon, Y. J. Production of Bioplastic through Food Waste Valorization. *Environ. Int.* **2019,** *127*, 625–644.
Wallace Hayes, A.; Sahu, S. C. Nanotechnology in the Food Industry: A Short Review—Food Safety Magazine. 2017, March.
Wang, Y.; Feng, Y.; Zhang, M.; Huang, C.; Yao, J. A Green Strategy for Preparing Durable Underwater Superoleophobic Calcium Alginate Hydrogel Coated-Meshes for Oil/Water Separation. *Int. J. Biol. Macromol.* **2019,** *136*, 13–19.
Weiss, J.; Takhistov, P.; McClements, D. J. Functional Materials in Food Nanotechnology, *J. Food Sci.* **2006,** *71* (9), R107–R116.
Wicochea-Rodríguez, J. D.; Chalier, P.; Ruiz, T.; Gastaldi, E. Active Food Packaging Based on Biopolymers and Aroma Compounds: How to Design and Control the Release. *Front. Chem.* **2019,** *7* (JUN), 398.

CHAPTER 8

Edible Nanocoatings and Films for Preservation of Food Matrices

SHIJI MATHEW and E. K. RADHAKRISHNAN*

School of Biosciences, Mahatma Gandhi University, Kottayam 686560, Kerala, India

Corresponding author. E-mail: radhakrishnanek@mgu.ac.in

ABSTRACT

Edible packagings in the form of coatings and thin films are one among the leading interesting and attractive primary packaging approaches for optimization of food quality. Edible coatings/thin films are prepared from renewable natural biomaterials, such as polysaccharides, lipids, and proteins which can be applied directly on food products to improve their quality and shelf life. Edible packaging possesses the unique advantages of being edible along with the packed food, biodegradable, eco-friendly, and washable. Advanced research has shown that the incorporation of nanomaterials in edible films can make this venture more promising and efficient. Nanomaterials or nanocomposite-based edible films/coatings can offer better encapsulation of bioactive agents and confer controlled release of antioxidants, antimicrobials, nutraceuticals, and flavoring agents. Besides, edible nanocoatings/films can also lead to improvement of food functional aspects like sensory attributes, maintain natural appearance, as well as provide protection from microbial spoilage thereby preserving the food freshness. This chapter discusses about the general aspects of edible coating and edible nanocoating, various types of nano-based edible coating biomaterials used, methods of applying edible nanocoatings on fresh and processed foods, and the recent developments and successful applications in this area.

8.1 INTRODUCTION

The food from the producer has to be protected from various environmental factors until it reaches the consumer. Hence, food-packaging materials are considered to have a vital role in the food chain supplies. Edible packagings are thin layer of primary packaging composed of biopolymeric material that is used to coat/wrap the food directly to extend its shelf life without changing the original architecture of the food (Dehghani et al., 2018). These coatings form an integral part of the food such that it can be consumed along with the food. Edible packagings generally can be applied by two different ways: in the liquid form such that it forms a coat around the food or as thin solid film laminates which can be used to wrap the food (Falguera et al., 2011).

Edible coatings have been used since the 12th and 13th centuries in China in the form of waxes which were applied on lemons and oranges for increasing its shelf life (Zeuthen and Bøgh-Sørensen, 2003). Afterwards, edible coatings have received great attraction in various countries owing to their advantages that they can be consumed along with the food, leaving behind no waste to be discarded and also thereby lessening the environmental impact. Moreover, being simple process in application, this technology greatly influenced the consumer demands and new market requirements. With the incorporation of active agents (antimicrobials, antioxidants, and natural extracts), edible coatings can also be converted to active edible coatings which besides extending the product's shelf life, can also release the active agents in a controlled manner to improve the physical, chemical, sensory, and organoleptic properties of the food (Santos and Melo, 2020).

As we all are aware, in recent times, the study and application of nanotechnology in various fields, primarily in the food sector has increased with promising results. Due to their unique characteristics, nanomaterials when added to edible coatings can have a great impact, resulting in an upgrading of the functionalities and applications of edible coatings (González-Reza et al., 2018). Numerous studies have shown that the application of a bionanocomposite or nanomaterial-based edible coating/film over highly perishable fresh produces and processed foods can prevent their rapid spoilage, maintain freshness without affecting their original architecture, and also provides an extended shelf life.

This chapter provides an in-depth knowledge on the various advancements and practical aspects of nanomaterials-based edible packaging systems. The first section of this chapter mainly focuses on the important functions attributed by nano-based edible coatings/films as an ideal food packaging. This is followed by a detailed portion on the main strategies used for the preparation and application of edible nanocoatings or films on food matrices. Then, the different biomaterials used for the fabrication of nanostructure-based edible packaging materials are briefed with suitable examples. The later section covers the practical role and effect of edible nanocoatings/films in packaging of varied fresh and processed food products which are explained with appropriate examples.

8.2 MULTIFUNCTIONAL EDIBLE NANOCOATINGS

Edible nanocoats/films offer many advantages which make them attractive and convenient compared with synthetic packaging materials. Edible nano-based packagings are considered as a one-step solution which offers all the essential functions required for an ideal packaging. The multiple functions performed by nanostructure-based edible coatings are described below.

Edible: As the name indicates, the main advantage and attractive feature of using edible nano-based package is that it forms an integral part of the food and can be consumed along with the packaged product (Janjarasskul and Krochta, 2010) as these packagings are mainly composed of bio-based renewable and edible ingredients. **Protection:** Edible nanocoatings/films can give physical protection to foods and can also prevent the draining of liquid from it. They can also extend the food's shelf life by providing effective barrier to gases, light, and water. Moreover, these coatings can reduce the loss of moisture, aromas, and solutes from the food and allows selective and controlled exchange of gases involved in food respiration, such as O_2, CO_2, ethylene, etc. (Huber and Embuscado, 2009). **Eco-friendly:** Since these edible packages do not create any waste material of their own, they also play a major role in reducing the serious environmental impact of white pollution. **Preserves organoleptic properties:** Nanocomposites or nanostructured edible coating matrices act as a good medium for adding various additives and functional groups which can play a vital role in maintaining the original texture, taste, and odor of the packed foods (Bharti et al., 2020). **Antimicrobial performance:** The incorporation of biocidal nanomaterials in the coatings can impart antimicrobial ability and hence can function in greatly reducing microbial contamination of foods. **Antioxidant property:** With the inclusion of antioxidants, the edible coatings can have antioxidant ability which can result in the preservation of fresh produces and prevent them from browning. **Role as carrier:** Edible nanocoatings can act as carrier of many bioactive natural agents, such as antimicrobials, antioxidants, flavoring agents, dyes, plant/animal extracts, or prebiotic and probiotic agents. Such encapsulated nanocoatings can cause controlled release of these agents which can further enhance the nutritional value and boost up the sensory attributes of food. **Washable:** Some nanocoatings are even washable, so that they form an intact protective covering over the food surface and can be easily removed by washing prior to consumption of the coated food. **Sensorial ability:** Edible coatings can also be incorporated with certain nanosensors which can rapidly detect any possible spoilage or chemical changes on the coated fruit and immediately notify it to the consumer. Figure 8.1 summarizes the multiple functions of nano-based edible coatings.

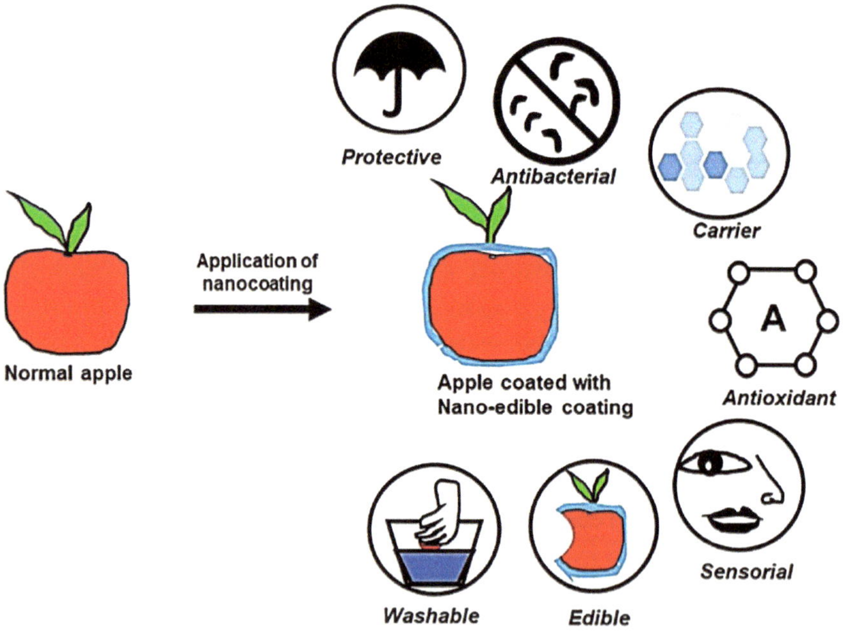

FIGURE 8.1 Multiple functions performed by edible nanocoatings on food materials

On the other side, certain limitations are also associated with the implementation of nanotechnological applications in foods. Regulation of the use and safety aspects of nano-based edible coatings is still a controversial issue. Currently, there are many uncertainties regarding their response to public health, impact on environment and occupational risks associated with their manufacturing (Zambrano-Zaragoza et al., 2018).

8.3 METHODS OF APPLYING EDIBLE NANOCOATINS/FILMS ON FOOD

Different methods are available for preparing nanolaminate food wraps and application of edible nanocoatings on the surface of food. The major methods for applying nanocoatings include dipping/drenching or immersing, coating (spray, spread, or spin rotation method), fluidized bed, and panning (Suhag et al., 2020; Yanyun Zhao, 2011). Similarly, the main methods that are followed for preparing nanolaminates or films include casting, extrusion, electrospinning, etc. In addition, multilayered nano-based edible coating can be prepared by dipping, layer-by-layer method, and co-extrusion technology. The selection of these methods depends upon the nature of the food to be coated, objective of coating process, and surface attributes of coating process, such as surface tension, density, and viscosity (Andrade et al., 2012). The major steps involved in these procedures are illustrated in Figure 8.2. Dipping, spraying, and solvent casting methods are discussed in detail in the following section.

8.3.1 DIPPING OR IMMERSING

Dipping/drenching or immersing method of nanocoating aids in the formation of uniform coatings on the food surface (Lu et al., 2010). The first step in this process involves the development of a pH adjusted homogenous nanocoating solution/dispersion. Next step is the preparation of the food sample to be coated which includes its cleaning, cutting, drying, and weighing. After this step, the food is dipped/immersed in the nanocoating solution/dispersion for a desired time (30 s–30 min). Then the coating deposited food is taken out, the excess solution is drained off and then the solvent is allowed to dry by evaporation. Finally, the coated food can be stored in plastic or any other containers. The wetting capacity of

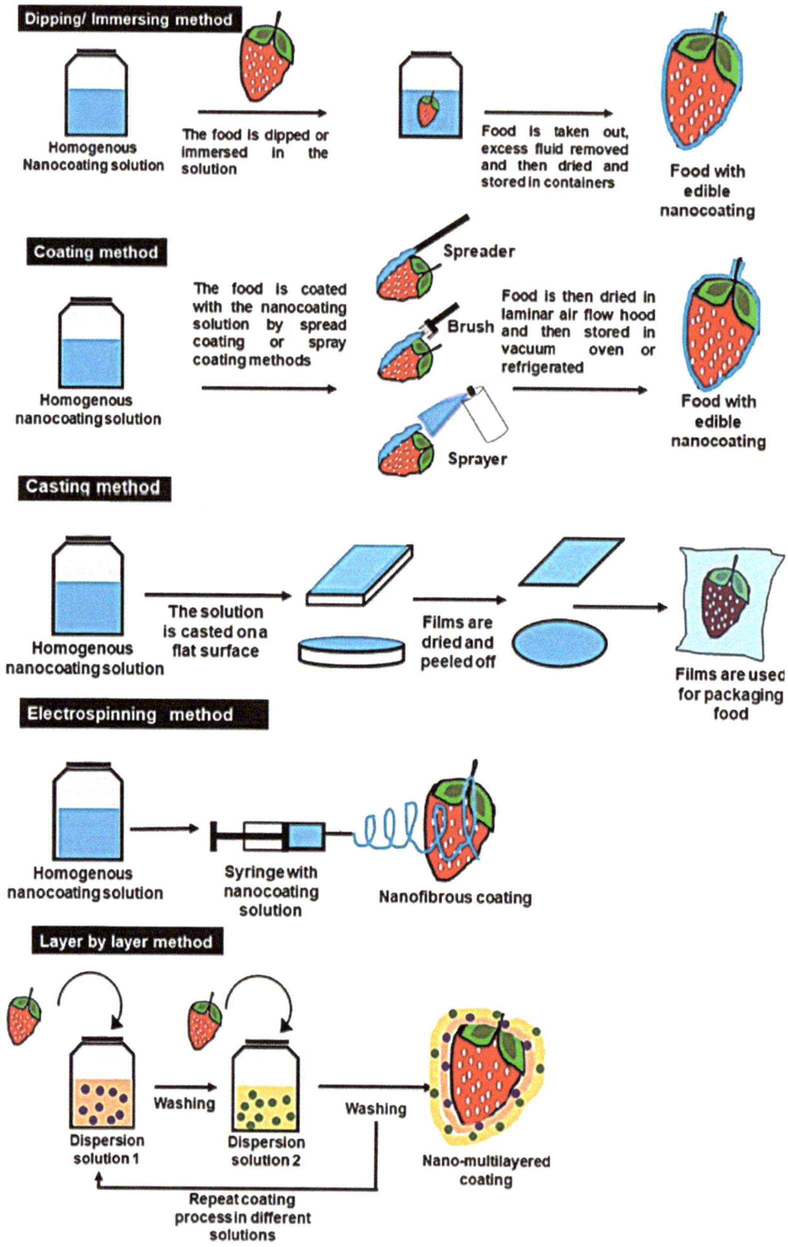

FIGURE 8.2 Different strategies used for application of edible nanocoatings on food surface. *Source:* Adapted and modified from Gheorghita (Puscaselu) et al. (2020), Zambrano-Zaragoza et al. (2020).

the food surface, processing, and draining time plays an important role in determining the success of this process. Immersion or dipping process is a simple procedure which provides better preservation of the coated foods, especially fresh produces. This method can also be employed to generate multiple layers of coating on the food, which may provide even better performance (Wang et al., 2018).

8.3.2 COATING

Coating is a method of applying the nano-based coats either directly on the surface of the food, such as fruits, meat, vegetables or indirectly on the surface of the packaging material. This can be done by three different ways: spread coating, spray coating, and spin coating.

Spread coating: Direct Spread coating on food surface can be done by the use of sterile tools, such as spreader, brush, or spatula. This type of coating is a potentially effective method in preserving the quality of the food and limiting the microbial growth. These direct spread coatings can also influence the gas permeability and thereby extend the shelf life of the coated products. Indirect spread coating of nanoformulations on packaging material can also offer better antibacterial property and can also be used as a method of fabricating multilayer sheets of coating on the packaging material (Guo et al., 2014).

Spray coating: Spraying can be done with a set of nozzles which forms droplets that can be dispensed on food surfaces. Spray coating can be attained with the use of tools, such as knapsack sprayer, compressed air-assisted sprayer, or copper backpack. Three types of spraying techniques for coating on foods include air spray atomization, air-assisted airless atomization, and pressure atomization (Gheorghita (Puscaselu) et al., 2020; Suhag et al., 2020).

8.3.3 CASTING

Casting or solvent casting is the most common method of preparing edible nanolaminates or films. This process involves three steps: (1) Solubilization process where the selected polymer and chosen nano-additives are dissolved or dispersed in a suitable solvent separately. (2) Casting process, in this process, the dissolved/dispersed polymer/solvent/additive mixture is poured onto a flat surface such as a petridish/Teflon plate. This is followed by the (3) drying process where the solvent gets removed by evaporation, leaving behind a thin film of the nanocomposite (Suhag et al., 2020). Finally, this film can be used to wrap the food.

8.4 MATRICES USED FOR EDIBLE NANOCOATINGS

Biopolymeric and easily available and renewable sources, such as polysaccharides, proteins, and lipids are generally employed for the fabrication of nanocomposite or nanostructured edible matrices. The following section discusses on the use of these important matrices for the development of a nano-based edible coating/film with certain examples.

8.4.1 POLYSACCHARIDES-BASED EDIBLE NANOCOATINGS

Being one of the most widely available biopolymers, polysaccharides of animal, plant, and marine origin are the extensively researched matrix for the development of nanomaterial-based edible packagings (Aguirre-Joya et al., 2018). The structure and properties of these polysaccharide favors the effective incorporation, as well as biological and molecular functioning of nanomaterials in this matrix (Zheng et al., 2015). The common polysaccharides exploited for the fabrication of nano-based edible coatings include pectin (Sucheta et al., 2019), starch (Escamilla-García et al., 2018), alginate (Emamifar and Bavaisi, 2020), chitosan (Ortiz-Duarte et al., 2019), gum xanthan (K. S. et al., 2020), and carrageenan. Definitely, starch-based edible nanocoatings/films are the most researched ones which are found to protect and act as efficient carriers of bioactive agents for controlled delivery.

Polysaccharide-based edible nanocoatings can be fabricated in two important ways: (1) exploitation of any polysaccharide matrix for the incorporation of various functional nanoparticles or (2) by nanostructuring of the polysaccharides itself, for example, use of cellulose nanocrystals. The nanomaterials often used as additives in polysaccharide matrix to provide functional characteristics include metallic nanoparticles, such as silver (Ortiz-Duarte et al., 2019), zinc oxide (Koushesh Saba and Amini,

2017), and titanium dioxide (Mozhgan Nasiri et al., 2019), solid lipid nanoparticles (Zambrano-Zaragoza et al., 2013), polymeric nanoparticles, such as nanochitosan (Mohammadi et al., 2016) or cellulose nanocrystals (Souza Vieira da Silva et al., 2019), essential oil nanoparticles (Zhang et al., 2019).

Many reports are available on the development and application of polysaccharide-based coatings/thin films for food packaging purpose. In a very recent work, a novel approach was reported where structured oil nanoparticles (SONs) fabricated from sunflower oil was used to improve the hydrophobicity of polysaccharide-based edible films (Ghiasi et al., 2020). Here, farsi gum (FG) was used as the polysaccharide matrix. Comparatively, the physicochemical, mechanical, and thermal properties of FG films with 0.5% SONs showed better performance over FG films with SONs with 0 and 1% (w/w) and sunflower oil-incorporated FG control films.

8.4.2 PROTEINS-BASED EDIBLE NANOCOATINGS

Recently, proteins derived from both animal and plant sources have been greatly used for the preparation of nano-based edible coatings/films as they offer hydrophilic surfaces providing improved water and gas barrier properties. The common protein matrices used for the preparation of nanomaterial-based edible food coatings include whey (Wang et al., 2020) and casein proteins (Bora and Mishra, 2016) from milk, zein from corn (Hager et al., 2019), gluten from wheat (Tanada-Palmu and Grosso, 2005), and gelatine from animal tissues (Zhang et al., 2017) and egg proteins.

Protein-based edible nanocoatings can be developed by three ways: (1) either by using an animal or plant-based protein matrix onto which various functional nanoparticles can be incorporated or (2) by using nanostructured form of proteins as an essential component of coatings (such as nanostructured zein, gelatin and bovine serum albumin), or (3) by the application of protein-based nanofibers such as electrospun whey nanofiber in edible coatings. Nanostructured zein is a highly attractive nanostructured protein candidate which is highly stable and an efficient vehicle for entrapping and controlled delivering of bioactive substances.

In a recent work, an interesting edible and washable poly (albumen)-based multifunctional bionanocomposite coating composed of egg white protein, glycerol, egg yolk, curcumin extract, and cellulose nanocrystals was reported by Jung and colleagues. For its practical application, perishable fruits, such as papaya, banana, strawberries, and avocado were coated by dipping method and the results showed that this nanocoating would preserve the cosmetic appearance of fruits as well as reduce microbial growth, dehydration, and respiration of fruits, thereby contributing to an extended fruit shelf life (Jung et al., 2020). Table 8.1 shows the list of protein-based edible nanocoatings and their practical applications in food preservation.

8.4.3 LIPIDS-BASED EDIBLE NANOCOATINGS

Many natural sources, such as plants, animals, and insects are rich in lipids. Recently, researchers have focused more on to the usage of lipids as edible coatings/thin films as a matrix for the inclusion of functional nanoparticles which in addition to act as efficient edible packaging, can also provide gloss, reduce moisture loss, and lessens the complexity and cost of packaging among others. The important lipids used for this purpose include oils, fats, waxes, essential oils, plasticisers, resins, and emulsifiers. In addition, nanostructured lipid matrices possess high encapsulation ability and impart controlled delivery of bioactive substances. Nanoemulsions of essential oils from various spices, nuts, and fruits are important examples of nanostructured lipid which forms an important ingredient in active nano-based edible coatings due to its potential antimicrobial and flavor enhancing capability. Table 8.2 summarizes various examples of edible coatings with nanoemulsions of essential oil and their potential applications.

8.5 DIRECT APPLICATION OF EDIBLE NANOCOATINGS ON FOODS

One of the most important applications of edible coatings/films comes in the case of perishable food stuffs, such as fresh produces like vegetables and fruits, meat and poultry, fish and marine foods, and eggs and dairy products. The section below details on the use of various nano-based edible coatings/thin films that have been applied for preservation and shelf life extension of both processed and fresh foods.

TABLE 8.1 List of Protein-Based Edible Nanocoatings and Their Practical Application.

Components of edible nanocoating	Preparation method	Functions performed	References
Hazelnut industry waste with nanoemulsions of clove essential oil	Ultrasonication	Provided better mechanical, antibacterial, and antioxidant properties	Gul et al. (2018)
Gluten films containing chitosan–gelatin nanofibers	Nozzle-less electrospinning	Improved mechanical properties	Ebrahimi et al. (2019)
Gamma-aminobutyric acid-rich edible films with soy fermented protein and chitosan	Casting method	Antimicrobial and antioxidant	Zareie et al. (2020)
Soy protein SiOx nanocomposite film	Casting method	Extended the shelf life of apples	Liu et al. (2017)
Whey protein concentrate-corn oil-TiO_2 nanoparticles	Casting method	Extended shelf life of cheese	Montes-de-Oca-Ávalos et al. (2020)
Chitosan–whey protein	Casting method	Chestnut preservation, antimicrobial properties	Huang et al. (2020)
Semolina protein with zinc oxide nanoparticles	Casting method	Improved mechanical and antibacterial properties	Jafarzadeh et al. (2017)

TABLE 8.2 List of Edible Food Coatings Composed of Nanoemulsions of Essential Oil for Food Packaging Application.

Edible coating matrix	Essential oil present	Preparation method	Functions performed	References
Whey protein films	Nanoemulsion of *Grammosciadium ptrocarpum* Bioss. essential oil	Casting	Lowered water permeability, improved mechanical, and antimicrobial performance	Ghadetaj et al. (2018)
Chitosan coatings	Nano-encapsulated *Paulownia tomentosa* essential oil	Casting	Improved shelf life of ready-to-cook pork chops	Aguilar-Sánchez et al. (2019)
Chitosan–gelatin coating	Nanoencapsulated tarragon essential oil	Ionic gelation	Preservation of pork slices	Zhang et al. (2020)
Pullulan films	Nanoemulsions of cinnamon essential oil	Ultrasonication	Antibacterial performance	Chu et al. (2020)
Whey protein isolate/chitosan	Nanoencapsulated garlic essential oil	Casting	Extended shelf life of vacuum-packed sausages	H. Esmaeili et al. (2020)
Sodium caseinate films reinforced with ZnO nanoparticles	Microencapsulated *Melissa officinalis* essential oil	Casting	High antioxidant and antibacterial properties	Sani et al. (2021)
Large mouth mass Fish sarcoplasmic protein–chitosan	Nanoemulsions of ginger essential oil	Casting	Extended shelf life of red sea bream fillets	Cai et al. (2020)
Chitosan films	Nanoemulsion of cumin essential oil	Casting	Improved quality of beef loins	Dini et al. (2020)
Banana starch films	Nanoemulsions of lemon grass and rosemary essential oils	Casting	Nanoemulsions increased the plasticity and decreased the water barrier properties of the films	Restrepo et al. (2018)

8.5.1 FRUITS AND VEGETABLES

Fruits and vegetables are one among the highly perishable food items wherein 40–50% of products are wasted every year. The factors which attribute to the rapid decay of fruits and vegetables during the postharvest management processes, such as handling, transportation, and storage include the presence of insects, microorganisms, loss of water, respiration, texture deterioration, transpiration, and senescence (Ali et al., 2010; Jung et al., 2020).

Many methods are adopted for preserving the freshness and to extend their shelf life of fruits and vegetables. Refrigeration is considered as the most common, traditional, and effective method to preserve fresh produces, but the process itself is not found to be sufficient enough. Other methods include the use of modified atmosphere packaging (MAP) and active coatings based on paraffin wax. Unfortunately, all these techniques possess the limitations of being expensive, time-consuming, and can also result in altering the texture and organoleptic properties of these produces. Another common method is waxing which involves coating fruit with preservatives containing weak acids and their derivatives, but is often associated with adverse effects in human body following its consumption. In this context, the use of bionanocomposite or nanostructure-based coatings is considered to be a promising alternative without causing any alteration to the physiological or physiochemical characteristics of the coated food. Numerous literatures have reported the preparation and application of nano-based edible coatings for the preservation of fruits and vegetables (Table 8.3). Few are discussed below.

In a recent study, different formulations based on chitosan nanoparticles and chitosan thyme essential oil (15%, 30%, and 45%) was used to develop natural edible nanocoatings by Correa-Pacheco et al., 2021. Then these formulations named CS15, CS30, CS45, TEO15, TEO30, and TEO45 were used to coat on fresh green bell peppers. Later, the quality and physiological parameters of coated and uncoated bell peppers inoculated and uninoculated with a common phytopathogen, *Pectobacterium carotovorum* was assessed for 12 days. Of all the formulations tested, preparation containing 15% chitosan nanoparticles (CS15) was considered as the highly efficient coating which showed lowest CO_2 production, reduced incidence of *P. carotovorum* infection. Figure 8.3 shows the severity of *P. carotovorum* infection on bell peppers. As indicated by this figure, no

TABLE 8.3 A Summary of Various Nano-Based Edible Coatings Used for Preservation of Fruits and Vegetables.

Components	Fruit/vegetable treated	Effect introduced	References
Carrageenan and ZnO nanoparticles	Mango	Reduced the total acidity, maintained firmness and delayed discoloration and decay and provided protection against microbes	Meindrawan et al. (2018)
Alginate-based limonene liposomes	Strawberry	The coated fruit showed lower respiration rates, pH and higher anthocyanin content	Dhital et al. (2018)
Beeswax solid lipid nanoparticles in xanthan gum and propylene glycol	Strawberry	The coated fruits showed less weight loss and decay and exhibited better firmness	Zambrano-Zaragoza et al. (2020)
Chitosan thyme essential oil nanocoating	Green bell pepper	Coated bell peppers showed lower CO_2 production, maintained firmness and weight loss and protected against bacteria *P. carotovorum*	Correa-Pacheco et al. (2021)
Chitosan and propolis nanocoatings	Fig fruit	Reduced weight loss and increased antioxidant capacity and antifungal activity	Aparicio-García et al. (2021)
Chitosan-based nano-TiO_2 and nano-SiO_2 coatings	Blueberry	Extended shelf life, delayed ripening and controlled spoilage organisms	Li et al. (2021)
Nanolaminate coatings based on alginate, chitosan, and antimicrobial extract of *Flournesia cernua*	Tomato	Improved WVP and O_2 permeabilities, inhibited microbes, extended shelf life and decreased weight loss	Salas-Méndez et al. (2019)
Chitosan, nisin, silicon dioxide nanocomposite coating films	Blueberry	Maintained fruit texture and acted as antimicrobial agent	Eldib et al. (2020)
Chitosan and alginate-based coatings with cinnamon essential oil microcapsules	Mango	Improvement in pH value, fruit firmness, vitamin C content, soluble solid contents and acid content were noticed	Yin et al. (2019)

TABLE 8.3 *(Continued)*

Components	Fruit/vegetable treated	Effect introduced	References
Chitosan nanoparticles and thyme essential oil nanocoating	Green Bell pepper	Extension of fruit shelf life without altering cellular or physicochemical properties with antibacterial effect against *Pectobacterium carotovorum*	Correa-Pacheco et al. (2021)
Alginate films with cellulose nanofibrils from cocoa by products	Wild Andean blueberries	Decreased water vapor permeability and transparency of films. decreased the weight loss, respiration rate, and improved the firmness of blueberries	Medina-Jaramillo et al. (2020)

infection was observed in the case of uninoculated control fruit. At the same time, the inoculated-uncoated fruit showed complete infection. The lowest incidence of infection was observed for CS15 coated fruit when compared with TEO30 coated ones.

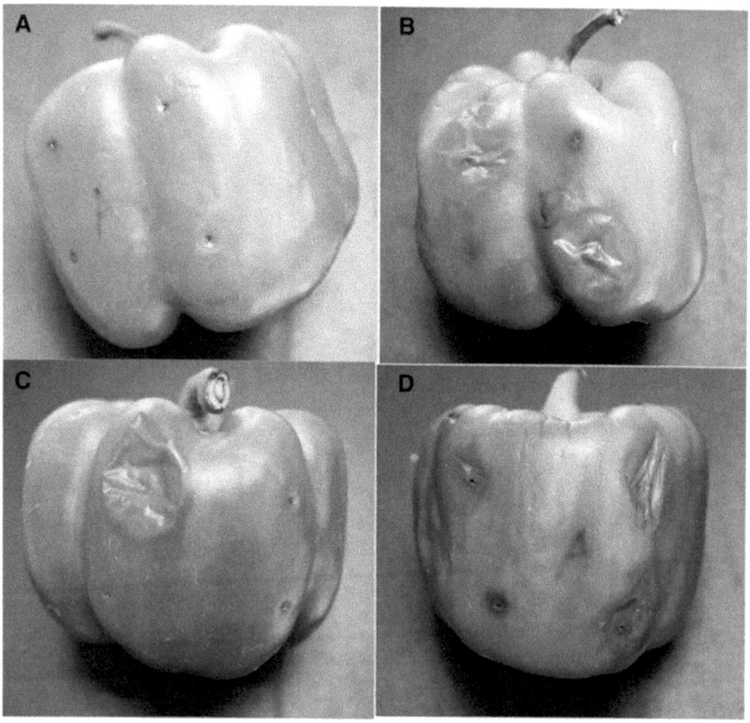

FIGURE 8.3 Severity of *P. carotovorum* infection in (A) uninoculated control, (B) inoculated control, (C) CS15 coated, and (D) TEO30-coated bell pepper.

Source: Reprinted with permission from Correa-Pacheco et al. (2021). Copyrights @ 2020 Institute of Food Technologists.

Also, in recent years, there has been a growing consumer demand on fresh-cut vegetables and fruits. But the peeling and cutting process hastens the metabolic activities of plant tissue and then make them more perishable than the intact fruits and vegetables (Chiumarelli and Hubinger, 2012). In such cases, coating of fresh cut fruits or vegetables with edible nanocoatings has been used to prevent their quicker spoilage. Recently, Saravanakumar and coworkers developed nanocoatings composed of both biogenic and chemical silver nanoparticles polyvinylpyrrolidone-based glycerosomes (G/C-PVP–AgNPs) and compared their role in augmenting the shelf life of fresh cut yellow and red bell peppers (Saravanakumar et al., 2020). The biogenic silver nanoparticle-based coating showed more superior properties than the chemically synthesized one in terms of antibacterial and physico-chemical attributes. Figure 8.4 shows the comparative changes in texture, color, moisture content, total dissolved solids, and antimicrobial analysis. After storing the fruits at 15°C for 15 days, G-PVP-AgNPs-coated fruits exhibited high efficiency in preventing the growth of gray molds when compared with control and C-PVP–AgNPs-coated fruits (Fig. 8.4a). At the same time, no detectable difference in controlling mould growth was seen in the case of red fresh cut bell peppers in all cases (Fig. 8.4b). In addition, the fruits coated with G-PVP-AgNPs maintained better firmness and played a significant role in extending the shelf life of coated fruits without altering the cellular and physicochemical nature of the fruits. Table below shows the development of various nanocomposite edible coatings for the fruits.

8.5.2 EGG AND DAIRY PRODUCTS

Egg

Eggs are worldwide available and acceptable food very high in nutritional and calorific values which are also easy to cook and ready to eat. As eggshell is very brittle and breathable material, it permits the entry of CO_2 and moisture, which can lead to significant weight loss of egg yolk. In such situations, when such pores can be sealed using an efficient edible coating, the internal changes occurring in the egg can be minimized, as well as the unexpected damage of eggshell can be avoided (Saeed et al., 2017).

In a study, an edible nanocoating based on whey protein isolate nanofibers with antibacterial agent carvacrol and glycerol as a plasticizer (WPNFs-CA/Gly) was developed to study its role in maintaining the texture of salted duck egg yolk (SDEY), a traditional pickled egg product (Wang et al., 2020). The WPNFs-CA/Gly-coated SDEY showed the lowest weight loss after 10 days of storage, which could be attributed to the unique dense structures and hydrophobicity of WPNFs (Fig. 8.5a). Moreover, WPNFs-CA/Gly coating resulted in reducing the sensory score of egg yolk in terms of hardiness, chewiness, and springiness. It also prevented the formation of cracks on the yolk as seen in Figure 8.5b.

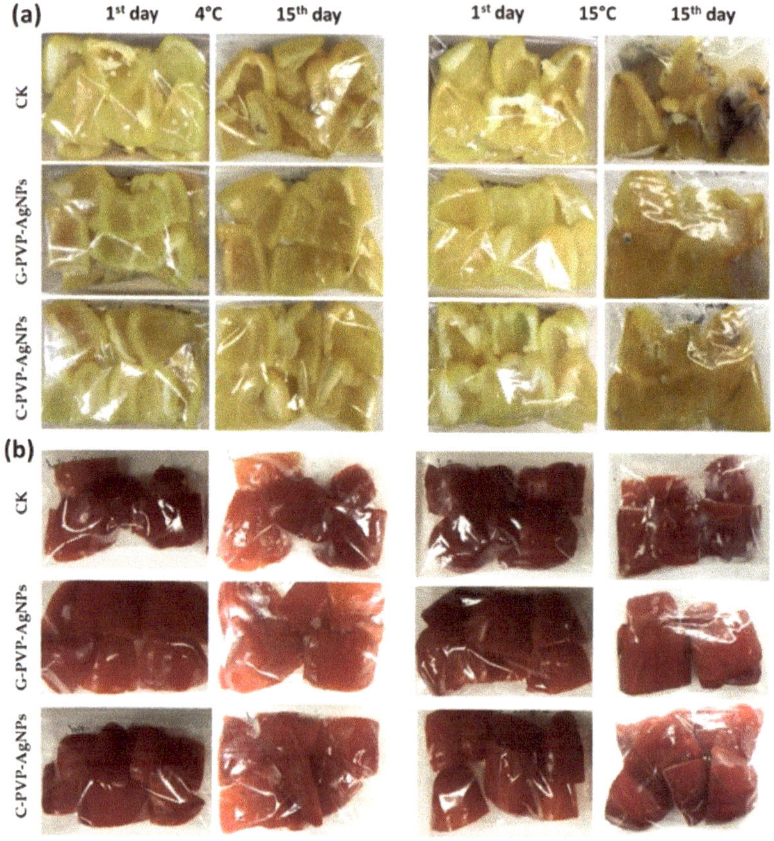

FIGURE 8.4 Effect of G–PVP-AgNPs and C-PVP–AgNPs-based nanocoatings on the shelf life of fresh cut bell peppers compared on 1st and 15th day.

Source: Reused with permission from Saravanakumar et al. (2020). Copyrights @ 2019 Elsevier B. V.

Dairy Products/Cheese

Among the varied dairy products, cheese is rich in protein, calcium, minerals, and vitamins and is one the most commonly consumed and regular part of human diet. Cheese gets spoiled rapidly by microbes when stored under inappropriate conditions and hence its safe packaging is of great importance. Edible coatings can be functionalized with nanostructured active agents to render them antibacterial, as a protective coating for cheese spoilage. The type of coatings/films used and the method of application must be selected depending on the type of cheese and storage conditions. The varieties of cheeses available are hard, soft, fresh, and pasteurized/processed cheeses (Ramos et al., 2016).

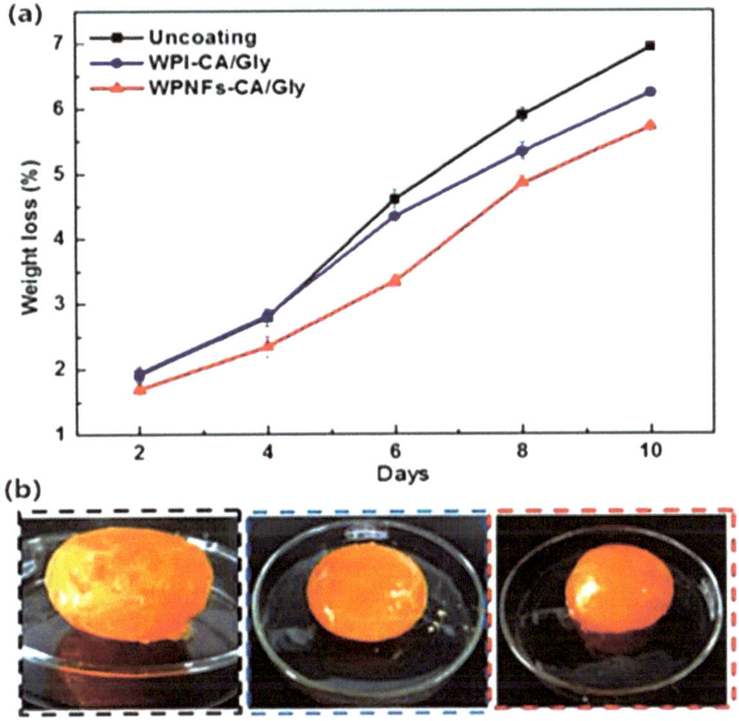

FIGURE 8.5 (a) Graph showing weight loss of uncoated and coated SDEYs stored for 10 days at 4°C, (b) digital image of SDEYS (starting from left to right) uncoated SDEY, WPI-CA/Gly-coated SDEY and WPNFs-CA/Gly-coated SDEY.

Source: Reprinted with permission from Wang et al. (2020), Open access.

Youssef et al. (2019) in their work developed a bionanocomposite coating composed of chitosan/ PVA/ TiO_2 nanoparticles and used it for protective coating on Ras cheese. Three different formulations were prepared using 0.5%, 1% and 2% (w/v) concentration of TiO_2 nanoparticle. The results showed that the coating with 2% TiO_2 nanoparticles showed the lowest weight loss and prevented the growth of mold on the cheese surface (Fig. 8.6). This study proved the efficiency of TiO_2-based edible nanocoating in extending the shelf life and limiting microbial contamination of cheese products.

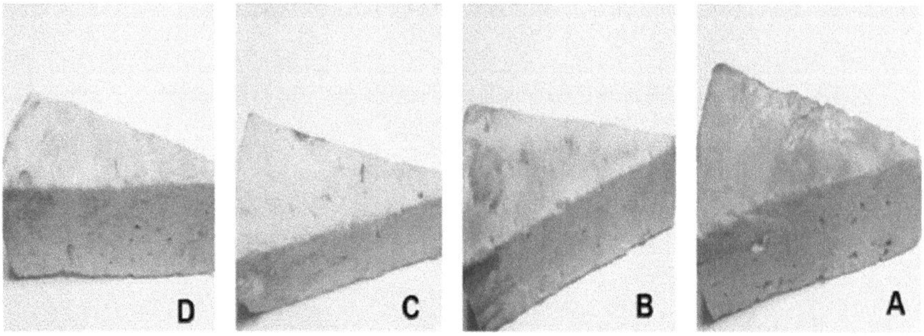

FIGURE 8.6 Changes observed on Ras cheese after coating with bionanocomposite containing TiO_2 nanoparticle concentrations (A) 0%, (B) 0.5%, (C) 1, and (D) 2%.
Source: Reprinted with permission from (Youssef et al., 2019). Copyrights @ 2018 Elsevier Ltd.

Recently, in a similar study, Ligaj et al. observed the impact of polyolefin foil with PLA coating and antibacterial nanocoating composed of polyolefin foil/PLA/zero-valent iron nanoparticle on inhibiting microbial growth on goat cream cheese (Ligaj et al., 2020). Figure 8.7a shows that after 5 weeks storage at chilled temperature, the cheese sample packed with polyolefin foil with PLA showed the presence of microorganisms, whereas the nanocoating film inhibited microbial growth both on the package and the cheese surface (Fig. 8.7b).

8.5.3 MEAT, POULTRY AND FISHERY PRODUCTS

Meat or muscle foods and fishery products form an important source of animal protein and hence they are influential ingredient of healthy and well-balanced diet. Meat and fish products are considered as the highly vulnerable to rapid microbial and oxidative deterioration during the pre-slaughter handling processes. Hence, muscle foods must be packaged in most appropriate manner that can prevent or delay its undesirable spoilage. From centuries onwards, edible coating techniques have been applied on meat and fishes to prevent shrinkage, discoloration, off-flavors, and microbial spoilage.

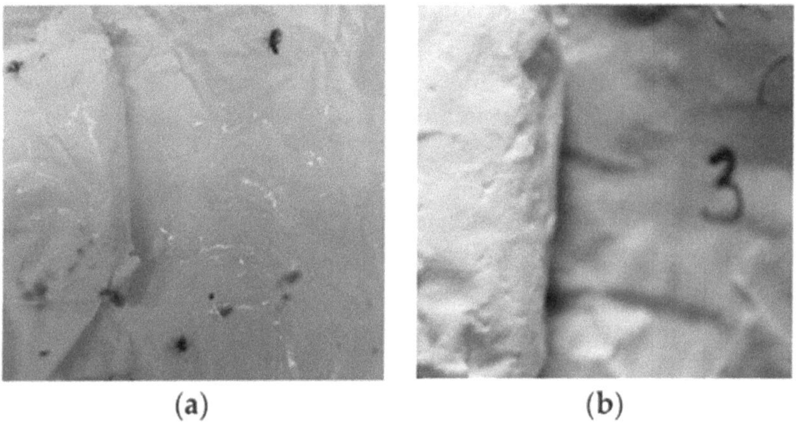

FIGURE 8.7 Goat cream cheese wrapped with (a) polyolefin foil with PLA coating and (b) polyolefin foil with ZVI/PLA nanocoating.
Source: Reprinted with permission from Ligaj et al. (2020), Open Access Journal.

Meindrawan et al. reported the development of a bionanocomposite edible nanocoating made of gelatin–ZnO nanoparticle for preserving broiler chicken fillets (Meindrawan et al., 2020). Based on the concentration of ZnO nanoparticles, four different formulations of the coatings were prepared and were designated as F1 (0% w/w), F2 (0.024% w/w), F3 (0.048% w/w), and F4 (0.096% w/w). Figure 8.8 shows the physical appearance of chicken fillets uncoated and coated with gelatin–ZnO bionanocomposite. From the figure, it is clear that the fillets coated with F3 and F4 films maintained the freshness of the chicken as indicated by the reddish color of the meat. On the contrary, the uncoated, F1 and F2 formulation-coated samples showed yellow discoloration due to oxidative or microbial spoilage. The study concluded that the F3 film formulation was the best which provided lowest microbial count relatively high fillet firmness, and lowest weight loss. A handful of chapters have reported the preparation and application of nano-based edible coatings for preservation of meat and fishery products (Table 8.4).

8.6 CONCLUSION

Inclusion of nanomaterials in edible coating matrices is one of the most promising and advancing avenues in food nanotechnology. Nanomaterials in edible coatings/films not only improve the food shelf life, ensure food safety, but also can cause the enhancement of food quality and nutritive value. The main role of edible nanocoatings application is to preserve highly perishable food commodities like fresh and cut pieces of fruits and vegetables, meat, poultry, fish, egg and dairy products. Also, it is worth to mention that the selection of nanomaterials for use in edible coatings has to be made wisely as it should not impart any toxicity and do not interact with the functionalities of any natural additive compounds present in the coating.

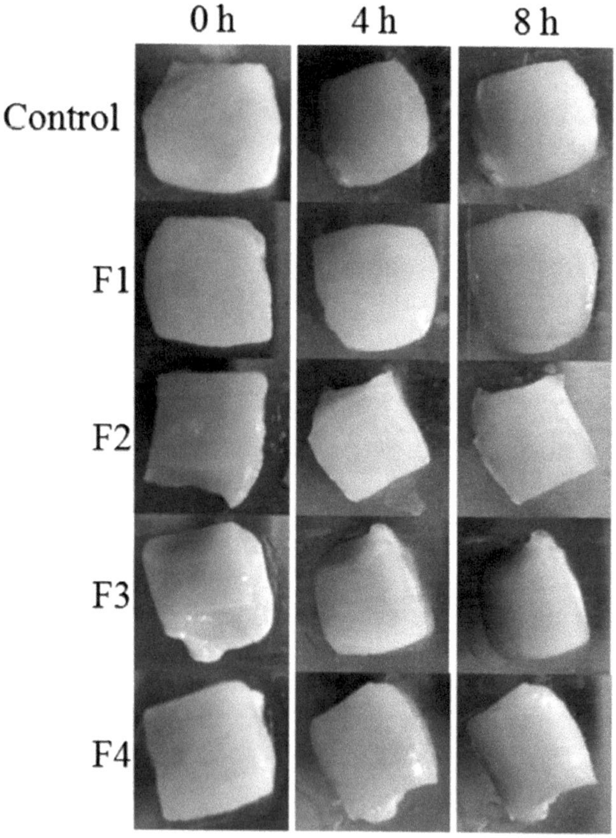

FIGURE 8.8 Physical appearance of chicken fillets coated with different formulations of gelatin–ZnO bionanocomposite coating.

Source: Reprinted with permission from Meindrawan et al. (2020). Copyrights @ 2020 WILEY VCH Verlag GmbH & Co. KGaA, Weinheim.

TABLE 8.4 List of Edible Nanocoatings Developed to Preserve Meat and Fishery Products.

Components	Meat treated	Effect introduced	References
Jujube gum and nettle oil-loaded nanoemulsion coatings	Beluga sturgeon fillets	Extended the fillet shelf life	Gharibzahedi and Mohammadnabi (2017)
Chitosan-based nanoliposome coating incorporated with *Satureja* plant essential oil	Lamb meat	Prolonged antimicrobial and antioxidant activity	Pabast et al. (2018)
Chitosan–montmorillonite-based nanocoatings with α tocopherol	Sliced dry-cured ham	Coated meat retained stronger radical-scavenging activity and showed lower thiobarbituric acid reactive substances	Yan et al. (2019)
Chitosan-*Lepidium sativum* sea gum nanocoating	Beef	Delayed microbial and oxidative spoilage of beef at three different storage temperatures	M. Esmaeili et al. (2020)
Nanochitosan incorporated with nanoliposome cumin	Sardine fish	antimicrobial and antioxidant activity	Homayounpour et al. (2020)
Chitosan zein coating with nanoencapsulated with *Pulicaria gnaphalodes (Vent.) Boiss*. aqueous extract	Rainbow trout fish	Showed lower thiobarbituric acid reactive substances and peroxide value	Mehdizadeh et al. (2021)
Low density polyethylene films with silver nanoparticles	Chicken breast fillets	Enhanced antimicrobial and antioxidant activity	Azlin-Hasim et al. (2016)
Thymol-loaded chitosan nanofiber coating	Gilthead sea bream fillets	Delayed chemical deterioration	Ceylan et al. (2017)
Whey protein nanofibrils with TiO_2 nanotubes	Chilled meat	Limited lipid peroxidation and promoted antioxidant activity in beef	Feng et al. (2019)

KEYWORDS

- **edible films**
- **nanocoatings**
- **food preservation**
- **solvent casting**
- **dipping method**
- **extension of food shelf life**
- **nanomaterials**

REFERENCES

Aguilar-Sánchez, R.; Munguía-Pérez, R.; Reyes-Jurado, F.; Navarro-Cruz, A. R.; Cid-Pérez, T. S.; Hernández-Carranza, P.; Beristain-Bauza, S. del C.; Ochoa-Velasco, C. E.; Avila-Sosa, R. Structural, Physical, and Antifungal Characterization of Starch Edible Films Added with Nanocomposites and Mexican Oregano (Lippia berlandieri Schauer) Essential Oil. *Molecules* **2019,** *24*, 2340.

Aguirre-Joya, J. A.; De Leon-Zapata, M. A.; Alvarez-Perez, O. B.; Torres-León, C.; Nieto-Oropeza, D. E.; Ventura-Sobrevilla, J. M.; Aguilar, M. A.; Ruelas-Chacón, X.; Rojas, R., Ramos-Aguiñaga, M. E.; Aguilar, C. N. Basic and Applied Concepts of Edible Packaging for Foods. In *Food Packaging and Preservation*; Elsevier, 2018; pp 1–61.

Ali, A.; Maqbool, M.; Ramachandran, S.; Alderson, P. G. Gum Arabic as a Novel Edible Coating for Enhancing Shelf-Life and Improving Postharvest Quality of Tomato (*Solanum lycopersicum* L.) Fruit. *Postharvest Biol. Technol.* **2010,** *58*, 42–47.

Andrade, R. D.; Skurtys, O.; Osorio, F. A. Atomizing Spray Systems for Application of Edible Coatings. *Compr. Rev. Food Sci. Food Saf.* **2012,** *11*, 323–337.

Aparicio-García, P. F.; Ventura-Aguilar, R. I.; del Río-García, J. C.; Hernández-López, M.; Guillén-Sánchez, D.; Salazar-Piña, D. A.; Ramos-García, M. de L.; Bautista-Baños, S. Edible Chitosan/Propolis Coatings and Their Effect on Ripening, Development of Aspergillus flavus, and Sensory Quality in Fig Fruit, during Controlled Storage. *Plants* **2021,** *10*, 112.

Azlin-Hasim, S.; Cruz-Romero, M. C.; Morris, M. A.; Padmanabhan, S. C.; Cummins, E.; Kerry, J. P. The Potential Application of Antimicrobial Silver Polyvinyl Chloride Nanocomposite Films to Extend the Shelf-Life of Chicken Breast Fillets. *Food Bioprocess Technol.* **2016,** *9*, 1661–1673.

Bharti, S. K.; Pathak, V.; Alam, T.; Arya, A.; Basak, G.; Awasthi, M. G. Materiality of Edible Film Packaging in Muscle Foods: A Worthwhile Conception. *J. Packag. Technol. Res.* **2020,** *4*, 117–132.

Bora, A.; Mishra, P. Characterization of Casein and Casein-Silver Conjugated Nanoparticle Containing Multifunctional (Pectin–Sodium Alginate/Casein) Bilayer Film. *J. Food Sci. Technol.* **2016,** *53*, 3704–3714.

Cai, L.; Wang, Y.; Cao, A. The Physiochemical and Preservation Properties of Fish Sarcoplasmic Protein/Chitosan Composite Films Containing Ginger Essential Oil Emulsions. *J. Food Process Eng.* **2020,** *43*.

Ceylan, Z.; Sengor, G. F. U.; Yilmaz, M. T. A Novel Approach to Limit Chemical Deterioration of Gilthead Sea Bream *(Sparus aurata)* Fillets: Coating with Electrospun Nanofibers as Characterized by Molecular, Thermal, and Microstructural Properties: Coating with Electrospun Nanofiber. *J. Food Sci.* **2017,** *82*, 1163–1170.

Chiumarelli, M.; Hubinger, M. D. Stability, Solubility, Mechanical and Barrier Properties of Cassava Starch—Carnauba Wax Edible Coatings to Preserve Fresh-Cut Apples. *Food Hydrocoll.* **2012,** *28*, 59–67.

Chu, Y.; Cheng, W.; Feng, X.; Gao, C.; Wu, D.; Meng, L.; Zhang, Y.; Tang, X. Fabrication, Structure and Properties of Pullulan-Based Active Films Incorporated with Ultrasound-Assisted Cinnamon Essential Oil Nanoemulsions. *Food Packag. Shelf Life* **2020,** *25*, 100547.

Correa-Pacheco, Z. N.; Corona-Rangel, M. L.; Bautista-Baños, S.; Ventura-Aguilar, R. I. Application of Natural-Based Nanocoatings for Extending the Shelf Life of Green Bell Pepper Fruit. *J. Food Sci.* **2021,** *86*, 95–102.

Dehghani, S.; Hosseini, S. V.; Regenstein, J. M. Edible Films and Coatings in Seafood Preservation: A Review. *Food Chem.* **2018,** *240*, 505–513.

Dhital, R.; Mora, N. B.; Watson, D. G.; Kohli, P.; Choudhary, R. Efficacy of Limonene Nano Coatings on Post-Harvest Shelf Life of Strawberries. *LWT* **2018,** *97*, 124–134.

Dini, H.; Fallah, A. A.; Bonyadian, M.; Abbasvali, M.; Soleimani, M. Effect of Edible Composite Film Based on Chitosan and Cumin Essential Oil-Loaded Nanoemulsion Combined with Low-Dose Gamma Irradiation on Microbiological Safety and Quality of Beef Loins during Refrigerated Storage. *Int. J. Biol. Macromol.* **2020,** *164*, 1501–1509.

Ebrahimi, S.; Fathi, M.; Kadivar, M. Production and Characterization of Chitosan-Gelatin Nanofibers by Nozzle-Less Electrospinning and Their Application to Enhance Edible Film's Properties. *Food Packag. Shelf Life* **2019,** *22*, 100387.

Eldib, R.; Khojah, E.; Elhakem, A.; Benajiba, N.; Helal, M. Chitosan, Nisin, Silicon Dioxide Nanoparticles Coating Films Effects on Blueberry (*Vaccinium myrtillus*) Quality. *Coatings* **2020,** 10, 962.

Emamifar, A.; Bavaisi, S. Nanocomposite Coating Based on Sodium Alginate and Nano-ZnO for Extending the Storage Life of Fresh Strawberries (Fragaria × ananassa Duch.). *J. Food Meas. Charact.* **2020,** *14*, 1012–1024.

Escamilla-García, M.; Rodríguez-Hernández, M.; Hernández-Hernández, H.; Delgado-Sánchez, L.; García-Almendárez, B.; Amaro-Reyes, A.; Regalado-González, C. Effect of an Edible Coating Based on Chitosan and Oxidized Starch on Shelf Life of *Carica papaya L.*, and Its Physicochemical and Antimicrobial Properties. *Coatings* **2018,** *8*, 318.

Esmaeili, H.; Cheraghi, N.; Khanjari, A.; Rezaeigolestani, M.; Basti, A. A.; Kamkar, A.; Aghaee, E. M. Incorporation of Nanoencapsulated Garlic Essential Oil into Edible Films: A Novel Approach for Extending Shelf Life of Vacuum-Packed Sausages. *Meat Sci.* **2020,** *166*, 108135.

Esmaeili, M.; Ariaii, P.; Nasiraie, L. R.; Pour, M. Y. Comparison of Coating and Nano-Coating of Chitosan- Lepidium Sativum Seed Gum Composites on Quality and Shelf Life of Beef. *J. Food Meas. Charact.* **2020**.

Falguera, V.; Quintero, J. P.; Jiménez, A.; Muñoz, J. A.; Ibarz, A. Edible Films and Coatings: Structures, Active Functions and Trends in Their Use. *Trends Food Sci. Technol.* **2011,** *22*, 292–303.

Feng, Z.; Li, L.; Wang, Q.; Wu, G.; Liu, C.; Jiang, B.; Xu, J. Effect of Antioxidant and Antimicrobial Coating based on Whey Protein Nanofibrils with TiO_2 Nanotubes on the Quality and Shelf Life of Chilled Meat. *Int. J. Mol. Sci.* **2019,** *20*, 1184.

Ghadetaj, A.; Almasi, H.; Mehryar, L. Development and Characterization of Whey Protein Isolate Active Films Containing Nanoemulsions of Grammosciadium Ptrocarpum Bioss. Essential Oil. *Food Packag. Shelf Life* **2018,** *16*, 31–40.

Gharibzahedi, S. M. T.; Mohammadnabi, S. Effect of Novel Bioactive Edible Coatings Based on Jujube Gum and Nettle Oil-Loaded Nanoemulsions on the Shelf-Life of Beluga Sturgeon Fillets. *Int. J. Biol. Macromol.* **2017,** *95*, 769–777.

Gheorghita (Puscaselu), R.; Gutt, G.; Amariei, S. The Use of Edible Films Based on Sodium Alginate in Meat Product Packaging: An Eco-Friendly Alternative to Conventional Plastic Materials. *Coatings* **2020,** *10*, 166.

Ghiasi, F.; Golmakani, M.-T.; Eskandari, M. H.; Hosseini, S. M. H. A New Approach in the Hydrophobic Modification of Polysaccharide-Based Edible Films Using Structured Oil Nanoparticles. *Ind. Crops Prod.* **2020,** *154*, 112679.

González-Reza, R.; García-Betanzos, C.; Sánchez-Valdes, L.; Quintanar-Guerrero, D.; Cornejo-Villegas, M.; Zambrano-Zaragoza, M. The Functionalization of Nanostructures and Their Potential Applications in Edible Coatings. *Coatings* **2018,** *8*, 160.

Gul, O.; Saricaoglu, F. T.; Besir, A.; Atalar, I.; Yazici, F. Effect of Ultrasound Treatment on the Properties of Nano-Emulsion Films Obtained from Hazelnut Meal Protein and Clove Essential Oil. *Ultrason. Sonochem.* **2018,** *41*, 466–474.

Guo, M.; Jin, T. Z.; Wang, L.; Scullen, O. J.; Sommers, C. H. Antimicrobial Films and Coatings for Inactivation of Listeria Innocua on Ready-to-Eat Deli Turkey Meat. *Food Control* **2014,** *40*, 64–70.

Hager, J. V.; Rawles, S. D.; Xiong, Y. L.; Newman, M. C.; Webster, C. D. Edible Corn-zein-based Coating Incorporated with Nisin or Lemongrass Essential Oil Inhibits *Listeria monocytogenes* on Cultured Hybrid Striped Bass, *Morone chrysops* × *Morone saxatilis*, Fillets during Refrigerated and Frozen Storage: Lemongrass Essential Oil Inhibits *Listeria*. *J. World Aquac. Soc.* **2019,** *50*, 204–218.

Homayounpour, P.; Jalali, H.; Shariatifar, N.; Amanlou, M.; Khanjari, A. Protective Effect of Nanochitosan Incorporated with Free/nanoliposome Cumin (*Cuminum cyminum* L.) Aqueous Extract on Sardine Fish. *J. Aquat. Food Prod. Technol.* **2020,** *29*, 949–961.

Huang, Y.; Gu, C.; He, S.; Zhu, D.; Liu, X.; Chen, Z. Development and Characterization of an Edible Chitosan–Whey Protein Nano Composite Film for Chestnut (*Castanea mollissima* Bl.) Preservation. *J. Food Sci.* **2020,** *85*, 2114–2123.

Huber, K. C.; Embuscado, M. E. *Edible Films and Coatings for Food Applications*; Springer: Dordrecht ; London, 2009.

Jafarzadeh, S.; Alias, A.; Ariffin, F.; Mahmud, S. Characterization of Semolina Protein Film with Incorporated Zinc Oxide Nano Rod Intended for Food Packaging. *Pol. J. Food Nutr. Sci.* **2017,** *67*, 183–190.

Janjarasskul, T.; Krochta, J. M. Edible Packaging Materials. *Annu. Rev. Food Sci. Technol.* **2010,** *1*, 415–448.
Jung, S.; Cui, Y.; Barnes, M.; Satam, C.; Zhang, S.; Chowdhury, R. A.; Adumbumkulath, A.; Sahin, O.; Miller, C.; Sajadi, S. M.; Sassi, L. M.; Ji, Y.; Bennett, M. R.; Yu, M.; Friguglietti, J.; Merchant, F. A.; Verduzco, R.; Roy, S.; Vajtai, R.; Meredith, J. C.; Youngblood, J. P.; Koratkar, N.; Rahman, M. M.; Ajayan, P. M. Multifunctional Bio-Nanocomposite Coatings for Perishable Fruits. *Adv. Mater.* **2020,** *32*, 1908291.
K. S., J.; Jose, J.; Li, T.; Thomas, M.; Shankregowda, A. M.; Sreekumaran, S.; Kalarikkal, N.; Thomas, S. Application of Novel Zinc Oxide Reinforced Xanthan Gum Hybrid System for Edible Coatings. *Int. J. Biol. Macromol.* **2020,** *151*, 806–813.
Koushesh Saba, M.; Amini, R. Nano-ZnO/Carboxymethyl Cellulose-Based Active Coating Impact on Ready-to-Use Pomegranate during Cold Storage. *Food Chem.* **2017,** *232*, 721–726.
Li, Y.; Rokayya, S.; Jia, F.; Nie, X.; Xu, J.; Elhakem, A.; Almatrafi, M.; Benajiba, N.; Helal, M. Shelf-Life, Quality, Safety Evaluations of Blueberry Fruits Coated with Chitosan Nano-Material Films. *Sci. Rep.* **2021,** *11*, 55.
Ligaj, M.; Tichoniuk, M.; Cierpiszewski, R.; Foltynowicz, Z. Efficiency of Novel Antimicrobial Coating Based on Iron Nanoparticles for Dairy Products' Packaging. *Coatings* **2020,** *10*, 156.
Liu, R.; Liu, D.; Liu, Y.; Song, Y.; Wu, T.; Zhang, M. Using Soy Protein SiOx Nanocomposite Film Coating to Extend the Shelf Life of Apple Fruit. *Int. J. Food Sci. Technol.* **2017,** *52*, 2018–2030.
Lu, F.; Ding, Y.; Ye, X.; Liu, D. Cinnamon and Nisin in Alginate–Calcium Coating Maintain Quality of Fresh Northern Snakehead Fish Fillets. *LWT—Food Sci. Technol.* **2010,** *43*, 1331–1335.
Medina-Jaramillo, C.; Quintero-Pimiento, C.; Gómez-Hoyos, C.; Zuluaga-Gallego, R.; López-Córdoba, A. Alginate-Edible Coatings for Application on Wild Andean Blueberries (Vaccinium meridionale Swartz): Effect of the Addition of Nanofibrils Isolated from Cocoa By-Products. *Polymers* **2020,** *12*, 824.
Mehdizadeh, A.; Shahidi, S.-A.; Shariatifar, N.; Shiran, M.; Ghorbani-HasanSaraei, A. Evaluation of Chitosan-zein Coating Containing Free and Nano-encapsulated *Pulicaria gnaphalodes (Vent.) Boiss*. Extract on Quality Attributes of Rainbow Trout. *J. Aquat. Food Prod. Technol.* **2021,** *30*, 62–75.
Meindrawan, B.; Putri, S.; Susanto, C. S.; Ofe, O.; Mangindaan, D.; Ayman, A.; Kasih, T. P. Bionanocomposite of Gelatin–ZnO Nanoparticles as Potential Edible Coating for Broiler Chicken Fillet. *Macromol. Symp.* **2020,** *391*, 1900165.
Meindrawan, B.; Suyatma, N. E.; Wardana, A. A.; Pamela, V. Y. Nanocomposite Coating Based on Carrageenan and ZnO Nanoparticles to Maintain the Storage Quality of Mango. *Food Packag. Shelf Life* **2018,** *18*, 140–146.
Mohammadi, A.; Hashemi, M.; Hosseini, S. M. Postharvest Treatment of Nanochitosan-Based Coating Loaded with Zataria Multiflora Essential Oil Improves Antioxidant Activity and Extends Shelf-Life of Cucumber. *Innov. Food Sci. Emerg. Technol.* **2016,** *33*, 580–588.
Montes-de-Oca-Ávalos, J. M.; Altamura, D.; Herrera, M. L.; Huck-Iriart, C.; Scattarella, F.; Siliqi, D.; Giannini, C.; Candal, R. J. Physical and Structural Properties of Whey Protein

Concentrate—Corn Oil—TiO_2 Nanocomposite Films for Edible Food-Packaging. *Food Packag. Shelf Life* **2020,** *26*, 100590.

Nasiri, M.; Sani, A. M.; Hakimzadeh, V.; Shahidi, M. Antimicrobial Effects of Edible Nano-Composite Based on Bean Pod Shell Gum, Nano-TiO_2, and Mentha Pulegium Essential Oil. *J. Appl. Biol. Biotechnol.* **2019,** *7*, 75–78.

Ortiz-Duarte, G.; Pérez-Cabrera, L. E.; Artés-Hernández, F.; Martínez-Hernández, G. B. Ag-Chitosan Nanocomposites in Edible Coatings Affect the Quality of Fresh-Cut Melon. *Postharvest Biol. Technol.* **2019,** *147*, 174–184.

Pabast, M.; Shariatifar, N.; Beikzadeh, S.; Jahed, G. Effects of Chitosan Coatings Incorporating with Free or Nano-Encapsulated Satureja Plant Essential Oil on Quality Characteristics of Lamb Meat. *Food Control* **2018,** *91*, 185–192.

Ramos, O.; Pereira, R. N. C.; Martins, J.; Malcata, F. Edible Packaging for Dairy Products. 2016.

Restrepo, A. E.; Rojas, J. D.; García, O. R.; Sánchez, L. T.; Pinzón, M. I.; Villa, C. C. Mechanical, Barrier, and Color Properties of Banana Starch Edible Films Incorporated with Nanoemulsions of Lemongrass (*Cymbopogon citratus*) and Rosemary (*Rosmarinus officinalis*) Essential Oils. *Food Sci. Technol. Int.* **2018,** *24*, 705–712.

Saeed, F.; Javaid, A.; Ahmed, N.; Nadeem, M. T.; Arshad, M. S.; Imran, A.; Sohaib, M.; Khan, A. U. Influence of Edible Coating Techniques on Quality Characteristics of Eggs: Influence of Edible Coating Techniques on Eggs. *J. Food Process. Preserv.* **2017,** *41*, e12815.

Salas-Méndez, E. de J.; Vicente, A.; Pinheiro, A. C.; Ballesteros, L. F.; Silva, P.; Rodríguez-García, R.; Hernández-Castillo, F. D.; Díaz-Jiménez, M. de L. V.; Flores-López, M. L.; Villarreal-Quintanilla, J. Á.; Peña-Ramos, F. M.; Carrillo-Lomelí, D. A.; Jasso de Rodríguez, D. Application of Edible Nanolaminate Coatings with Antimicrobial Extract of Flourensia Cernua to Extend the Shelf-Life of Tomato (*Solanum lycopersicum L.*) Fruit. *Postharvest Biol. Technol.* **2019,** *150*, 19–27.

Sani, I. K.; Marand, S. A.; Alizadeh, M.; Amiri, S.; Asdagh, A. Thermal, Mechanical, Microstructural and Inhibitory Characteristics of Sodium Caseinate Based Bioactive Films Reinforced by ZnONPs/Encapsulated *Melissa officinalis* Essential Oil. *J. Inorg. Organomet. Polym. Mater.* **2021,** *31*, 261–271.

Santos, A. M. P.; Melo, E. de A. Application of Edible Biopolymer Coatings to Extend the Storage Life of Fresh Fruits and Vegetables. In *Biopolymer Membranes and Films*; Elsevier, 2020; pp 505–513.

Saravanakumar, K.; Hu, X.; Chelliah, R.; Oh, D.-H.; Kathiresan, K.; Wang, M.-H. Biogenic Silver Nanoparticles-Polyvinylpyrrolidone Based Glycerosomes Coating to Expand the Shelf Life of Fresh-Cut Bell Pepper (*Capsicum annuum* L. var. grossum (L.) Sendt). *Postharvest Biol. Technol.* **2020,** *160*, 111039.

Souza Vieira da Silva, I.; Soares Prado, N.; Gontijo de Melo, P.; Campion Arantes, D.; Zeni Andrade, M.; Otaguro, H.; Pasquini, D. Edible Coatings Based on Apple Pectin, Cellulose Nanocrystals, and Essential Oil of Lemongrass: Improving the Quality and Shelf Life of Strawberries (Fragaria Ananassa). *J. Renew. Mater.* **2019,** *7*, 73–87.

Sucheta, Chaturvedi, K.; Sharma, N.; Yadav, S. K. Composite Edible Coatings from Commercial Pectin, Corn Flour and Beetroot Powder Minimize Post-Harvest Decay, Reduces Ripening and Improves Sensory Liking of Tomatoes. *Int. J. Biol. Macromol.* **2019,** *133*, 284–293.

Suhag, R.; Kumar, N.; Petkoska, A. T.; Upadhyay, A. Film Formation and Deposition Methods of Edible Coating on Food Products: A Review. *Food Res. Int.* **2020,** *136,* 109582.

Tanada-Palmu, P. S.; Grosso, C. R. F. Effect of Edible Wheat Gluten-Based Films and Coatings on Refrigerated Strawberry (*Fragaria ananassa*) Quality. *Postharvest Biol. Technol.* **2005,** *36,* 199–208.

Wang, H.; Qian, J.; Ding, F. Emerging Chitosan-Based Films for Food Packaging Applications. *J. Agric. Food Chem.* **2018,** *66,* 395–413.

Wang, Q.; Liu, W.; Tian, B.; Li, D.; Liu, C.; Jiang, B.; Feng, Z. Preparation and Characterization of Coating Based on Protein Nanofibers and Polyphenol and Application for Salted Duck Egg Yolks. *Foods* **2020,** *9,* 449.

Yan, W.; Chen, W.; Muhammad, U.; Zhang, J.; Zhuang, H.; Zhou, G.; Preparation of α-tocopherol-Chitosan Nanoparticles/Chitosan/Montmorillonite Film and the Antioxidant Efficiency on Sliced Dry-Cured Ham. *Food Control* **2019,** *104,* 132–138.

Yin, C.; Huang, C.; Wang, J.; Liu, Y.; Lu, P.; Huang, L. Effect of Chitosan- and Alginate-Based Coatings Enriched with Cinnamon Essential Oil Microcapsules to Improve the Postharvest Quality of Mangoes. *Materials* **2019,** *12,* 2039.

Youssef, A. M.; Assem, F. M.; Abdel-Aziz, M. E.; Elaaser, M.; Ibrahim, O. A.; Mahmoud, M.; Abd El-Salam, M. H. Development of Bionanocomposite Materials and Its Use in Coating of Ras Cheese. *Food Chem.* **2019,** *270,* 467–475.

Zambrano-Zaragoza, M. L.; Mercado-Silva, E.; Ramirez-Zamorano, P.; Cornejo-Villegas, M. A. Gutiérrez-Cortez, E.; Quintanar-Guerrero, D. Use of Solid Lipid Nanoparticles (SLNs) in Edible Coatings to Increase Guava (*Psidium guajava* L.) Shelf-Life. *Food Res. Int.* **2013,** *51,* 946–953.

Zambrano-Zaragoza, M. L.; Quintanar-Guerrero, D.; Del Real, A.; González-Reza, R. M.; Cornejo-Villegas, M. A.; Gutiérrez-Cortez, E. Effect of Nano-Edible Coating Based on Beeswax Solid Lipid Nanoparticles on Strawberry's Preservation. *Coatings* **2020,** *10,* 253.

Zambrano-Zaragoza, M.; González-Reza, R.; Mendoza-Muñoz, N.; Miranda-Linares, V.; Bernal-Couoh, T.; Mendoza-Elvira, S.; Quintanar-Guerrero, D. Nanosystems in Edible Coatings: A Novel Strategy for Food Preservation. *Int. J. Mol. Sci.* **2018,** *19,* 705.

Zareie, Z.; Tabatabaei Yazdi, F.; Mortazavi, S. A. Development and Characterization of Antioxidant and Antimicrobial Edible Films Based on Chitosan and Gamma-Aminobutyric Acid-Rich Fermented Soy Protein. *Carbohydr. Polym.* **2020,** *244,* 116491.

Zeuthen, P.; Bøgh-Sørensen, L., Eds. *Food Preservation Techniques, Woodhead Publishing in Food Science and Technology*; CRC Press; Woodhead: Boca Raton : Cambridge, 2003.

Zhang, H.; Li, X.; Kang, H. Chitosan Coatings Incorporated with Free or Nano-Encapsulated Paulownia Tomentosa Essential Oil to Improve Shelf-Life of Ready-to-Cook Pork Chops. *LWT* **2019,** *116,* 108580.

Zhang, H.; Liang, Y.; Li, X.; Kang, H. Effect of Chitosan-Gelatin Coating Containing Nano-Encapsulated Tarragon Essential Oil on the Preservation of Pork Slices. *Meat Sci.* **2020,** *166,* 108137.

Zhang, L.; Liu, A.; Wang, W.; Ye, R.; Liu, Y.; Xiao, J.; Wang, K. Characterisation of Microemulsion Nanofilms Based on Tilapia Fish Skin Gelatine and ZnO Nanoparticles

Incorporated with Ginger Essential Oil: Meat Packaging Application. *Int. J. Food Sci. Technol.* **2017,** *52*, 1670–1679.

Zhao, Y. *Application of Commercial Coatings, Edible Coatings and Films to Improve Food Quality*; CRC Press, 2011.

Zheng, Y.; Monty, J.; Linhardt, R. J. Polysaccharide-Based Nanocomposites and Their Applications. *Carbohydr. Res.* **2015,** *405*, 23–32.

CHAPTER 9

Health and Safety Issues of Nanotechnology in Food Applications

SHIJI MATHEW and E. K. RADHAKRISHNAN*

School of Biosciences, Mahatma Gandhi University, Kottayam 686560, Kerala, India

**Corresponding author. E-mail: radhakrishnanek@mgu.ac.in*

ABSTRACT

Based on the current scenario, it is clear that the advanced and upcoming nanoinnovations in food applications can further open new possibilities in the betterment and improvement of food quality and preservation. Both inorganic (metal and metal oxides) and organic (carbohydrates, proteins, lipids) nanoparticles are applied in nanofoods and packagings for offering improved shelf life, providing antimicrobial protection and maintaining fresh quality of food. Nevertheless, many in vitro and in vivo studies have demonstrated that the inorganic nanoparticles pose potential threats to human health as they are exposed in higher concentrations for longer periods and are degraded and excreted in lower amounts. Today, researchers are more focused on the speedy development and applications of nanotechnology due to its attractive impact in every field. However, considering the sake of consumers, it is mandatory to provide a comprehensive information regarding the interface between nanoparticles and cells, tissues, and organisms, particularly in relation to possible hazards to human health. This chapter details on the possible routes of human exposure to nanoparticles, their potential adverse effects on human and environment, and the different mechanisms involved behind nanotoxicity. The chapter also provides information on the conventional and advanced in vitro and in vivo nanotoxicity risk assessment tests currently practiced.

9.1 INTRODUCTION

The introduction of nanotechnology has definitely improved our life quality, economy, and environmental status. The food industry has been beneficially revolutionized with the introduction of "Nanofoods"; the foods that are cultivated, manufactured, processed, and packaged using nanotechnology (Joseph et al., 2006). Today, in the market, a variety of nanofoods are available which include nanotechnology-based processed food, nanomaterial-containing packagings and containers, and health supplements incorporated with nanoparticles. Table 9.1 provides a list of currently available nanofoods in the market. The highly attractive nano-technological application in food industry includes nanoencapsulation for food fortification, nanoinnovations for smart, functional and improved packaging, and nanomaterial containing health adjuvating products ensuring controlled delivery of nutraceuticals (Acosta, 2009). But the excessive use of organic and inorganic nanomaterials in daily use products of agri-food industry is linked with increased alarm, uncertainties and unforeseen toxic impacts on human health and environmental (Amenta et al., 2015). Despite the attractive advantages of nanotechnology applications in all sectors, their toxic effects to human and environment still remains unclear. Previous studies suggest that the inclusion of nanomaterials in foods can possibly cause acute or chronic toxicity. In the case of foods, as low level of nanoparticles is consumed for prolonged periods, it is found to be associated with chronic type of toxicity.

The three main routes of nanomaterial exposure in human beings include dermal, inhalation, and ingestion. In all the three cases, the nanomaterials can enter into body, pass cell barriers, penetrate into various organs and tissues, and finally get accumulated in different body parts. The type and physiochemical nature of the nanoparticle determines the reaction and final consequences on the body. Organic nanoparticles (carbohydrates, proteins, lipids, and vitamins) are generally harmless as they are digestible. On the other hand, inorganic metal nanoparticles (silver, titanium dioxide, silicon dioxide, iron, and zinc oxide) are indigestible. The ingested nanoparticles can either cause cellular or organ damage within the gastrointestinal tract or on the sites on to which nanoparticles are adsorbed. Moreover, the nanoparticles can also interfere with the resident microflora of the gastrointestinal tract, leading to serious health impairment (Buzea et al., 2007; Fröhlich and Fröhlich, 2016).

TABLE 9.1 Commercially Available Nanofoods in Market.

Application	Product name	Type of product	Manufacturing company	Nanoparticle	Product and Function	References
Food processing	Nanoceuticals Slim Shake Chocolate	Supplement drink	RBC Life Sciences/ USA	Cocoa nanoclusters	Flavor enhancement	www.nanotechproject.tech/cpi/products/nanoceuticalstm-slim-shake-chocolate/ (2007)
	Canola active oil	Cooking	Shemen Industries, Israel	Nanomicelles	Transport of nanoingredients	www.nanotechproject.tech/cpi/products/canola-active-oil/, (2007)
	Maternal water	Baby product	La Posta del Aguila, Argentina	Nanocolloid Silver	Antimicrobial protection	www.nanotechproject.tech/cpi/products/maternal-water/ (2009)
	Nanotea		Shenzhen Become Industry & Trade Co., Ltd., China	Selenium	Antimicrobial protection, increased bioavailability	www.nanotechproject.tech/cpi/products/nanotea/ (2007)
	Eclipse Spearmint Chewy Mints	Chewing gum	Mars Wrigley, USA	TiO_2 nanoparticles	Coloring E171	https://nanodb.dk/en/product/?pid=4171 (2015)
	Old El Paso Taco	Seasoning Mix	Old el Paso, USA	Silicon dioxide	Anticaking agent E551	https://product.statnano.com/product/6089/old-el-paso-taco-seasoning-mix (2016)
	Dickinson's Coconut Curd		Dickinson's, USA	TiO_2 nanoparticles	To provide natural flavor	https://product.statnano.com/product/6093/dickinson%27s-coconut-curd (2016)
Containers and food packaging	Kinetic Go Green	Food container set	Kinetic, USA	Silver	Antimicrobial protection, maintains freshness of food	https://product.statnano.com/product/6075 (2016)

TABLE 9.1 *(Continued)*

Application	Product name	Type of product	Manufacturing company	Nanoparticle	Product and Function	References
	FresherLonger Miracle	Food storage bags	FresherLonger, UK	Silver	Antimicrobial protection	"FresherLonger™ Miracle Food Storage" (2007)
	Gourmet Trends Original Always Fresh	Container	Always Fresh, USA	Silver	Antimicrobial protection	https://nanodb.dk/en/product/?pid=3447 (2007)
	Silver–nano Noble GS	New born baby nursing bottle	Baby Dream Co. Ltd, South Korea	Silver	Antimicrobial protection	https://product.statnano.com/product/6841/silver-nano-noble-gs-nursing-bottle-(newborn-baby-use) (2016)
	Nanosilver Storage Box		Quan Zhou Hu Zheng Nano Technology Co., Ltd, China	Silver	Antimicrobial protection	"Quan Zhou Hu Zheng Nano Technology Co., Ltd.® Nano-silver Storage Box (Baoxianhe)" (2007)
Nutritional supplements	Liquid ionic magnesium	Fitness mineral supplements	Inno Tech Nutrition, Canada	Nano-magnesium	Enhances bioavailability	https://nanodb.dk/en/product/?pid=5057 (2007)
	Liquid ionic iron	Mineral supplements	Good State, USA	Nano-iron	Increases bioavailability	https://goodstate.com/products/good-state-liquid-ionic-iron-48-servings-at-10mg-plus-2-mg-fulvic-acid-8-fl-oz (2013)
	Liposomal vitamin C	Vitamin supplements	NanoNutra, Denmark	Nanoliposomes	Enhances Bioavailability	https://www.nanonutrausa.com/collections/all/products/liposomal-vitamin-c (2017)

Source: Adapted and modified with permission from Tarhan (2020). Copyrights @ 2020 Elsevier Inc.

Hence, a comprehensive examination on the potential adverse effects and toxicity issues of nanomaterials in food has to be made before its commercialization. In regard to this, various regulations, rules, legislations, policies, guidance, and assessment tests are being introduced by various prestigious legal authorities which categorize the numerous nanomaterials used in agro-food sector and indicate their toxic levels for ensuring their safe handling and consumption. The important legal authorities include the European Commission (EC), European Food Safety Agency (EFSA), Environmental Protection Agency (EPA), Organization for Economic Corporation and Development (OECD), Food and Drug Administration (FDA), World Health Organization (WHO) and International Standard Organization (ISO). It is expected that the uncertainties and lack of knowledge about the safety of nanomaterials in food products can be eliminated if these guidelines and legislations are properly practiced and followed worldwide.

This chapter mainly discusses about the health and safety issues connected with nanotechnological applications in foods. Here, we provide a comprehensive information on the main routes of nanoparticle exposure into human body, the physicochemical factors promoting nanoparticle toxicity and the possible mechanism of nanotoxicity. The chapter also details on the different analytical and imaginary techniques used for the identification of nanomaterials in food and also discusses about the in vivo and in vitro assessment tests for determining toxicity-related information. Later sections will also include a note on the public perception regarding the nanomaterials used in food applications.

9.2 POSSIBLE ROUTES OF NANOPARTICLE ENTRY INTO HUMAN BODY

Many food items may naturally contain some amounts of particles in nanoscale, which may be harmless due to minute concentration. The same is not applicable in the case of a specific nanoingredient enriched food product, and can end up in overexposure, allergy, and toxicity (Tarhan, 2020). Numerous studies have shown that nanomaterials from food can enter into our tissues and organs mainly through three potential routes; through skin via dermal contact, lungs through inhalation, and the gastrointestinal tract during ingestion (Maisanaba et al., 2015). Figure 9.1 represents the three main routes of nanoparticle exposure in human beings. The following section discusses the three main route of entry of nanoparticles into human body.

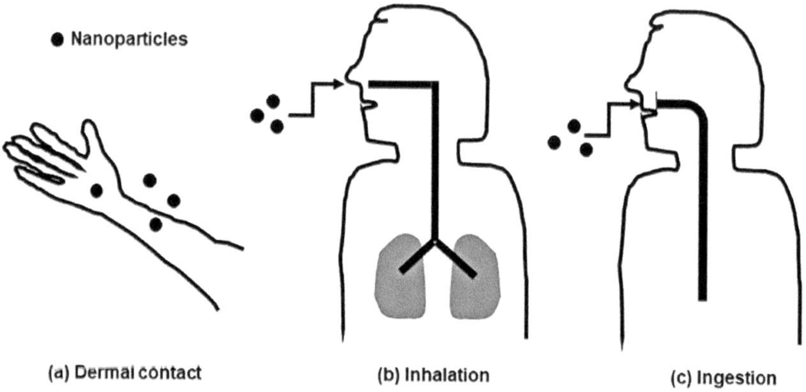

FIGURE 9.1 The three main routes of nanoparticles exposure in human body (a) dermal contact, (b) inhalation, and (c) ingestion.

9.2.1 DERMAL EXPOSURE

Nanomaterials have the ability to penetrate through skin following dermal exposure. Normally, the nanomaterials present in cosmetics and medical goods are expected to enter into our body through skin. In recent years, there is a great focus on the skin as a potential route of absorption of nanoparticles due to increased consumption of cosmetics and sunscreens. In general, an intact skin is found to be a protective barrier and prevents the entry of nanoparticles. But, it may not be true in all cases, as many studies have shown that some fine nanoparticles may gain entry through skin and get translocated and deposited in other sites. Nanoparticles are able to penetrate through the outer layers of the skin and there is little information on the hazard which they might cause (Maynard, 2006). Some evidences suggest that the penetration of inorganic nanoparticles through skin can cause harmful impacts such as oxidative damage due to the generation of free radicals (Kreilgaard, 2002) and elicitation of an immune response (Wakefield et al., 2004). Furthermore, the entrance through skin penetration is a matter of high concern of occupational hazard, especially for workers, professionals, and consumers coming in direct and regular contact with nanoparticles (Youssef, 2013). Once they penetrate into the biological environment, nanoparticles will inevitably come into contact with a huge variety of biological macromolecules (proteins, sugars, and lipids) which are dissolved in our body fluids such as the interstitial fluid between cells, lymph, or blood (Farhoodi, 2016).

9.2.2 INHALATION

Entry of nanoparticles through lungs via inhalation of nonfood and agri-food products is considered as a relevant route of exposure. Nanoparticles mainly get entry via inhalation during its production, such as handling, aerosols get generated while vortexing, weighing, mixing, sonication, and blending (Oberdörster et al., 2005). The important physicochemical characteristics of nanoparticles, such as particle size, shape, distribution, chemical structure, mass, and rate of accumulation are highly crucial in determining the potential hazards it may cause after human body exposure (Chau et al., 2007). Studies have shown that nanoparticles with size <10 nm can pass from the nasal cavity into lungs and then penetrate the alveolar region. Hence, low mass nanoparticles are potentially dangerous compared with larger ones. Toxicity issues related to nanoparticles through inhalation is generally considered to occur as an occupational hazard. Titanium dioxide, a commonly used antimicrobial nanomaterial in food products is often reported to trigger pneumonia and pulmonary infections due to its accumulation in the lungs (Nel, 2006). Much supporting evidence is still lacking on the hazards caused by inhalation of nanoingredients from food which needs to be evaluated by individual case studies (Chau et al., 2007).

9.2.3 INGESTION

Currently, researchers have observed that the exposure to nanomaterials occurs through ingestion in higher concentration can cause negative effects in humans. Exposure to nanoparticles through the gastrointestinal tract (GIT) occurring via direct oral uptake or ingestion of nanofood is the most common and frequent route. Nanoparticles may also accidentally arrive in GIT contact via leaching/migration from nanomaterial-based packaging to food commodities (He and Hwang, 2016). By the addition of nanostructured materials to food, water, and drugs, nanoparticles can be absorbed from the intestine, and may even enter the circulatory system. The interaction of nanoparticles with the microvilli present in the intestine plays an important role in the further absorption of nanomaterials and transport through mucous barrier to other tissues and organs (Chen et al., 2006). Liver and spleen are mainly responsible for the distribution of nanoparticles, mediating their passage from intestine to the blood circulation (Dimitrijevic et al., 2015). Nanoparticles that find their way to blood stream may influence the blood vessel lining and their function, may lead to blood clotting, or may even contributes toward cardiovascular diseases (Dimitrijevic et al., 2015).

Many studies reported that the presence of nanoparticles can elicit pathophysiological changes in the human body even leading to cancer (Gatti, 2004), cardiovascular diseases (Duan et al., 2013), neurodegenerative disorders, such as Alzheimer's and Parkinson's disease (Migliore et al., 2015), respiratory tract discomfort (Qiao et al., 2015), and other manifestations. Several nanoparticles have been reported to cause cellular damage to biological systems when they accumulate within the system. At times, they also disrupt the normal working of the cellular components within the biological system as they attach to cellular receptors of the cells of the immune system and confound them (Jordan et al., 2005). The nanoparticles also at times get coated with proteins and this leads to the degradation of the protein and hence the normal cellular mechanism gets disrupted. The smaller nanoparticles are found to be absorbed by the GIT more readily than the larger ones. Also, longer the duration of nanoparticles remaining in the intestine, higher will be the exposure. The reduced size of the nanoparticles also allows them to cross the blood-brain barrier and transit via blood and lymphatics to various organs and tissues and cause systemic damage (Maness et al., 1999).

9.3 FACTORS INFLUENCING NANOPARTICLE TOXICITY

The propensity of nanoparticles to cause toxicity depends upon its varied physicochemical characteristics, such as its dimensions as size, differential shape, high surface-to-volume ratio, well-ordered composition, solubility, hydrophobic or hydrophilic nature, surface and core composition, surface charge, aggregation state, presence of biochemical moieties on surface, and interfacial properties. All these properties individually or in combination can affect is the fate of toxicity of nanoscale material. Herein, we discuss on the various characteristics of nanoparticles that can lead to toxicity.

9.3.1 SIZE AND SHAPE

Depending upon the food matrix, the dimensions of the nanoparticles incorporated in it may vary from few nanometers to hundred nanometers. The dimensions of particles greatly determine the fate and toxicity of nanoparticles in food. The small size of nanoparticles provides an opportunity for augmented uptake and increased interaction with the biological tissues (Nel, 2006). It is documented that the smaller nanoparticles are found to be easily dissolved or digested by GIT than larger nanoparticles of the same composition. Moreover, the small size offers them more surface area for the GIT components to interact with them and can also penetrate into the mucous layer coating the epithelial cells (Bellmann et al., 2015; McClements and Xiao, 2017). This is because when the particle size reduces, the surface area-to-volume ratio increases, which again may promote increased display of their atoms/molecules on the surface thereby enabling increased interaction of nanomaterials with biological entities, culminating in adverse effects. Cho and colleagues evaluated the size-dependant cytotoxicity effect by observing the changes occurring in the acute toxicity of intraperitoneally administered AgNPs of 10, 60, and 100 nm sizes on female BALB/c mice. The results showed that the mice treated with the smallest size (10 nm) AgNPs showed histopathological changes including spleen congestion, vacuolation, liver necrosis, and apoptosis in thymus cortex. These findings were absent in the case of mice treated with 60 and 100 nm AgNPs (Cho et al., 2018). Similar like size, the shape of nanoparticles can also have influence on determining their toxicity. This is an area which is under investigation and only few information is available. The shape of a nanoparticle also contributes significantly to the surface area. For example, an octagonal particle will have different surface area compared with a spherical particle with similar size (Navya and Daima, 2016). In this milieu, Mahmoudi et al. demonstrated that biological macromolecules exhibited more effective rate of uptake in the case of spherical nanoparticles compared with the nanorods (Mahmoudi et al., 2011).

9.3.2 COMPOSITION

The composition of nanoparticles included in food is an important factor which determines its toxicity nature and fate in the GIT. Organic nanoparticles consisting of carbohydrates, proteins, lipids, and vitamins can be digested by proteases, amylases and lipases present in mouth, stomach, and upper GIT. But, at the same time, organic nanoparticles of mineral oil and dietary fibers cannot be digested by the GIT, which can turn out to be hazardous later. Inorganic nanoparticles including carbon-based, metallic, and metal oxide nanoparticles, such as carbon nanotubes, silver, titanium dioxide, silicon dioxide, iron oxide, and zinc oxide, do not get digested by the upper GIT and is associated with adverse effects. These nanoparticles can relocate to the lower GIT and cause detrimental effects, such as the release of ROS and can also alter the normal microbiome (Fröhlich and Fröhlich, 2016; McClements and Xiao, 2017). Table 9.2 lists the possible adverse effects caused by organic and inorganic nanoparticles after exposure in humans.

9.3.3 SURFACE CHARGE

Surface charge of nanomaterials play a crucial role in deciding the toxic nature of nanoparticles after its interaction with biological macromolecules. Nanoparticles can have positive, negative, or neutral charge. Reports suggest that positively charged nanomaterials can have more toxic effects on living things, than compared with negatively charged ones. The reason behind this is found to be associated with the interaction of positively charged nanoparticles with negatively charged genetic material (DNA, RNA), triggering genotoxicity. On the other hand, negatively charged nanoparticles are found to get repelled on the surfaces of biological entities possessing alike charges (Wiesner et al., 2006).

9.3.4 AGGREGATION STATE AND PROTEIN CORONA

Depending upon the pH, ionic strength, mechanical forces, and ingredient interactions present in the reactive environment, nanoparticles can exist as single individual particles and can also exist in clustered or aggregated form. Even though we can fabricate nanoparticles with controlled size in a laboratory set-up, there are chances for them to aggregate into larger clusters, when being in realistic environment. The aggregated state of

TABLE 9.2 Adverse Effects Associated with Inorganic and Organic Nanoparticles.

Nanoparticles		Purpose	Route of entry	Estimated human exposure	Risks	References
Inorganic nanoparticles	Silver nanoparticles (AgNPs)	To provide antimicrobial effect to food packagings and as edible food coatings	Oral uptake from foods or migration from food packagings	20–80 μg/day	Associated with a. Cytotoxicity, causing damage to cellular membrane, nucleus and organelles due to generation of ROS b. alteration of gut microbiota and abnormal mucus content in intestine ad lymphocyte infiltration c. penetration into organs like brain, kidney, intestine, liver, spleen and stomach	Cattò et al. (2019), Fröhlich and Fröhlich (2016), Hendrickson et al. (2016), Mercier-Bonin et al. (2018), van den Brule et al. (2015), and Williams et al. (2015, 2016)
	Titanium dioxide (TiO_2NPs)	Lightening agents used in milk powders, desserts, gums, and candies	Oral uptake through incorporate foods	United Kingdom: 1.1 mg/kg (bw/day) Unites states: 2.2 mg/kg	a. Fate in GIT varies according to the size, shape, crystal form, and aggregation state. b. accumulation in liver, kidney, spleen, lungs causing damage to blood, and cardiovascular system	Ali et al. (2019), Banaee et al. (2019), Baranowska-Wójcik et al. (2020), Chen et al. (2019), Mu et al. (2019) Shabbir et al. (2021), and Zhang et al. (2020)
	Zinc oxide (ZnO NPs)	Antimicrobial agents used in foods and packaging materials and as UV absorbers to protect food, nutritional supplements	Oral uptake from foods and migration from packaging materials		Cell damage due to ROS generation, injuries to liver, kidney, intestine and lungs, dysfunction of reproductive organs	Aboulhoda et al. (2020), Du et al. (2018), Duan et al. (2013), Kuang et al. (2020), Srivastav et al. (2019), Wang et al. (2017), and Yang et al. (2017)

TABLE 9.2 *(Continued)*

Nanoparticles		Purpose	Route of entry	Estimated human exposure	Risks	References
	Silica nanoparticles (SiO_2 NPs)	Anticaking agents in milk powder, salt, spices, icing sugar	Oral uptake from incorporated foods	20–50 mg/day	Cytotoxicity due to ROS, accumulation in liver	Guo et al. (2021, 2018), Hempt et al. (2020), Que et al. (2020), Sohal et al. (2020), and Yan et al. (2020)
	Iron nanoparticles (Fe_2O_3 NPs)	Colorant in foods such as sausages and dietary supplements	Oral uptake from foods	10–23 mg/day	Cytotoxicity by generation of RCS	De Matteis (2017), Garcia-Fernandez et al. (2020), Laffon et al. (2018), Shah and Dobrovolskaia (2018) and Wang et al. (2020)
Organic nanoparticles	Carbohydrate nanoparticles	Carrier for nutraceuticals and additives	Oral uptake from incorporated foods		Potential toxicity not yet reported. Fate in GIT depends on its digestibility, alteration of gut microbiota and bioactivity and bioavailability of the supplemented food	Liu et al. (2019), McClements and Xiao (2017), Toragall et al. (2020), and Yang et al. (2020)
	Protein nanoparticles	Helps in transport and delivery of nutrients, colorants, preservatives and other bioactive agents	Oral uptake from incorporated foods		Potential toxicity not yet reported. Fate in GIT depends on environmental conditions	Cao et al. (2019), Han et al. (2019), Hong et al. (2020), Tarhini et al. (2017), and Verma et al. (2018)
	Lipid nanoparticles	Delivery system for antimicrobials, antioxidants, nutrients, colorants	Oral uptake from incorporated foods		Potential toxicity not yet reported. Fate in GIT depends on environmental conditions	Campos et al. (2020), Rahman et al. (2021), and Talegaonkar and Bhattacharyya (2019)

Source: Adapted and modified with permission from Tarhan (2020). Copyrights @ 2020 Elsevier Inc.

nanoparticles may contain particles of varied shape, size, and strength, which are held together by Van der waal, hydrophobic, and electrostatic forces. The nanoparticle aggregates are found to have larger dimensions compared with single particles, which thereby determines their fate and toxic propensity. So, it is very vital to study the stability of nanoparticles in experimental set-up or physiological condition to recognize the tendency of nanoparticles to form colloidal aggregates.

Another important factor that promotes the toxicity of nanoparticles is the formation of surface protein corona. The proteins present in biological fluids can bind to nanoparticles and generate a surface coating known as protein corona. This can have adverse effects on the normal functioning of biological systems (Liu et al., 2020; Monopoli et al., 2012; Tenzer et al., 2013). So, it is imperative to investigate the possibility of such corona formation before the application of nanoparticles in food commodities. Figure 9.2 summarizes the different physicochemical factors of nanoparticles that influence their toxic behavior. All these evidences show that for effective inclusion in food products, these physicochemical parameters of nanoparticles must be considered seriously and hence rationally engineered for keeping their toxicity issues in count.

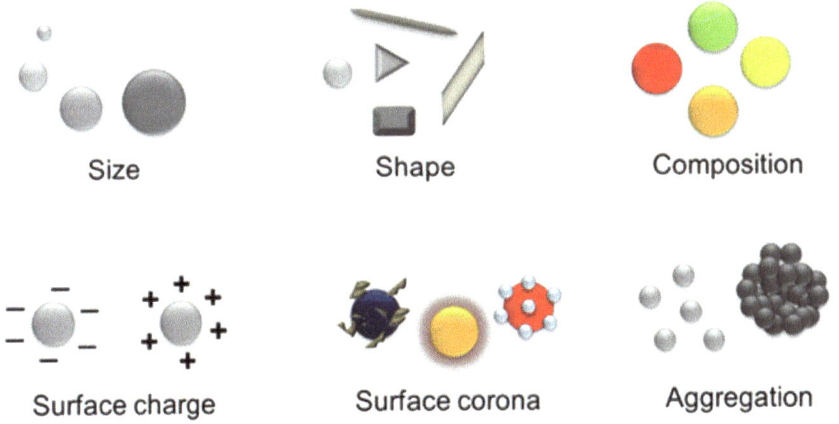

FIGURE 9.2 Illustration representing physicochemical properties of nanoparticles that influence its toxicity.

Source: Reprinted with permission from Navya and Daima (2016), Open access article. (http://creativecommons.org/licenses/by/4.0/).

9.4 MECHANISMS OF NANOPARTICLE TOXICITY

The nanoparticles that enter into human body in over concentration and those persist for longer duration can lead to toxicity due to their peculiar physicochemical and physiological mechanisms. This section briefs some of the important mechanisms of nanoparticle toxicity. As mentioned earlier, the inorganic nanoparticles incorporated in nanofoods are mainly responsible for inducing toxicity.

9.4.1 INTERFERENCE WITH NORMAL GIT FUNCTION

The normal functions of the GIT can be altered due to the presence of nanoparticles. The interference of nanoparticles on GIT function can be of many ways. First, the smaller size of the nanoparticles enables them to have a high specific surface area which offers large area for the adsorption of surface-active components, such as metabolic enzymes and globular proteins. Such adsorption can interfere with their normal catalytic functions. Secondly, the high levels of nanoparticles can reduce the rate of digestion of starch, lipids, and proteins normally present within GIT (McClements and Xiao, 2017). Thirdly, the presence of nanoparticles can physically disrupt the important structures of GIT such as the microvilli or the tight junctions, thereby interrupting the normal nutrient absorption and protective functions of the cells (Fröhlich and Fröhlich, 2016). Lastly, the nanoparticles present in the GIT can also elicit an immune response, creating adverse effects in human health (Őrfi and Szebeni, 2016).

9.4.2 ACCUMULATION IN ORGANS AND TISSUES

The ingested nanoparticles have shown evidences of being adsorbed in the GIT and accumulated in other specific tissues. Depending upon the above-mentioned nanoparticle characteristics, they can have various fates after entry into the human body. It is presumed that the nanoparticles can travel along the mucus layer of the GIT and get adsorbed actively or passively into the cells. After being adsorbed, the nanoparticles can either be metabolized, transferred out of cells to other sites, or get accumulated within other organs. When the nanoparticles get accumulated in specific tissues and exceeds the accumulation threshold level, long-term problems may arise. This type of toxicity is often seen associated with inorganic nanoparticles.

9.4.3 CELLULAR MALFUNCTION AND CYTOTOXICITY

Depending upon their composition and structure, nanoparticles can cause cytotoxicity and malfunctioning of cells. Nanoparticles are known to generate reactive oxygen species (ROS), such as singlet oxygen, hydrogen peroxide, hydroxyl radicals, and superoxides. These ROS released can interact with proteins, lipids and nucleic acids present in bodily fluids and can thereby damage the cell membrane, nucleus, and other organelles. This mechanism of nanoparticle cytotoxicity is found to be mostly associated with inorganic nanoparticles.

9.4.4 ALTERATION IN TARGETED RELEASE

Sometimes, nanoparticles encapsulated with bioactive agents intended for targeted release in mouth, stomach, or small intestine may undergo alteration from their targeted location and get deposited in other sites such as the GIT. As a result of this, the physiological impact that the bioactive agent exerts in the GIT would create adverse effects. Also, our GIT is normally inhabited by various microflora which play an important role in healthy digestion and well-being. In some situations, antimicrobial agents encapsulated within nanoparticles similarly can also undergo altered release in GIT, which may alter the viability and eliminate the normal flora of the colon, hence resulting in serious health condition (McClements and Xiao, 2017). The proper and exact mechanism of how this targeted location is altered has to be well researched on a case-by-case basis.

9.4.5 INCREASES PESTICIDE BIOAVAILABILITY

Nanoparticles have the ability to enhance the oral bioavailability of hydrophobic substances, such as pesticides and hormones present in foods, bringing up health hazards. For instance, foods, such as beverages, salads, dressings, or cream may have the presence of incorporated lipid nanoparticles which thereby can increase the oral bioavailability of pesticides present on the surface of the fruits or vegetables. The actual information behind this mechanism is still unknown and is often related to nanoemulsions of lipid nanoparticles in food.

9.5 METHODS OF NANOTECHNOLOGY RISK ASSESSMENT

As seen above, it is obvious that the nanoparticles must be investigated efficiently to check if they possess certain necessary criteria such as specific properties to be devoid of hazardous consequences. However, it is sure that humans and other living things often come to exposure to these nanoparticles, which can enter into our body through various routes and can also interact with the biological system. The impact of nanoparticles on human health and environment is an intense area requiring serious research. However, no comprehensive data is yet available on the potential risks caused due to the exposure of nanoparticles and its interaction with the biological system. Several in vitro and in vivo approaches are available for determining the toxicity related to nanoparticle containing food products. The EFSA committee has published certain important guidelines regarding the safe handling and the risk assessment of nanotechnology in food commodities and food chain (EFSA Scientific Committee et al., 2018).

9.5.1 CONVENTIONAL IN VITRO AND IN VIVO ASSAYS

The conventional methods employed for assessing the risk of nanomaterials incorporated in food include cytotoxic studies by in vitro cell culture assay and histopathological studies by in vivo animal experiments. The in vitro studies include cell-based assays with phenotypic readout parameters, such as oxidative stress, apoptosis, membrane integrity, cell morphology, and proliferation. The conventional cytotoxicity testing when compared with animal testing methods is devoid of ethical problems, and is cost-effective and faster in evaluation, at the expense of lower predictive value. Another approach of assessing the risk of nanoparticles is by animal experiments commonly using rodent models. This includes exposure of rodents to the particular nanoparticle and examining the histopathological effects on organs, such as liver, kidney, spleen, lung, brain, gastrointestinal tract, and blood analysis. But conducting this analysis alone is not sufficient to assess the risk factor of nanoparticles. Many in vitro and in vivo studies have demonstrated the hazardous effect of nanoparticles exposure, but the predictive value of these tests in human body remains unclear. This happens because in experiments, unrealistic higher doses are tested and moreover, the anatomical and physiological difference between humans and animals also accounts.

9.5.2 ADVANCED METHODS

9.5.2.1 EX VIVO METHODS

Today, nanotoxicity testing is conducted using advanced cellular models in order to overcome the limitations of conventional animal testing. In vivo animal experiments are considered as the gold standard for nano-toxicity testing, but in recent years, this had been replaced with ex vivo and in vitro studies. The complete replacement of animal studies is not possible but still new initiatives in toxicity testing has been introduced, called the ex vivo methods which includes the testing of tissue explants, tissue sections, and advanced in vitro methods, such as coculture, culture in 3D systems, and flow and mechanical stimulation (Fröhlich, 2018). One of the best alternatives to replace the nanoparticle exposure and testing on animals is the application of ex vivo methods. Depending on the exposure route and the tissue under investigation, their use varies. Ex vivo samples are artificially reconstructed commercially available tissues which include protective epithelia and receptive barriers and even ready to use model systems of organs, such as liver, kidney, intestine, etc.

9.5.2.1.1 Nanoriskcat

Nanomaterial risk categorization concept, NanoRiskCat (NRC), is a conceptual decision support tool introduced by EPA to identify, assess, rank, communicate, and determine the toxicity and hazardous nature of the nanoparticle containing consumer products (Hansen et al., 2014, 2011). According to this approach, on the inventory of each product, a short title that describes the use of nanomaterial and a color code comprising of five dots is assigned. After the evaluation process, five-color-coded rank is given to the product out of which, the first three refer to the exposure potential to professional end-users, consumers, and the environment, in that order. The last two dots will refer to the potential effect on human and the environment. Based on the color indicators, four main categories of risk include high, medium, low, and unknown which are signified by red, yellow, green, and grey colors, respectively (Figure 9.3). The evaluation procedure takes into account the aspect ratio, its adverse impact in the bulk form, and factors and indicators that determine genotoxicity, mutagenicity, carcinogenicity, neurotoxicity, respiratory, and cardiovascular threats. For each case, the result is interpreted along with the scientific data and literature review. After processing of all the combined information, NanoRiskCat finally categorizes the nanoparticles based on all the collected data. NanoRiskCat approach possesses positive and negative sides. The negative side is that for the assessment, expert judgment panel, and the inclusion of latest data is required.

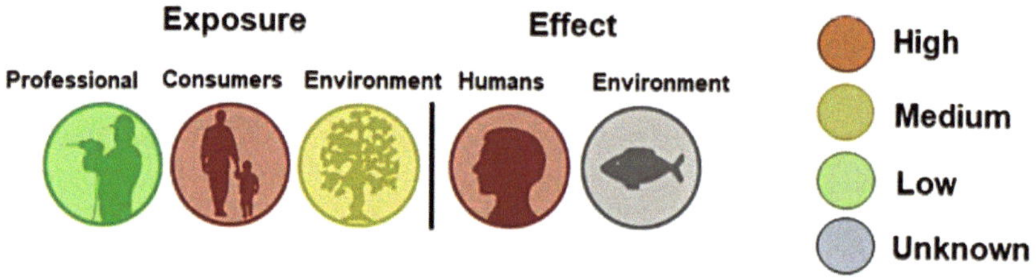

FIGURE 9.3 NanoRiskCat evaluation of a sample consumer product containing nanoparticles. The first three dots represent the potential exposure during intended use to professional, consumers, and environment. The last two dots refer to the hazardous effect on humans and environment respectively. The colored dots refer to specific risk category; red (high), yellow (medium), green (low), and grey (unknown) respectively.

Source: Adapted and modified with permission from Hansen et al. (2014).

9.5.2.2 SYSTEMS TOXICOLOGY

This is a novel strategy which avoids relying on high phenotypic responses in animals (Yang et al., 2010). This technique was initially termed as "systems biology by Ideker et al. to describe the integrated study of biological systems at molecular level, involving perturbation of systems, monitoring of molecular expressions, integrating response data, and modeling the systems molecular structure and network functions (Ideker et al., 2001). Later, Waters et al. used the term "systems toxicology" in a similar manner to describe toxicogenomic evaluation of biological systems (Waters, 2003). It is a high throughput molecular biology strategy and includes the "omics" techniques which includes genomics, epigenomics,

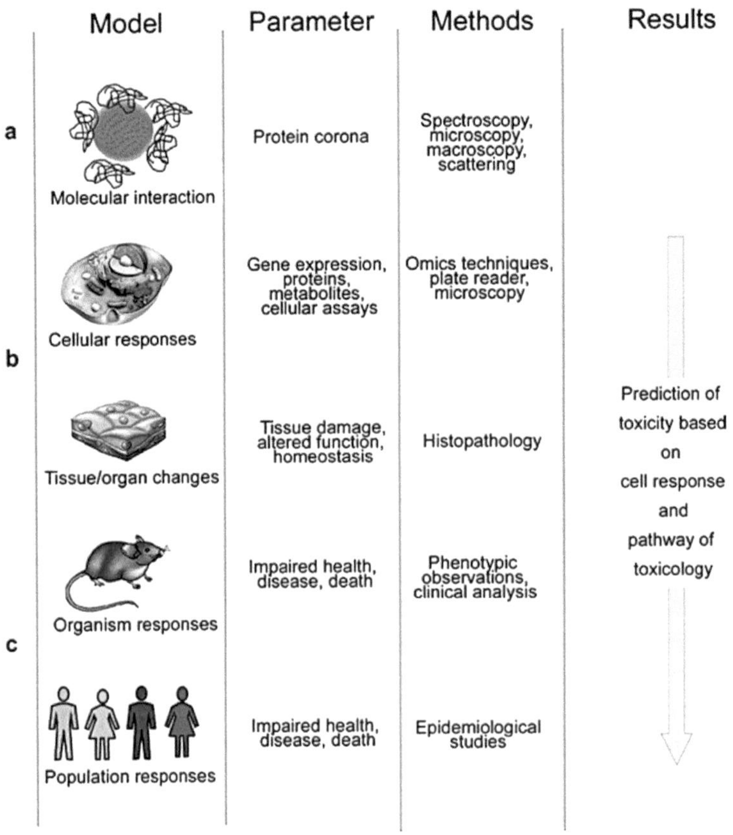

FIGURE 9.4 The models, readout parameters and methods adopted in systems toxicology strategy. (a) Represents the various spectroscopical analytical techniques used for the characterization of nanoparticles in food. The analytical techniques include UV-visible spectroscopy, infrared absorption, raman spectroscopy, photoluminescence, circular dichroism spectroscopy, electron paramagnetic spectroscopy, and fluorescence spectroscopy, (b) various biological assays utilize these technologies and also rely on readers based on luminescence, absorbance and fluorescence, image analysis and on other separation and detection platforms, such as mass spectrometry, gas chromatography, nuclear magnetic resonance spectroscopy, electrophoresis, and high-pressure liquid chromatography etc. (c) then the analysis of nanoparticle effects on (1) organs are done mainly by histopathology and staining techniques, (2) whole animal can be done by imaging techniques such as ultrasound imaging, magnetic resonance, radiography, computed tomography, photoacoustic tomography, positron emission tomography, single photon emission computed tomography, thermography, etc., and (3) as well as observation of changes in behavior, appearance, deterioration to health and death. The predicted value of nanoparticles to human toxicology increases from top to bottom.

Source: Reprinted with permission from Fröhlich (2017), Open access article. (http://creativecommons.org/licenses/by/4.0/).

proteomics, metabolomics, and transcriptomics data to achieve a better mechanistic understanding of the toxicological response (Yang et al., 2010). Figure 9.4 depicts the models, readout parameters, and methods applied in systems toxicology. One of the important features of omics technique is that it provides mechanistic information and identifies the toxicity pathway which is not possible in conventional assessment methods. Another specialty of this technique is that it helps to identify the adaptive responses elicited by cells or organisms under stress to toxic nanoparticles even exposed to lower levels, without the onset of phenotypic changes.

The advantages of using this method include the identification of new targets, markers, and molecular mechanisms for nanoparticle toxicity. The peculiarity of omics strategy is that it can be used both for in vitro and in vivo evaluation of nanoparticles. Another advantage includes low interference of omics techniques with nanoparticles. But the disadvantages include the requirement of skilled personnel and expensive infrastructure for the sample preparation and data analysis. We advise the readers to read other literatures for detailed information on the different techniques employed in systems toxicology (Costa and Fadeel, 2016; Fadeel et al., 2018; Fröhlich, 2017; Yang et al., 2010).

9.6 PUBLIC AWARENESS AND PERCEPTION

Public perception and acceptance are very vital for the commercial success of any food product. Misinformation or no information on the use of nanomaterials in food can significantly affect the public preferences. Public concerns about health hazards are often unpredictable, dictated more by fashion than by sober consideration of available facts. Many studies in this regard showed that public response toward the awareness of nanotechnology is neutral, but there is more deflection toward those applications where nanoscience is not directly applied to food. Giles et al. (2015) suggested that the use of nanomaterials in food packages is more likely to be accepted by public rather than the direct incorporation of nanoparticles in food.

In the marketing world, public awareness and acceptance are important but often ignored by a food manufacturer. Most of the food manufacturers would prefer to keep their new product development "underground" and would not like to share with the public (maybe partly due to the competition and trade secret) (Chun, 2009). This can be a conflict with the fact that the public intend to know what and why the food manufacturer is marketing a new product. Consumers are expected to purchase products which are low-priced and offer more benefits, but reluctance was also seen toward the use of nanotechnology in food (Siegrist et al., 2007; Spence and Townsend, 2006). In this context, the European Commission has introduced a regulation (EU, 2011, no 1169/2011) which imposes proper labeling of any nano-ingredient food so that the public is aware of what they consume (http://data.europa.eu/eli/reg/2011/1169/oj, 2011). According to this regulation, it is the manufacturer's responsibility to keep the product's information transparent and available to the public. However, this regulation was not successfully applied in all aspects. There were many cases reported where nanoparticles were detected in products which had no such labels (Ropers, 2019). So in general, people are less aware about the ingredients in their food.

To determine public perception of nanofoods and nanomaterials containing food and containers, sensory analysis and surveys seeking information about consumer attitudes, feelings, beliefs, motivation, and reasons can be done. In a literature review, Siegrist et al. conducted a survey to evaluate the public perception of different types of food materials, which included an antibacterial food packaging material, a nanocoating on tomatoes which protects from humidity and oxygen, bread containing nanocapsules of omega-3 fatty acids, and a juice with vitamin A encapsulated in starch (Siegrist et al., 2007). The survey data obtained from 153 people revealed that public acceptance to nanomaterial-incorporated packaging was more compared with their view to nanotechnology-engineered foods. These results also supported the hypothesis that nanotechnology inside a food is perceived as less acceptable than being on the outside (i.e., in the food packaging). A survey study conducted in Singapore has already proved that unawareness of nanotechnology and the adverse effects of nanotechnology increase the negative perception among the public (George et al., 2014). Another survey conducted on the island of Ireland reported that even the agri-food organization (stakeholders) also has a very low awareness with regards to nanotechnology (Handford et al., 2015). According to Cushen et al. on the public acceptance of nanotechnology in the United States, consumers were found to have limited knowledge about this technology, but had an optimistic perception (Cushen et al., 2012). Studies in Europe showed less positive results (Gaskell et al., 2004), while results in Taiwan were more positive where the perception of benefits was higher (Chen et al., 2013).

Therefore, these survey reports show that proper maintenance of the regulations, mandatory labeling, and by strictly conducting the risk assessment tests, are expected to shape the public perception on nanofoods. Improving the knowledge and social trust among consumers could improve the public perception regarding the use of nanotechnology in food (Siegrist et al., 2007). Furthermore, in order to improve the public acceptance of food nanotechnology, major efforts to ensure the safety of nanofoods need to be undertaken by governments, manufacturers, and concerned authorities (Chen et al., 2013).

9.7 CONCLUSION

Different nanoinnovations in food technology have benefitted the producers, suppliers and consumers to a great extent. On the other side, the wider applications of such innovations have evoked serious concern and uncertainties due to their exposure, interaction, and deposition of nanomaterials with biological macromolecules in human body and environment. Adding to this, it is worth to mention that majority of the public is unaware about the general aspects regarding the inclusion of nanoparticles in food and food packagings. In this context, it is the prime responsibility of the producers/manufacturers to inform the public about the nanoingredient involved in their product and detail the potential function as well as its hazardous effect on the human health and environment. Public acceptance and perception are always essential for every product to become successfully marketed. For bringing this out, certain authorities have introduced legislations, regulations, and guidelines for the safe handling and efficient application of nanomaterials in food industry. With the implementation of such guidelines, the inclusion of highly hazardous nanoparticles and uncontrolled use can be excluded. If these guidelines and regulations are to be followed globally, then it would be easy to implement the safe use of nanomaterial containing products. Moreover, further detailed investigations on each nano-based product introduced in market are necessary to unravel the possible health and environmental complications associated with it.

KEYWORDS

- **nanotoxicity**
- **inorganic nanoparticles**
- **organic nanoparticles**
- **route of exposure**
- **risk assessment**
- **in vitro and in vivo assays**

REFERENCES

Aboulhoda, B. E.; Abdeltawab, D. A.; Rashed, L. A.; Abd Alla, M. F.; Yassa, H. D. Hepatotoxic Effect of Oral Zinc Oxide Nanoparticles and the Ameliorating Role of Selenium in Rats: A Histological, Immunohistochemical and Molecular Study. *Tissue Cell* **2020,** *67*, 101441.

Acosta, E. Bioavailability of Nanoparticles in Nutrient and Nutraceutical Delivery. *Curr. Opin. Colloid Interf. Sci.* **2009,** *14*, 3–15.

Ali, S. A.; Rizk, M. Z.; Hamed, M. A.; Aboul-Ela, E. I.; El-Rigal, N. S.; Aly, H. F.; Abdel-Hamid, A.-H. Z. Assessment of Titanium Dioxide Nanoparticles Toxicity via Oral Exposure in Mice: Effect of Dose and Particle Size. *Biomarkers* **2019,** *24*, 492–498.

Amenta, V.; Aschberger, K.; Arena, M.; Bouwmeester, H.; Botelho Moniz, F.; Brandhoff, P.; Gottardo, S.; Marvin, H. J. P.; Mech, A.; Quiros Pesudo, L.; Rauscher, H.; Schoonjans, R.; Vettori, M. V.; Weigel, S.; Peters, R. J. Regulatory Aspects of Nanotechnology in the Agri/Feed/Food Sector in EU and Non-EU Countries. *Regul. Toxicol. Pharmacol.* **2015,** *73*, 463–476.

Banaee, M.; Tahery, S.; Nematdoost Haghi, B.; Shahafve, S.; Vaziriyan, M. Blood Biochemical Changes in Common Carp (*Cyprinus carpio*) Upon Co-Exposure to Titanium Dioxide Nanoparticles and Paraquat. *Iran. J. Fish. Sci.* **2019,** *18*.

Baranowska-Wójcik, E.; Szwajgier, D.; Oleszczuk, P.; Winiarska-Mieczan, A. Effects of Titanium Dioxide Nanoparticles Exposure on Human Health—a Review. *Biol. Trace Elem. Res.* **2020,** *193*, 118–129.

Bellmann, S.; Carlander, D.; Fasano, A.; Momcilovic, D.; Scimeca, J. A.; Waldman, W. J.; Gombau, L.; Tsytsikova, L.; Canady, R.; Pereira, D. I. A.; Lefebvre, D. E. Mammalian Gastrointestinal Tract Parameters Modulating the Integrity, Surface Properties, and Absorption of Food-Relevant Nanomaterials: Nanomaterial Gastrointestinal Absorption. Wiley *Interdiscip. Rev. Nanomed. Nanobiotechnol.* **2015,** *7*, 609–622.

Buzea, C.; Pacheco, I. I.; Robbie, K. Nanomaterials and Nanoparticles: Sources and Toxicity. *Biointerphases* **2007,** *2*, MR17–MR71.

Campos, J. R.; Severino, P.; Santini, A.; Silva, A. M.; Shegokar, R.; Souto, S. B.; Souto, E. B. Solid Lipid Nanoparticles (SLN). In *Nanopharmaceuticals*; Elsevier, 2020; pp 1–15.

Cao, S.; Xu, S.; Wang, H.; Ling, Y.; Dong, J.; Xia, R.; Sun, X. Nanoparticles: Oral Delivery for Protein and Peptide Drugs. *AAPS PharmSciTech* **2019,** *20*, 190.

Cattò, C.; Garuglieri, E.; Borruso, L.; Erba, D.; Casiraghi, M. C.; Cappitelli, F.; Villa, F.; Zecchin, S.; Zanchi, R. Impacts of Dietary Silver Nanoparticles and Probiotic Administration on the Microbiota of an in-Vitro Gut Model. *Environ. Pollut.* **2019,** *245*, 754–763.

Chau, C.-F.; Wu, S.-H.; Yen, G.-C. The Development of Regulations for Food Nanotechnology. *Trends Food Sci. Technol.* **2007,** *18*, 269–280.

Chen, L.; Remondetto, G. E.; Subirade, M. Food Protein-Based Materials as Nutraceutical Delivery Systems. *Trends Food Sci. Technol.* **2006,** *17*, 272–283.

Chen, M.-F.; Lin, Y.-P.; Cheng, T.-J. Public Attitudes toward Nanotechnology Applications in Taiwan. *Technovation* **2013,** *33*, 88–96.

Chen, Z.; Zhou, D.; Han, S.; Zhou, S.; Jia, G. Hepatotoxicity and the Role of the Gut-Liver Axis in Rats after Oral Administration of Titanium Dioxide Nanoparticles. *Part. Fibre Toxicol.* **2019,** *16*, 48.

Cho, Y.-M.; Mizuta, Y.; Akagi, J.; Toyoda, T.; Sone, M.; Ogawa, K. Size-Dependent Acute Toxicity of Silver Nanoparticles in Mice. *J. Toxicol. Pathol.* **2018,** *31*, 73–80.

Chun, A. L. Will the Public Swallow Nanofood? *Nat. Nanotechnol.* **2009,** *4*, 790–791.

Costa, P. M.; Fadeel, B. Emerging Systems Biology Approaches in Nanotoxicology: Towards a Mechanism-Based Understanding of Nanomaterial Hazard and Risk. *Toxicol. Appl. Pharmacol.* **2016,** *299*, 101–111.

Cushen, M.; Kerry, J.; Morris, M.; Cruz-Romero, M.; Cummins, E. Nanotechnologies in the Food Industry—Recent Developments, Risks and Regulation. *Trends Food Sci. Technol.* **2012,** *24*, 30–46.

De Matteis, V. Exposure to Inorganic Nanoparticles: Routes of Entry, Immune Response, Biodistribution and in Vitro/in Vivo Toxicity Evaluation. *Toxics* **2017,** *5*, 29.

Dimitrijevic, M.; Karabasil, N.; Boskovic, M.; Teodorovic, V.; Vasilev, D.; Djordjevic, V.; Kilibarda, N.; Cobanovic, N. Safety Aspects of Nanotechnology Applications in Food Packaging. *Procedia Food Sci.* **2015,** *5*, 57–60.

Du, L.-J.; Xiang, K.; Liu, J.-H.; Song, Z.-M.; Liu, Y.; Cao, A.; Wang, H.; Intestinal Injury Alters Tissue Distribution and Toxicity of ZnO Nanoparticles in Mice. *Toxicol. Lett.* **2018,** *295*, 74–85.

Duan, J.; Yu, Yongbo, Li, Y.; Yu, Yang, Sun, Z. Cardiovascular Toxicity Evaluation of Silica Nanoparticles in Endothelial Cells and Zebrafish Model. *Biomaterials* **2013,** *34*, 5853–5862.

EFSA Scientific Committee, Hardy, A.; Benford, D.; Halldorsson, T.; Jeger, M. J.; Knutsen, H. K.; More, S.; Naegeli, H.; Noteborn, H.; Ockleford, C.; Ricci, A.; Rychen, G.; Schlatter, J. R.; Silano, V.; Solecki, R.; Turck, D.; Younes, M.; Chaudhry, Q.; Cubadda, F.; Gott, D.; Oomen, A.; Weigel, S.; Karamitrou, M.; Schoonjans, R.; Mortensen, A.; Guidance on Risk Assessment of the Application of Nanoscience and Nanotechnologies in the Food and Feed Chain: Part 1, Human and Animal Health. *EFSA J.* **2018,** *16*.

Fadeel, B.; Farcal, L.; Hardy, B.; Vázquez-Campos, S.; Hristozov, D.; Marcomini, A.; Lynch, I.; Valsami-Jones, E.; Alenius, H.; Savolainen, K. Advanced Tools for the Safety Assessment of Nanomaterials. *Nat. Nanotechnol.* **2018,** *13*, 537–543.

Farhoodi, M.; Nanocomposite Materials for Food Packaging Applications: Characterization and Safety Evaluation. *Food Eng. Rev.* **2016,** *8*, 35–51.
FresherLonger™ Miracle Food Storage [WWW Document], 2007. https://www.nanotech-project.tech/cpi/products/fresherlongertm-miracle-food-storage/(accessed Feb 10, 2021).
Fröhlich, E. Comparison of Conventional and Advanced in Vitro Models in the Toxicity Testing of Nanoparticles. *Artif. Cells Nanomed. Biotechnol.* **2018,** *46*, 1091–1107.
Fröhlich, E. Role of Omics Techniques in the Toxicity Testing of Nanoparticles. *J. Nanobiotechnology* **2017,** *15*, 84.
Fröhlich, Esther; Fröhlich, Eleonore. Cytotoxicity of Nanoparticles Contained in Food on Intestinal Cells and the Gut Microbiota. *Int. J. Mol. Sci.* **2016,** *17*, 509.
Garcia-Fernandez, J.; Turiel, D.; Bettmer, J.; Jakubowski, N.; Panne, U.; Rivas García, L.; Llopis, J.; Sánchez González, C.; Montes-Bayón, M. In Vitro and in Situ Experiments to Evaluate the Biodistribution and Cellular Toxicity of Ultrasmall Iron Oxide Nanoparticles Potentially Used as Oral Iron Supplements. *Nanotoxicology* **2020,** *14*, 388–403.
Gaskell, G.; Eyck, T. T.; Jackson, J.; Veltri, G. Public Attitudes to Nanotech in Europe and the United States. *Nat. Mater.* **2004,** *3*, 496–496.
Gatti, A. M. Biocompatibility of Micro- and Nano-Particles in the Colon. Part II. *Biomaterials* **2004,** *25*, 385–392.
George, S.; Kaptan, G.; Lee, J.; Frewer, L. Awareness on Adverse Effects of Nanotechnology Increases Negative Perception among Public: Survey Study from Singapore. *J. Nanoparticle Res.* **2014,** *16*, 2751.
Giles, E. L.; Kuznesof, S.; Clark, B.; Hubbard, C.; Frewer, L. J. Consumer Acceptance of and Willingness to Pay for Food Nanotechnology: A Systematic Review. *J. Nanopart. Res.* **2015,** *17*.
Guo, C.; Liu, Y.; Li, Y. Adverse Effects of Amorphous Silica Nanoparticles: Focus on Human Cardiovascular Health. *J. Hazard. Mater.* **2021,** *406*, 124626.
Guo, Z.; Martucci, N. J.; Liu, Y.; Yoo, E.; Tako, E.; Mahler, G. J. Silicon Dioxide Nanoparticle Exposure Affects Small Intestine Function in an in Vitro Model. *Nanotoxicology* **2018,** *12*, 485–508.
Han, Y.; Gao, Z.; Chen, L.; Kang, L.; Huang, W.; Jin, M.; Wang, Q.; Bae, Y. H. Multifunctional Oral Delivery Systems for Enhanced Bioavailability of Therapeutic Peptides/Proteins. *Acta Pharm. Sin.* **2019,** *B9*, 902–922.
Handford, C. E.; Dean, M.; Spence, M.; Henchion, M.; Elliott, C. T.; Campbell, K. Awareness and Attitudes towards the Emerging Use of Nanotechnology in the Agri-Food Sector. *Food Control* **2015,** *57*, 24–34.
Hansen, S. F.; Baun, A.; Alstrup-Jensen, K. NanoRiskCat: A Conceptual Decision Support Tool for Nanomaterials, 2011.
Hansen, S. F.; Jensen, K. A.; Baun, A. NanoRiskCat: A Conceptual Tool for Categorization and Communication of Exposure Potentials and Hazards of Nanomaterials in Consumer Products. *J. Nanopart. Res.* **2014,** *16*, 2195.
He, X.; Hwang, H.-M. Nanotechnology in Food Science: Functionality, Applicability, and Safety Assessment. *J. Food Drug Anal.* **2016,** *24*, 671–681.
Hempt, C.; Kaiser, J.-P.; Scholder, O.; Buerki-Thurnherr, T.; Hofmann, H.; Rippl, A.; Schuster, T. B.; Wick, P.; Hirsch, C. The Impact of Synthetic Amorphous Silica (E 551) on Differentiated Caco-2 Cells, a Model for the Human Intestinal Epithelium. *Toxicol. in Vitro* **2020,** *67*, 104903.

Hendrickson, O. D.; Klochkov, S. G.; Novikova, O. V.; Bravova, I. M.; Shevtsova, E. F.; Safenkova, I. V.; Zherdev, A. V.; Bachurin, S. O.; Dzantiev, B. B. Toxicity of Nanosilver in Intragastric Studies: Biodistribution and Metabolic Effects. *Toxicol. Lett.* **2016,** *241*, 184–192.

Hong, S.; Choi, D. W.; Kim, H. N.; Park, C. G.; Lee, W.; Park, H. H. Protein-Based Nanoparticles as Drug Delivery Systems. *Pharmaceutics* **2020,** *12*, 604.

http://data.europa.eu/eli/reg/2011/1169/oj, 2011. EUR-Lex–32011R1169–EN–EUR-Lex [WWWDocument].https://eur-lex.europa.eu/legal-content/EN/ALL/?uri=CELEX%3A 32011R1169 (accessed Feb 11, 2021).

https://goodstate.com/products/good-state-liquid-ionic-iron-48-servings-at-10mg-plus-2-mg-fulvic-acid-8-fl-oz, 2013. Best Iron Supplement | Formulated for Healthy Blood Cells—Good State [WWW Document]. https://goodstate.com/products/good-state-liquid-ionic-iron-48-servings-at-10mg-plus-2-mg-fulvic-acid-8-fl-oz (accessed Feb 10, 2021).

https://nanodb.dk/en/product/?pid=3447, 2007. Product—The Nanodatabase [WWW Document]. https://nanodb.dk/en/product/?pid=3447 (accessed Feb 2, 2021).

https://nanodb.dk/en/product/?pid=4171, 2015. Product—The Nanodatabase [WWW Document]. https://nanodb.dk/en/product/?pid=6748 (accessed Feb 6, 2021).

https://nanodb.dk/en/product/?pid=5057, 2007. Product—The Nanodatabase [WWW Document]. https://nanodb.dk/en/product/?pid=5057 (accessed Feb 10, 21).

https://product.statnano.com/product/6075, 2016. Kinetic Go Green 8 piece Food Container Set [WWW Document]. https://product.statnano.com/product/6075 (accessed Feb 6, 2021).

https://product.statnano.com/product/6089/old-el-paso-taco-seasoning-mix, 2016. Old el Paso Taco Seasoning Mix|Seasoning|Nanotechnology Products|NPD [WWW Document]. https://product.statnano.com/product/6089/old-el-paso-taco-seasoning-mix (accessed Feb 6, 2021).

https://product.statnano.com/product/6093/dickinson%27s-coconut-curd, 2016. Dickinson's Coconut Curd [WWW Document]. https://product.statnano.com/product/6093/dickinson%27s-coconut-curd (accessed Feb 6, 2021).

https://product.statnano.com/product/6841/silver-nano-noble-gs-nursing-bottle-(newborn-baby-use), 2016. Silver-nano Noble GS nursing bottle (newborn-baby use) [WWW Document]. https://product.statnano.com/product/6841/silver-nano-noble-gs-nursing-bottle-(newborn-baby-use) (accessed Feb 6, 2021).

https://www.nanonutrausa.com/collections/all/products/liposomal-vitamin-c, 2017. Liposomal Vitamin C—NanoNutra [WWW Document]. https://www.nanonutrausa.com/collections/all/products/liposomal-vitamin-c (accessed Feb 7, 2021).

Ideker, T.; Galitski, T.; Hood, L. A New Approach to Decoding Life : Systems Biology. *Annu. Rev. Genomics Hum. Genet.* **2001,** *2*, 343–372.

Jordan, J.; Jacob, K. I.; Tannenbaum, R.; Sharaf, M. A.; Jasiuk, I. Experimental Trends in Polymer Nanocomposites—a Review. *Mater. Sci. Eng.* **2005,** *A393*, 1–11.

Joseph, T.; Morrison, M. Nanoforum.org, Institute for Nanotechnology (European Commission), Nanotechnology in Agriculture and Food: A Nanoforum Report. Nanoforum.org, 2006.

Kreilgaard, M. Influence of Microemulsions on Cutaneous Drug Delivery. *Adv. Drug Deliv. Rev.* **2002,** *54*, S77–S98.

Kuang, H.; Zhang, W.; Yang, L.; Aguilar, Z. P.; Xu, H. Reproductive Organ Dysfunction and Gene Expression after Orally Administration of ZnO Nanoparticles in Murine. *Environ. Toxicol. tox.* **2020,** 23060.

Laffon, B.; Fernández-Bertólez, N.; Costa, C.; Brandão, F.; Teixeira, J. P.; Pásaro, E.; Valdiglesias, V. Cellular and Molecular Toxicity of Iron Oxide Nanoparticles. In *Cellular and Molecular Toxicology of Nanoparticles, Advances in Experimental Medicine and Biology*; Saquib, Q., Faisal, M., Al-Khedhairy, A. A., Alatar, A. A., Eds.; Springer International Publishing: Cham, 2018; pp 199–213.

Liu, N.; Tang, M.; Ding, J. The Interaction Between Nanoparticles-Protein Corona Complex and Cells and Its Toxic Effect on Cells. *Chemosphere* **2020,** *245*, 125624.

Liu, X.; Zhang, B.; Sohal, I. S.; Bello, D.; Chen, H. Is "Nano Safe to Eat or Not"? A Review of the State-of-the Art in Soft Engineered Nanoparticle (sENP) Formulation and Delivery in Foods. In *Advances in Food and Nutrition Research*; Elsevier, 2019; pp 299–335.

Mahmoudi, M.; Azadmanesh, K.; Shokrgozar, M. A.; Journeay, W. S.; Laurent, S. Effect of Nanoparticles on the Cell Life Cycle. *Chem. Rev.* **2011,** *111*, 3407–3432.

Maisanaba, S.; Pichardo, S.; Puerto, M.; Gutiérrez-Praena, D.; Cameán, A. M.; Jos, A. Toxicological Evaluation of Clay Minerals and Derived Nanocomposites: A Review. *Environ. Res.* **2015,** *138*, 233–254.

Maness, P. C.; Smolinski, S.; Blake, D. M.; Huang, Z.; Wolfrum, E. J.; Jacoby, W. A. Bactericidal Activity of Photocatalytic TiO(2) Reaction: Toward an Understanding of Its Killing Mechanism. *Appl. Environ. Microbiol.* **1999,** *65*, 4094–4098.

Maynard, A. D. Nanotechnology: Assessing the Risks. *Nano Today* **2006,** *1*, 22–33.

McClements, D. J.; Xiao, H. Is Nano Safe in Foods? Establishing the Factors Impacting the Gastrointestinal Fate and Toxicity of Organic and Inorganic Food-Grade Nanoparticles. *Npj Sci. Food* **2017,** *1*, 6.

Mercier-Bonin, M.; Despax, B.; Raynaud, P.; Houdeau, E.; Thomas, M. Mucus and Microbiota as Emerging Players in Gut Nanotoxicology: The Example of Dietary Silver and Titanium Dioxide Nanoparticles. *Crit. Rev. Food Sci. Nutr.* **2018,** *58*, 1023–1032.

Migliore, L.; Uboldi, C.; Di Bucchianico, S.; Coppedè, F. Nanomaterials and Neurodegeneration: Nanomaterials and Neurodegeneration. *Environ. Mol. Mutagen.* **2015,** *56*, 149–170.

Monopoli, M. P.; Åberg, C.; Salvati, A.; Dawson, K. A. Biomolecular Coronas Provide the Biological Identity of Nanosized Materials. *Nat. Nanotechnol.* **2012,** *7*, 779–786.

Mu, W.; Wang, Y.; Huang, C.; Fu, Y.; Li, J.; Wang, H.; Jia, X.; Ba, Q. Effect of Long-Term Intake of Dietary Titanium Dioxide Nanoparticles on Intestine Inflammation in Mice. *J. Agric. Food Chem.* **2019,** *67*, 9382–9389.

Navya, P. N.; Daima, H. K. Rational Engineering of Physicochemical Properties of Nanomaterials for Biomedical Applications with Nanotoxicological Perspectives. *Nano Converg.* **2016,** *3*, 1.

Nel, A. Toxic Potential of Materials at the Nanolevel. *Science* **2006,** *311*, 622–627.

Oberdörster, G.; Oberdörster, E.; Oberdörster, J. Nanotoxicology: An Emerging Discipline Evolving from Studies of Ultrafine Particles. *Environ. Health Perspect.* **2005,** *113*, 823–839.

Őrfi, E.; Szebeni, J. The Immune System of the Gut and Potential Adverse Effects of Oral Nanocarriers on Its Function. *Adv. Drug Deliv.* **2016,** *106*, 402–409.

Qiao, H.; Liu, W.; Gu, H.; Wang, D.; Wang, Y. The Transport and Deposition of Nanoparticles in Respiratory System by Inhalation. *J. Nanomater.*, **2015,** 1–8.

Quan Zhou Hu Zheng Nano Technology Co., Ltd.® Nano-silver Storage Box (Baoxianhe) [WWW Document], 2007. https://www.nanotechproject.tech/cpi/products/

quan-zhou-hu-zheng-nano-technology-co-ltd-r-nano-silver-storage-box-baoxianhe/ (accessed Feb 7, 2021).

Que, D. E.; Hou, W.-C.; Yap Ang, M. B. M.; Lin, C.-C. Toxic Effects of Hydroxyl- and Amine-functionalized Silica Nanoparticles (SiO_2 and NH_2-SiO_2 NPs) on Caenorhabditis Elegans. *Aerosol Air Qual. Res.* **2020,** *20*, 1987–2020.

Rahman, M.; Souto, E. B.; Alam, K.; Beg, S. Biopharmaceutical Challenges in Using Lipid Nanoparticles for Oral Chemotherapy. In Nanoformulation Strategies for Cancer Treatment. Elsevier, 2021; pp 53–64.

Ropers, M.-H. Nanomaterials and Food Security: The Next Challenge for Consumers, Food Industries and Policies. In: *Encyclopedia of Food Security and Sustainability*; Elsevier, 2019; pp 575–581.

Shabbir, S.; Kulyar, M. F.-A.; Bhutta, Z. A.; Boruah, P.; Asif, M. Toxicological Consequences of Titanium Dioxide Nanoparticles (TiO_2NPs) and Their Jeopardy to Human Population. *BioNanoScience* 2021.

Shah, A.; Dobrovolskaia, M. A. Immunological Effects of Iron Oxide Nanoparticles and Iron-Based Complex Drug Formulations: Therapeutic Benefits, Toxicity, Mechanistic Insights, and Translational Considerations. *Nanomed. Nanotechnol. Biol. Med.* **2018,** *14*, 977–990.

Siegrist, M.; Cousin, M.-E.; Kastenholz, H.; Wiek, A. Public Acceptance of Nanotechnology Foods and Food Packaging: The Influence of Affect and Trust. *Appetite* **2007,** *49*, 459–466.

Sohal, I. S.; DeLoid, G. M.; O'Fallon, K. S.; Gaines, P.; Demokritou, P.; Bello, D. Effects of Ingested Food-Grade Titanium Dioxide, Silicon Dioxide, Iron (III) Oxide and Zinc Oxide Nanoparticles on an in Vitro Model of Intestinal Epithelium: Comparison between Monoculture vs. a Mucus-Secreting Coculture Model. *NanoImpact* **2020,** *17*, 100209.

Spence, A.; Townsend, E. Examining Consumer Behavior Toward Genetically Modified (GM) Food in Britain. *Risk Anal.* **2006,** *26*, 657–670.

Srivastav, A. K.; Dhiman, N.; Tiwari, R.; Arjaria, N.; Prakash, J.; Jagdale, P.; Ayanur, A.; Singh, D.; Patnaik, S.; Kumar, M. Sub-Acute Oral Exposure of Zinc Oxide Nanoparticles Causes Alteration in Iron Homeostasis through Acute Phase Response: A Protective Effect by Surface Modification. *J. Trace Elem. Med. Biol.* **2019,** *52*, 270–287.

Talegaonkar, S.; Bhattacharyya, A. Potential of Lipid Nanoparticles (SLNs and NLCs) in Enhancing Oral Bioavailability of Drugs with Poor Intestinal Permeability. *AAPS PharmSciTech* **2019,** *20*, 121.

Tarhan, Ö. Safety and Regulatory Issues of Nanomaterials in Foods. In *Handbook of Food Nanotechnology*; Elsevier, 2020, pp 655–703.

Tarhini, M.; Greige-Gerges, H.; Elaissari, A. Protein-Based Nanoparticles: From Preparation to Encapsulation of Active Molecules. *Int. J. Pharm.* **2017,** *522*, 172–197.

Tenzer, S.; Docter, D.; Kuharev, J.; Musyanovych, A.; Fetz, V.; Hecht, R.; Schlenk, F.; Fischer, D.; Kiouptsi, K.; Reinhardt, C.; Landfester, K.; Schild, H.; Maskos, M.; Knauer, S. K.; Stauber, R. H. Rapid Formation of Plasma Protein Corona Critically Affects Nanoparticle Pathophysiology. *Nat. Nanotechnol.* **2013,** *8*, 772–781.

Toragall, V.; Jayapala, N.; S P, M.; Vallikanan, B. Biodegradable Chitosan-Sodium Alginate-Oleic Acid Nanocarrier Promotes Bioavailability and Target Delivery of Lutein in Rat Model with No Toxicity. *Food Chem.* **2020,** *330*, 127195.

van den Brule, S.; Ambroise, J.; Lecloux, H.; Levard, C.; Soulas, R.; De Temmerman, P.-J.; Palmai-Pallag, M.; Marbaix, E.; Lison, D. Dietary Silver Nanoparticles Can Disturb the Gut Microbiota in Mice. *Part. Fibre Toxicol.* **2015,** *13*, 38.

Verma, D.; Gulati, N.; Kaul, S.; Mukherjee, S.; Nagaich, U. Protein Based Nanostructures for Drug Delivery. *J. Pharm.* **2018,** 1–18.

Wakefield, G.; Green, M.; Lipscomb, S.; Flutter, B. Modified Titania Nanomaterials for Sunscreen Applications—Reducing Free Radical Generation and DNA Damage. *Mater. Sci. Technol.* **2004,** *20*, 985–988.

Wang, C.; Cheng, K.; Zhou, L.; He, J.; Zheng, X.; Zhang, L.; Zhong, X.; Wang, T. Evaluation of Long-Term Toxicity of Oral Zinc Oxide Nanoparticles and Zinc Sulfate in Mice. *Biol. Trace Elem. Res.* **2017,** *178*, 276–282.

Wang, X.; Gong, J.; Tan, W.; Hu, T.; Rong, R.; Gui, Z.; Nie, K.; Xu, X. Adsorption of Proteins on oral Zn^{2+} Doped Iron Oxide Nanoparticles in Mouse Stomach and in Vitro: Triggering Nanoparticle Aggregation. *Nanoscale* **2020,** *12*, 22754–22767.

Waters, M. Systems Toxicology and the Chemical Effects in Biological Systems (CEBS) Knowledge Base. *Environ. Health Perspect.* **2003,** *111*, 811–812.

Wiesner, M. R.; Lowry, G. V.; Alvarez, P.; Dionysiou, D.; Biswas, P. Assessing the Risks of Manufactured Nanomaterials. *Environ. Sci. Technol.* **2006,** *40*, 4336–4345.

Williams, K.; Milner, J.; Boudreau, M. D.; Gokulan, K.; Cerniglia, C. E.; Khare, S. Effects of Subchronic Exposure of Silver Nanoparticles on Intestinal Microbiota and Gut-Associated Immune Responses in the Ileum of Sprague-Dawley Rats. *Nanotoxicology* **2015,** *9*, 279–289.

Williams, K. M.; Gokulan, K.; Cerniglia, C. E.; Khare, S. Size and Dose Dependent Effects of Silver Nanoparticle Exposure on Intestinal Permeability in an in Vitro Model of the Human Gut Epithelium. *J. Nanobiotechnol.* **2016,** *14*, 62.

www.nanotechproject.tech/cpi/products/canola-active-oil/, 2007. Canola Active Oil [WWW Document]. https://www.nanotechproject.tech/cpi/products/canola-active-oil/ (accessedFeb 6, 2021).

www.nanotechproject.tech/cpi/products/maternal-water/, 2009. Maternal Water [WWW Document]. https://www.nanotechproject.tech/cpi/products/maternal-water/ (accessed Feb 6, 2021).

www.nanotechproject.tech/cpi/products/nanoceuticalstm-slim-shake-chocolate/,2007.Nanoceuticals™ Slim Shake Chocolate [WWW Document]. https://www.nanotechproject.tech/cpi/products/nanoceuticalstm-slim-shake-chocolate/ (accessed Feb 6, 2021).

www.nanotechproject.tech/cpi/products/nanotea/, 2007. Nanotea [WWW Document]. URL https://www.nanotechproject.tech/cpi/products/nanotea/ (accessed Feb 6, 2021).

Yan, J.; Wang, D.; Li, K.; Chen, Q.; Lai, W.; Tian, L.; Lin, B.; Tan, Y.; Liu, X.; Xi, Z. Toxic Effects of the Food Additives Titanium Dioxide and Silica on the Murine Intestinal Tract: Mechanisms Related to Intestinal Barrier Dysfunction Involved by Gut Microbiota. *Environ. Toxicol. Pharmacol.* **2020,** *80*, 103485.

Yang, B.; Wang, Q. J.; Lei, R.; Wu, C., Shi, C.; Wang, Q. X.; Yuan, Y.; Wang, Y.; Luo, Y.; Hu, Z.; Ma, H.; Liao, M. Systems Toxicology Used in Nanotoxicology: Mechanistic Insights into the Hepatotoxicity of Nano-Copper Particles from Toxicogenomics. *J. Nanosci. Nanotechnol.* **2010,** *10*, 8527–8537.

Yang, P.; Hong, W.; Zhou, P.; Chen, B.; Xu, H. Nano and Bulk ZnO Trigger Diverse Zn-Transport-Related Gene Transcription in Distinct Regions of the Small Intestine in Mice after Oral Exposure. *Biochem. Biophys. Res. Commun.* **2017,** *493*, 1364–1369.

Yang, Y.; Liu, Y.; Chen, S.; Cheong, K.-L.; Teng, B. Carboxymethyl β-Cyclodextrin Grafted Carboxymethyl Chitosan Hydrogel-Based Microparticles for Oral Insulin Delivery. *Carbohydr. Polym.* **2020,** *246*, 116617.

Youssef, A. M. Polymer Nanocomposites as a New Trend for Packaging Applications. *Polym.-Plast. Technol. Eng.* **2013,** *52*, 635–660.

Zhang, S.; Jiang, X.; Cheng, S.; Fan, J.; Qin, X.; Wang, T.; Zhang, Y.; Zhang, J.; Qiu, Y.; Qiu, J.; Zou, Z.; Chen, C. Titanium Dioxide Nanoparticles via Oral Exposure Leads to Adverse Disturbance of Gut Microecology and Locomotor Activity in Adult Mice. *Arch. Toxicol.* **2020,** *94*, 1173–1190.

Index

S

T

X